AF699136

WITHDRAWN

AN INTRODUCTION TO SOLAR RADIATION

AN INTRODUCTION TO SOLAR RADIATION

Muhammad Iqbal
Department of Mechanical Engineering
The University of British Columbia
Vancouver, British Columbia, Canada

1983

ACADEMIC PRESS
A Subsidiary of Harcourt Brace Jovanovich, Publishers

Toronto New York London
Paris San Diego San Francisco São Paulo Sydney Tokyo

QC
911
.I63
1983

COPYRIGHT © 1983, BY ACADEMIC PRESS CANADA
ALL RIGHTS RESERVED.
NO PART OF THIS PUBLICATION MAY BE REPRODUCED OR TRANSMITTED IN ANY FORM OR BY ANY MEANS, ELECTRONIC OR MECHANICAL, INCLUDING PHOTOCOPY, RECORDING, OR ANY INFORMATION STORAGE AND RETRIEVAL SYSTEM, WITHOUT PERMISSION IN WRITING FROM THE PUBLISHER.

ACADEMIC PRESS CANADA
55 Barber Greene Road, Don Mills, Ontario M3C 2A1

United States Edition published by ACADEMIC PRESS, INC.
111 Fifth Avenue, New York, New York 10003

United Kingdom Edition published by ACADEMIC PRESS, INC. (LONDON) LTD.
24/28 Oval Road, London NW1 7DX

Library of Congress Cataloging in Publication Data

Iqbal, Muhammad
An introduction to solar radiation.

Includes index.
1. Solar radiation. I. Title.
QC911.I63 1983 551.5'271 82-7416
ISBN 0-12-373750-8
ISBN 0-12-373752-4 (pbk.)

Canadian Cataloguing in Publication Data

Iqbal, Muhammad
An introduction to solar radiation

Bibliography: p.
Includes index.
ISBN 0-12-373750-8
ISBN 0-12-373752-4 (pbk.)

1. Solar radiation. I. Title.

QC911.I72 551.5'271 C83-098054-7

83 84 85 86 9 8 7 6 5 4 3 2 1

To my daughter, Isabeau Anisa

CONTENTS

PREFACE

This book is written for those who must determine quantitatively the amount of solar radiation incident on a surface on the earth. The material has been derived from meteorological and engineering literature. The emphasis is on an understanding of the subject as well as on the methods of calculation.

The audience of this book can be divided into six groups: (1) *energy analysts* who are responsible for developing national programs on solar energy utilization and who are required to assess the potential of solar energy in their region; (2) *designers of thermal devices*, such as flat-plate collectors or concentrating devices, who need to calculate hourly and daily values of the incident radiation; (3) *photovoltaic engineers* who require an estimation of the spectral values of the direct plus diffuse solar radiation under cloudless skies; (4) *architects and engineers* who need to calculate thermal loads and natural illumination of buildings; (5) *agronomists* who need to estimate the radiative environment of plants; and (6) *hydrologists* who need to compute the amount of snowmelt and evaporation from reservoirs.

The required background for this book is some college-level elementary physics, chemistry, and calculus, and the book can form part of a course on solar energy utilization at senior–graduate levels or be used in an introductory course on solar radiation meteorology.

Chapter 1 introduces the trigonometric relationships between the sun–earth line and the position of an inclined surface. Because this material does not require a priori knowledge of any aspect of solar radiation, it was considered appropriate at the beginning. Chapter 2, on the characteristics of blackbody radiation, is written as a prelude to Chapter 3, which focuses on the solar constant and its spectral distribution. Chapter 4 deals with extraterrestrial radiation. Detailed formulations treating instantaneous, hourly, and daily quantities of radiation incident on inclined planes are presented.

A cloudless-sky atmosphere and its optics are treated in Chapter 5. The first part of this chapter describes earth's cloudless-sky atmosphere (U.S.S.A. 1976), concentration of the various molecules, vertical distribution of ozone and its seasonal variation, vertical distribution of water vapor, evaluation of the precipitable water through its partial pressure or dew-point temperature, and description of the aerosols. The second part deals with the calculation of optical masses for air, ozone, and water vapor.

Solar spectral radiation under cloudless skies is studied in Chapter 6. This chapter essentially has four parts. In the first part, the basic Bouguer equation is laid down; Rayleigh and aerosol scattering are also described, and aerosol attenuation is described through Ångström's turbidity formula. In the second part, molecular absorption is described, and tables of the monochromatic attenuation coefficients are given for ozone, uniformly mixed gases, and water vapor. In the third part, the formulation to compute direct spectral irradiance is presented. In the fourth part, a simple empirical approach to calculate the diffuse spectral irradiance is demonstrated; however, in this approach multiple scattering is ignored. The diffuse irradiance is considered to have three components: (1) Rayleigh-scattered diffuse radiation, (2) aerosol-scattered diffuse radiation, and (3) multiply reflected diffuse radiation.

Total (broadband) radiation under cloudless skies is the subject of Chapter 7, in which three parameterization models of calculating direct and diffuse irradiance are described and compared with each other. Finally, a comparison is made between one of the three models and the ASHRAE algorithm for clear-sky irradiance. A revised ASHRAE algorithm is recommended.

Chapter 8 treats the estimation of solar radiation arriving at horizontal surfaces on the earth through cloudy skies. Here, it is assumed that mean daily global (direct plus diffuse) radiation either is known from meteorological data or can be easily estimated through simple procedures such as the use of empirical correlations between global radiation and the number of bright sunshine hours or cloud cover. A number of such correlations are described, site dependent as well as site independent The rest of the chapter discusses methods of predicting diffuse radiation when the global value is known. A critical examination of the predictive correlations for global as well as diffuse radiation is presented.

In Chapter 9 the ground albedo and its spectral and angular variation are described, and tables of the albedo values for natural ground covers and man-made surfaces are presented.

Chapter 10 deals with calculation of angle factors. The angle factors between the ground and inclined surfaces and between the sky and inclined surfaces are developed.

The calculation of insolation on inclined surfaces is treated in Chapter 11. Direct, ground-reflected, and sky diffuse irradiance are taken up separately. Angular distribution of sky diffuse intensity is studied, and models of the anisotropic sky diffuse irradiance are presented.

Solar radiation measuring instruments are described and illustrated in Chapter 12. Development of the pyrheliometers beginning from Abbot water-flow and Ångström electrical compensation to the modern cavity-type radiometers is presented. The different types of pyranometers and sunshine recorders are discussed. Sources of errors and methods of calibration are presented.

The subject matter in this book is liberally illustrated by diagrams and worked examples. Wherever necessary, tables containing useful data are provided. At the end of each chapter, suggestions for further reading are given to encourage the enthusiast. With a few exceptions, SI units are used throughout. In Appendix A all basic constants are given in the SI as well as the conventional units. The nomenclature familiar to the community in this field is retained. A table of conversion factors is given in Appendix B.

ACKNOWLEDGMENTS

I have derived material from many research journals and am indebted to the authors of those publications and to these journals for granting permission to reproduce many useful diagrams and tables. A significant portion of this book is based on the research of several graduate students and assistants in this department who were supported by the Natural Sciences and Engineering Research Council of Canada. I am indebted to the Research Council for its indirect help. The Atmospheric Environment Service of Canada, which maintains an extensive radiation network and runs an excellent data collection and distribution system, was invaluable for development of this subject.

Several individuals have generously supplied me with unpublished reports, specific climatological data, and other useful information that has enriched this volume. I am particularly indebted to C. Fröhlich for supplying the table of extraterrestrial solar spectral irradiance and many other helpful reports from the World Radiation Center, to John Hay for allowing me early access to some of his research publications, to Marie Mermier and P. Fournier for some of the French data, and to Schieldrup Paulsen for the Norwegian data. The comments of Eldon C. Boes, who reviewed the manuscript, helped me improve this text.

Among my close collaborators, Cecilia Cameron carried out the initial computations, prepared diagrams and tables, and worked examples. Clement Ma reworked and extended the numerical procedures. I am indebted to Clement also for excellent proofreading of the final manuscript. Toby

Brine and Meinrad Mächler, through their theses and discussions, improved the contents of Chapters 6 and 7. David Morton prepared the albedo tables for Chapter 9.

Donna Schmid patiently typed and retyped several versions of the manuscript during a period of two years. Not only was her typing work excellent, but luckily she could also read my handwriting. My thanks are also due to Jean Horner for overseeing the typing of the manuscript. Richard Brun and Monica Gutirezz prepared the diagrams, and Irvine Devries edited my writing. Finally, I am indebted to my institution, which maintains and promotes a climate conducive for creative pursuits.

This book is based on my classroom notes, and I have kept, deliberately in spots, some informality in presentation.

Chapter 1

SUN–EARTH ASTRONOMICAL RELATIONSHIPS

1.1 Introduction

This book begins by considering the motion of the earth around the sun, the motion of the earth around its polar axis, and the angle between the earth's equator and the plane containing the sun–earth orbital system. Trigonometric equations relating the position of the sun to a horizontal or an inclined surface are presented. The main objective, however, is to write down the relevant astronomical relationships that do not require *a priori* knowledge of solar radiation.

1.2 Sun–Earth Distance *r*

The earth revolves around the sun in an elliptical orbit with the sun at one of the foci (Fig. 1.2.1). The amount of solar radiation reaching the earth is inversely proportional to the square of its distance from the sun; an accurate value of the sun–earth distance is, therefore, important. The mean sun–earth distance r_0 is called one astronomical unit:

$$1 \text{ AU} = 1.496 \times 10^8 \text{ km}$$

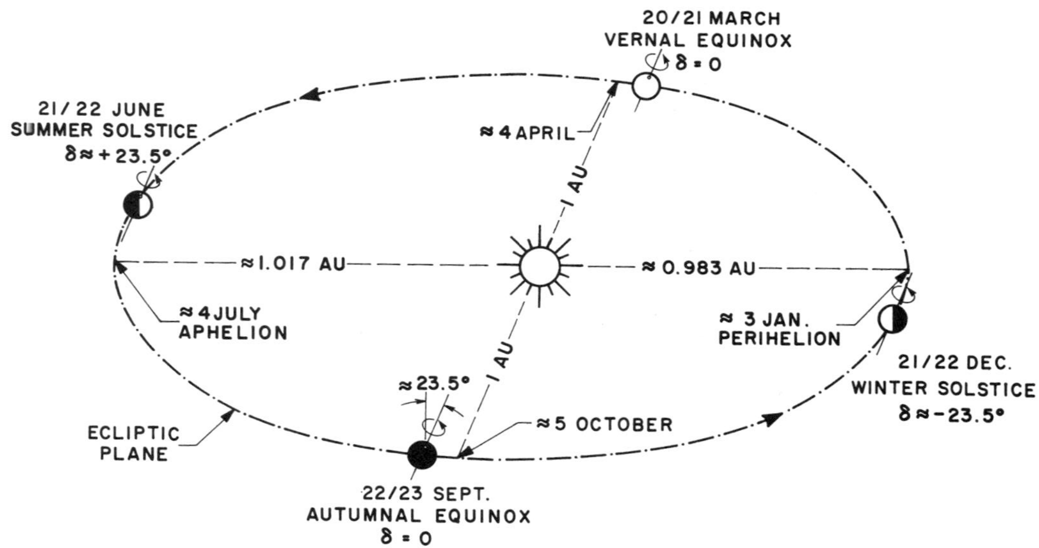

Figure 1.2.1 Motion of the earth around the sun.

or, more accurately, 149 597 890 ± 500 km.[1] The minimum sun–earth distance is about 0.983 AU, and the maximum approximately 1.017 AU. The earth is at its closest point to the sun (perihelion) on approximately 3 January and at its farthest point (aphelion) on approximately 4 July. The earth is at its mean distance from the sun on approximately 4 April and 5 October. In long-term cycles, these distances are influenced, however slightly, by other heavenly bodies and the leap year cycle. However, the sun–earth distance r for any day of any year is known with considerable accuracy. For precise information on a particular year, an ephemeris should be consulted, such as the "American Ephemeris and Nautical Almanac," published each year by the U.S. Naval Observatory.

It is more desirable, however, to have this distance expressed in a simple mathematical form: for this purpose, a number of mathematical expressions of varying complexities are available. Traditionally, the distance r is expressed in terms of a Fourier series type of expansion with a number of coefficients. With a maximum error of 0.0001, Spencer [1] developed the following expression for the reciprocal of the square of the radius vector of the earth, here called the eccentricity correction factor of the earth's orbit, E_0:

$$E_0 = (r_0/r)^2 = 1.000110 + 0.034221 \cos \Gamma + 0.001280 \sin \Gamma + 0.000719 \cos 2\Gamma + 0.000077 \sin 2\Gamma. \quad (1.2.1)$$

In this equation, Γ, in radians, is called the day angle. It is represented by

$$\Gamma = 2\pi(d_n - 1)/365, \quad (1.2.2)$$

where d_n is the day number of the year, ranging from 1 on 1 January to 365 on 31 December.[2] February is always assumed to have 28 days; because of the leap year cycle, the accuracy of Eq. (1.2.1) will vary slightly. For most engineering and technological applications, however, a very simple expression,

$$E_0 = (r_0/r)^2 = 1 + 0.033 \cos[(2\pi d_n/365)], \quad (1.2.3)$$

used by Duffie and Beckman [2] may be employed. A comparison of Eq. (1.2.3) with the Almanac values has shown that this equation can be safely employed for most engineering calculations. For greater accuracy and for use in digital machines, Eq. (1.2.1) is preferred. For individual calculations, Table 1.2.1 lists the eccentricity correction factor E_0 for all days of the calendar year.

[1] All the basic constants in this book are summarized in Appendix A, where their values are listed for SI and nonmetric units. Appendix B contains a table of conversion factors.

[2] A word of caution is necessary here. Spencer's original equation was expressed in the Julian system in which the day number is defined as the year day number, minus one. However, in many disciplines, including engineering, the day number is counted as defined above.

Table 1.2.1

Eccentricity Correction Factor $E_0 = (r_0/r)^2$ for Each Day of the Year[a]

Date	Jan	Feb	Mar	Apr	May	Jun	July	Aug	Sept	Oct	Nov	Dec
1	1.0350	1.0306	1.0190	1.0014	0.9845	0.9717	0.9666	0.9700	0.9814	0.9976	1.0155	1.0291
2	1.0351	1.0303	1.0185	1.0008	0.9840	0.9714	0.9666	0.9703	0.9819	0.9982	1.0161	1.0295
3	1.0351	1.0300	1.0180	1.0002	0.9835	0.9712	0.9666	0.9705	0.9823	0.9988	1.0166	1.0298
4	1.0351	1.0297	1.0174	0.9997	0.9830	0.9709	0.9666	0.9708	0.9828	0.9994	1.0171	1.0301
5	1.0351	1.0294	1.0169	0.9991	0.9825	0.9706	0.9666	0.9711	0.9833	0.9999	1.0177	1.0304
6	1.0350	1.0290	1.0164	0.9985	0.9821	0.9703	0.9666	0.9713	0.9838	1.0005	1.0182	1.0307
7	1.0350	1.0287	1.0158	0.9979	0.9816	0.9701	0.9666	0.9716	0.9843	1.0011	1.0187	1.0310
8	1.0350	1.0283	1.0153	0.9973	0.9811	0.9698	0.9666	0.9719	0.9848	1.0017	1.0192	1.0313
9	1.0349	1.0279	1.0147	0.9967	0.9806	0.9696	0.9667	0.9722	0.9854	1.0023	1.0197	1.0316
10	1.0348	1.0276	1.0142	0.9961	0.9802	0.9694	0.9667	0.9726	0.9859	1.0029	1.0202	1.0319
11	1.0347	1.0272	1.0136	0.9956	0.9797	0.9692	0.9668	0.9729	0.9864	1.0035	1.0207	1.0321
12	1.0347	1.0268	1.0131	0.9950	0.9793	0.9690	0.9668	0.9732	0.9869	1.0041	1.0212	1.0324
13	1.0346	1.0264	1.0125	0.9944	0.9788	0.9687	0.9669	0.9736	0.9875	1.0047	1.0217	1.0326
14	1.0344	1.0260	1.0119	0.9938	0.9784	0.9686	0.9670	0.9739	0.9880	1.0053	1.0222	1.0328

15	1.0343	1.0256	1.0114	0.9932	0.9780	0.9684	0.9671	0.9743	0.9885	1.0058	1.0226	1.0330
16	1.0342	1.0251	1.0108	0.9927	0.9775	0.9682	0.9672	0.9746	0.9891	1.0064	1.0231	1.0332
17	1.0340	1.0247	1.0102	0.9921	0.9771	0.9680	0.9673	0.9750	0.9896	1.0070	1.0235	1.0334
18	1.0339	1.0243	1.0097	0.9915	0.9767	0.9679	0.9674	0.9754	0.9902	1.0076	1.0240	1.0336
19	1.0337	1.0238	1.0091	0.9910	0.9763	0.9677	0.9675	0.9758	0.9907	1.0082	1.0244	1.0338
20	1.0335	1.0234	1.0085	0.9904	0.9759	0.9676	0.9677	0.9762	0.9913	1.0088	1.0249	1.0339
21	1.0334	1.0229	1.0079	0.9899	0.9755	0.9675	0.9678	0.9766	0.9918	1.0093	1.0253	1.0341
22	1.0332	1.0224	1.0073	0.9893	0.9751	0.9673	0.9680	0.9770	0.9924	1.0099	1.0257	1.0342
23	1.0330	1.0220	1.0067	0.9888	0.9748	0.9672	0.9681	0.9774	0.9930	1.0105	1.0261	1.0344
24	1.0327	1.0215	1.0062	0.9882	0.9744	0.9671	0.9683	0.9778	0.9935	1.0111	1.0265	1.0345
25	1.0325	1.0210	1.0056	0.9877	0.9740	0.9670	0.9685	0.9782	0.9941	1.0116	1.0269	1.0346
26	1.0323	1.0205	1.0050	0.9872	0.9737	0.9669	0.9687	0.9787	0.9947	1.0122	1.0273	1.0347
27	1.0320	1.0200	1.0044	0.9866	0.9733	0.9669	0.9689	0.9791	0.9953	1.0128	1.0277	1.0348
28	1.0318	1.0195	1.0038	0.9861	0.9730	0.9668	0.9691	0.9795	0.9959	1.0133	1.0281	1.0349
29	1.0315	0.0	1.0032	0.9856	0.9727	0.9667	0.9693	0.9800	0.9964	1.0139	1.0284	1.0349
30	1.0312	0.0	1.0026	0.9851	0.9724	0.9667	0.9695	0.9805	0.9970	1.0144	1.0288	1.0350
31	1.0309	0.0	1.0020	0.0	0.9720	0.0	0.9698	0.9809	0.0	1.0150	0.0	1.0350

[a] Calculated from Eq. (1.2.1).

☐ EXAMPLE 1.2.1. Calculation of the sun–earth distance r on 16 October.

Solution. On October 16 the day number $d_n = 289$ and the day angle $\Gamma = 2\pi(288)/365 = 4.958$ rad or $\Gamma = 284.16°$.

From Eq. (1.2.1)

$$\begin{aligned}(r_0/r)^2 &= 1.000110 + 0.034221(0.2428) + 0.00128(-0.9701) \\ &\quad + 0.000719(-0.8821) + 0.000077(-0.4710) \\ &= 1.0064.\end{aligned}$$

Therefore, $r = 0.9968$ AU.

From Eq. (1.2.3)

$$(r_0/r)^2 = 1 + 0.033\cos[2\pi(289)/365] = 1.0091,$$

or

$$r = 0.98202 \text{ AU}.$$ ☐

1.3 Solar Declination δ

The plane of revolution of the earth around the sun is called the *ecliptic plane*. The earth itself rotates around an axis called the polar axis, which is inclined at approximately $23\frac{1}{2}°$ from the normal to the ecliptic plane (Fig. 1.2.1). The earth's rotation around its axis causes the diurnal changes in radiation income; the position of this axis relative to the sun causes seasonal changes in solar radiation. The angle between the polar axis and the normal to the ecliptic plane, however, remains unchanged. The same is true of the angle between the earth's equatorial plane and the ecliptic plane. However, the angle between a line joining the centers of the sun and the earth to the equatorial plane changes every day, in fact, every instant. This angle is called the *solar declination* δ. It is zero at the *vernal* and *autumnal equinoxes* (literally, equal nights) and has a value of approximately $+23\frac{1}{2}°$ at the *summer solstice* and about $-23\frac{1}{2}°$ at the *winter solstice*. The four seasons pertain here to the northern hemisphere; the reverse is true in the southern hemisphere.

The actual dates of the equinoxes and solstices vary slightly from year to year. Because of this, one encounters slightly different dates for when the earth is in these four principal positions.

Another means of describing the solar declination is by drawing a celestial sphere with the earth at the center and the sun revolving around the earth (Fig. 1.3.1). In the celestial sphere, the celestial poles are the points at which the earth's polar axis, when produced, cuts the celestial sphere. Similarly, the celestial equator is an outward projection of the earth's equatorial plane on

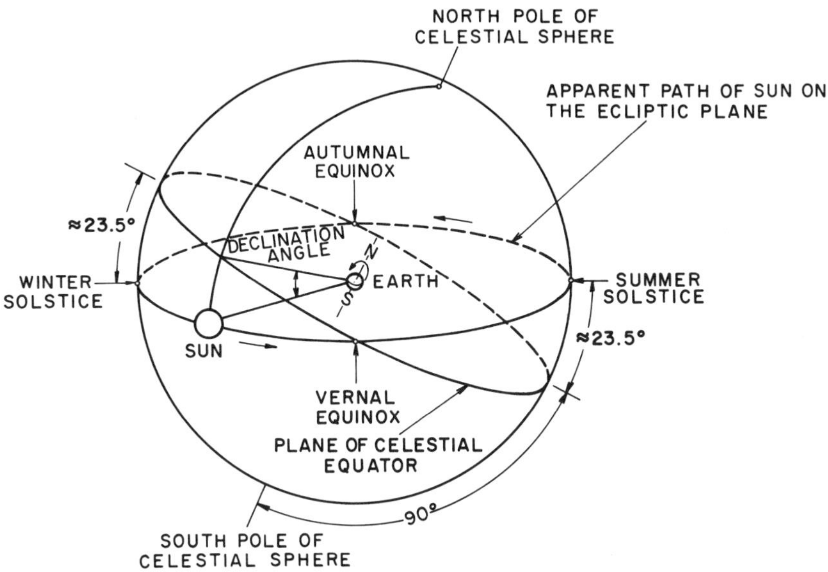

Figure 1.3.1 Celestial sphere showing apparent path of sun and sun's declination angle.

the celestial sphere. The intersection of the plane of the earth's equator with the plane of the sun's revolution, the ecliptic, makes an angle of approximately $23\frac{1}{2}°$. At any given time, the position of the sun relative to the plane of the celestial equator describes the declination angle. The main variations in the declination are due to the leap year cycle: during this four-year period, the declination may vary from the order of $\pm 10'$ at the equinoxes to less than $1'$ at the solstices [1].

In 24 h, the maximum change in declination (which occurs at the equinoxes) is less than $\frac{1}{2}°$. Therefore, if the declination is assumed constant for 24 h, a maximum error of $\frac{1}{2}°$ may occur in calculating the solar azimuth and zenith angles (to be described later). To obtain an accurate value of solar declination, an ephermeris should again be consulted. Expressions giving the approximate values of solar declination with varying degrees of accuracy have been developed by a number of authors. Spencer [1] presented the following expression for δ, in degrees:

$$\begin{aligned}\delta = (&0.006918 - 0.399912 \cos \Gamma + 0.070257 \sin \Gamma \\ &- 0.006758 \cos 2\Gamma + 0.000907 \sin 2\Gamma \\ &- 0.002697 \cos 3\Gamma + 0.00148 \sin 3\Gamma)(180/\pi). \qquad (1.3.1)\end{aligned}$$

This equation estimates δ with a maximum error of 0.0006 rad ($< 3'$) or, if the final two terms are omitted, with a maximum error of 0.0035 rad ($12'$).

Table 1.3.1

Declination for Each Day of the Year[a]

Date	Jan	Feb	Mar	Apr	May	Jun	Jul	Aug	Sept	Oct	Nov	Dec
1	−23.07	−17.28	−7.78	4.36	14.93	22.02	23.20	18.20	8.51	−2.95	−14.26	−21.74
2	−22.99	−17.00	−7.40	4.75	15.24	22.15	23.13	17.94	8.14	−3.33	−14.58	−21.90
3	−22.90	−16.71	−7.02	5.13	15.54	22.29	23.06	17.69	7.78	−3.72	−14.90	−22.05
4	−22.80	−16.41	−6.63	5.51	15.83	22.41	22.98	17.42	7.41	−4.11	−15.22	−22.19
5	−22.70	−16.11	−6.25	5.89	16.12	22.53	22.89	17.16	7.04	−4.50	−15.53	−22.32
6	−22.59	−15.81	−5.86	6.27	16.41	22.64	22.80	16.89	6.67	−4.88	−15.83	−22.45
7	−22.47	−15.50	−5.47	6.65	16.69	22.74	22.70	16.61	6.30	−5.27	−16.13	−22.57
8	−22.34	−15.18	−5.08	7.03	16.96	22.84	22.59	16.33	5.93	−5.65	−16.43	−22.68
9	−22.21	−14.87	−4.69	7.40	17.24	22.93	22.48	16.05	5.55	−6.03	−16.72	−22.79
10	−22.07	−14.54	−4.30	7.77	17.50	23.01	22.36	15.76	5.17	−6.41	−17.01	−22.28
11	−21.92	−14.22	−3.90	8.14	17.77	23.09	22.23	15.46	4.80	−6.79	−17.29	−22.98
12	−21.76	−13.89	−3.51	8.51	18.02	23.16	22.10	15.17	4.42	−7.17	−17.57	−23.06
13	−21.60	−13.55	−3.12	8.87	18.28	23.23	21.96	14.87	4.03	−7.55	−17.84	−23.13
14	−21.43	−13.22	−2.72	9.24	18.52	23.28	21.81	14.56	3.65	−7.92	−18.11	−23.20

15	−21.25	−12.87	−2.33	9.60	18.77	23.33	21.66	14.25	3.27	−8.30	−18.37	−23.26
16	−21.07	−12.53	−1.93	9.95	19.00	23.38	21.50	13.94	2.88	−8.67	−18.62	−23.31
17	−20.88	−12.18	−1.54	10.31	19.23	23.41	21.34	13.62	2.50	−9.04	−18.87	−23.36
18	−20.68	−11.83	−1.14	10.66	19.46	23.44	21.17	13.30	2.11	−9.40	−19.12	−23.39
19	−20.48	−11.47	−0.74	11.01	19.68	23.47	20.99	12.98	1.72	−9.77	−19.36	−23.42
20	−20.27	−11.12	−0.35	11.35	19.90	23.48	20.81	12.66	1.34	−10.13	−19.59	−23.44
21	−20.05	−10.76	0.05	11.70	20.10	23.49	20.63	12.33	0.95	−10.49	−19.82	−23.46
22	−19.83	−10.39	0.44	12.04	20.31	23.49	20.43	11.99	0.56	−10.85	−20.04	−23.46
23	−19.60	−10.03	0.84	12.37	20.51	23.49	20.23	11.66	0.17	−11.21	−20.25	−23.46
24	−19.37	−9.66	1.23	12.71	20.70	23.47	20.03	11.32	−0.22	−11.56	−20.46	−23.45
25	−19.13	−9.29	1.63	13.04	20.88	23.46	19.82	10.98	−0.61	−11.91	−20.67	−23.43
26	−18.88	−8.91	2.02	13.36	21.07	23.43	19.60	10.63	−1.00	−12.25	−20.86	−23.40
27	−18.63	−8.54	2.41	13.68	21.24	23.40	19.38	10.28	−1.39	−12.60	−21.05	−23.37
28	−18.37	−8.16	2.80	14.00	21.41	23.36	19.15	9.93	−1.78	−12.94	−21.23	−23.33
29	−18.11	0.0	3.19	14.32	21.57	23.31	18.92	9.58	−2.17	−13.27	−21.41	−23.28
30	−17.84	0.0	3.58	14.63	21.73	23.26	18.68	9.22	−2.56	−13.61	−21.58	−23.22
31	−17.56	0.0	3.97	0.0	21.87	0.0	18.44	8.87	0.0	−13.94	0.0	−23.16

[a] Calculated from Eq. (1.3.1).

Two other simple and commonly used formulas for declination are

$$\delta = \sin^{-1}\{0.4 \sin[\tfrac{360}{365}\ (d_n - 82)]\} \quad \text{in degrees} \tag{1.3.2}$$

obtained from Perrin de Brichambaut [4] and

$$\delta = 23.45 \sin[\tfrac{360}{365}\,(d_n + 284)] \quad \text{in degrees} \tag{1.3.3}$$

from Cooper [5].

The two simpler equations are in fact quite accurate. However, for greater accuracy and for use in digital machines, Eq. (1.3.1) is preferred. To facilitate rapid calculations for individual days, the values of the declination for all calendar days are given in Table 1.3.1.

□ EXAMPLE 1.3.1. Calculation of the solar declination on 16 October.

Solution. From Example 1.2.1, the day angle $\Gamma = 284.16°$.
From Eq. (1.3.1) the declination δ is

$$\begin{aligned}\delta = {} & [0.006918 - 0.399912(0.2428) + 0.07257(-0.9701) \\ & - 0.006758(-0.8821) + 0.000907(-0.4710) \\ & - 0.002697(-0.6711) + 0.00148(0.7414)](180/\pi) \\ = {} & -8.67°.\end{aligned}$$

From Eq. (1.3.2)

$$\begin{aligned}\delta &= \sin^{-1}\{0.4 \sin[\tfrac{360}{365}\,(289 - 82)]\} \\ &= -9.42°.\end{aligned}$$

From Eq. (1.3.3)

$$\begin{aligned}\delta &= 23.45 \sin[\tfrac{360}{365}\,(289 + 284)] \\ &= -9.97°.\end{aligned}$$

□

1.4 Equation of Time E_t

Solar time is based on the rotation of the earth about its polar axis and on its revolution around the sun. A solar day is the interval of time (not necessarily 24 h) as the sun appears to complete one cycle about a stationary observer on earth. The solar day varies in length through the year. The two principal factors for this variance are the following: (1) the earth sweeps out unequal areas on the ecliptic plane as it revolves around the sun, and (2) the earth's axis is tilted with respect to the ecliptic plane. In simple terms, this means that if an observer facing the equator today sets a clock (running at a uniform rate) at 12 noon, when the sun is directly over the local meridian, then after a month or so at 12 noon, clock time, the sun may not appear exactly over the local meridian. A discrepancy of as much as 16 min is possible. This discrepancy is called the equation of time and is measured relative to a perfectly uniform terrestrial motion. Again from Spencer [1], the following series gives the equation of time (in minutes):

$$E_t = (0.000075 + 0.001868 \cos \Gamma - 0.032077 \sin \Gamma - 0.014615 \cos 2\Gamma - 0.04089 \sin 2\Gamma)(229.18). \tag{1.4.1}$$

In this equation, the first right-hand-side term in parentheses represents E_t in radians and the multiplier 229.18 converts it into minutes. The maximum error[3] with this series is 0.0025 rad, equivalent to about 35 sec. For less accurate calculations, the equation of time in minutes may be obtained from Fig. 1.4.1.

Solar radiation data are often recorded in terms of local apparent time (LAT), also called true solar time (TST). On the other hand, some meteorological data such as temperature and wind speed are often recorded in terms of local clock time, which may be local standard time or daylight saving time. In many solar energy studies, it is usually necessary to obtain the radiation, temperature, and wind velocity data for the same instant or duration. Therefore, it is desirable to convert local standard time to local apparent time. It is necessary to add that it is not uncommon, as in the radiation network of the United States, for hourly or instantaneous data to be recorded against the local standard time. In order to compute the incidence angles, it is necessary to determine the corresponding local apparent time. To carry out this conversion, it is necessary to know the standard meridian for the local time zone. All international standard meridians are multiples of 15° east or west of Greenwich, England. Therefore, all standard times are hour multiples ahead of or behind the Greenwich mean time (GMT). The standard meridians for North and South America and Hawaii are given in Table 1.4.1. It may be added here that, although all international zone times are multiples of a whole hour ahead of or behind GMT, it is not uncommon to find standards with a half-hour time difference. For instance, Newfoundland standard time is a half hour ahead of Atlantic standard time, and consequently its standard meridian is $52\frac{1}{2}°$ W.

[3] The stated accuracies of Eqs. (1.2.1), (1.3.1), and (1.4.1) are all based on the year 1950.

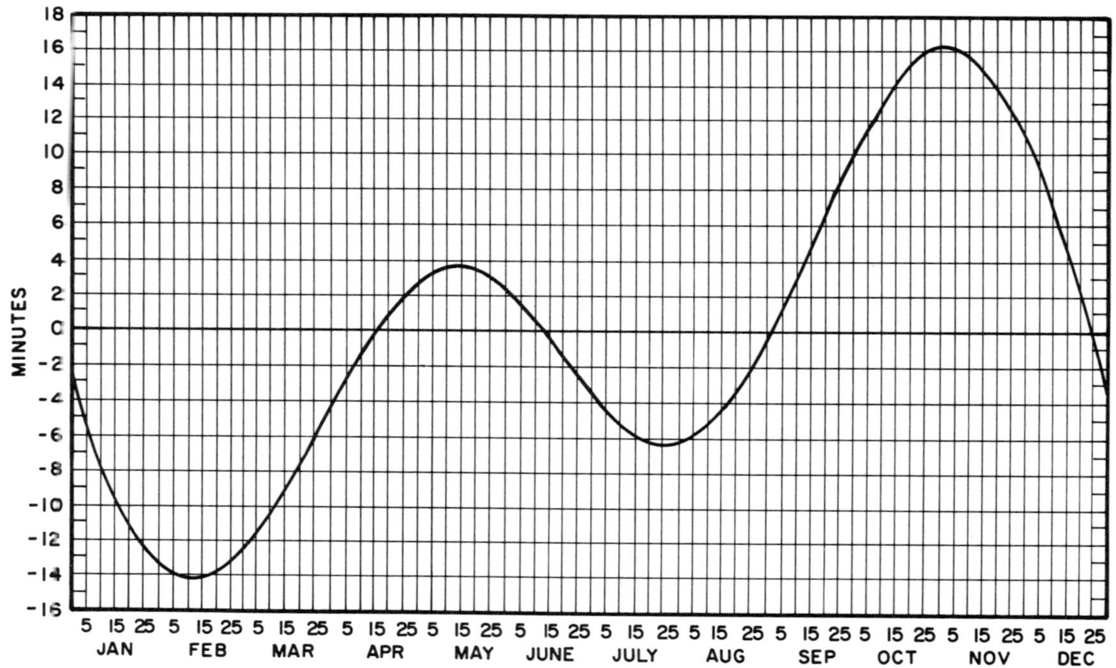

Figure 1.4.1 The yearly variation of the equation of time.

Table 1.4.1

International Zone Times for Several North and South American Time Zones

Time zone	Zone time (h)	Standard meridian
Atlantic standard time	4	60° W
Eastern standard time	5	75° W
Central standard time	6	90° W
Mountain standard time	7	105° W
Pacific standard time	8	120° W
Alaska standard time	9	135° W
Hawaii standard time	10	150° W

Local apparent time for a given standard time can now be written

$$\begin{aligned}\text{local apparent time} &= \text{local mean time} + \text{equation of time}\\ &= \text{local standard time} + \text{longitude correction}\\ &\quad + \text{equation of time}\\ &= \text{local standard time} + 4(L_s - L_e) + E_t, \qquad (1.4.2)\end{aligned}$$

where L_s is the standard longitude and L_e is the local longitude. The longitude correction, 4 min for every degree, accounts for the difference between the local and the standard meridians. It should be noted that the longitude correction is positive if the local meridian is *east* of the standard and is negative *west* of the standard meridian. The value of the equation of time E_t is added algebraically; it may be positive or negative. The correction for daylight saving time can be made appropriately.

□ EXAMPLE 1.4.1. Determination of the solar time (LAT) on 16 October at Buenos Aires (58°29′ W) when the standard time is 10:00.

Solution. From Example 1.2.1, on 16 October the day angle $\Gamma = 4.958$ rad, 284.16°.

From Eq. (1.4.1)

$$E_t = +14.62 \text{ min.}$$

Buenos Aires is 1°31′ east of the 60th meridian. The correction for latitude is

$$+4(60 - 58.48) = 6.07 \text{ min.}$$

Therefore,

$$\text{local apparent time} = 10{:}00 + (6.07 + 14.62) \text{ min} = 10{:}20{:}41. \qquad □$$

1.5 Position of the Sun Relative to Horizontal Surfaces

In order to calculate the solar radiation reaching a horizontal surface on the earth, it is necessary to write down the trigonometric relationships between the solar position in the sky and the surface coordinates on the earth. For instance, we can describe an observer on the earth by drawing again a celestial sphere with the earth as the center (see Fig. 1.5.1). At any given time, an observer on the earth's surface has a corresponding position in the celestial sphere called the observer's zenith; this is the point of intersection with the celestial sphere of a normal to the earth's surface at the observer's position. The point diagonally opposite to the local zenith is called the nadir. The observer's horizon is the great circle in the celestial sphere the plane of which passes through the center of the earth normal to the line joining the center of the earth and the zenith. The zenith angle θ_z (also called the zenith distance) is the angle between the local zenith and the line joining the observer and the

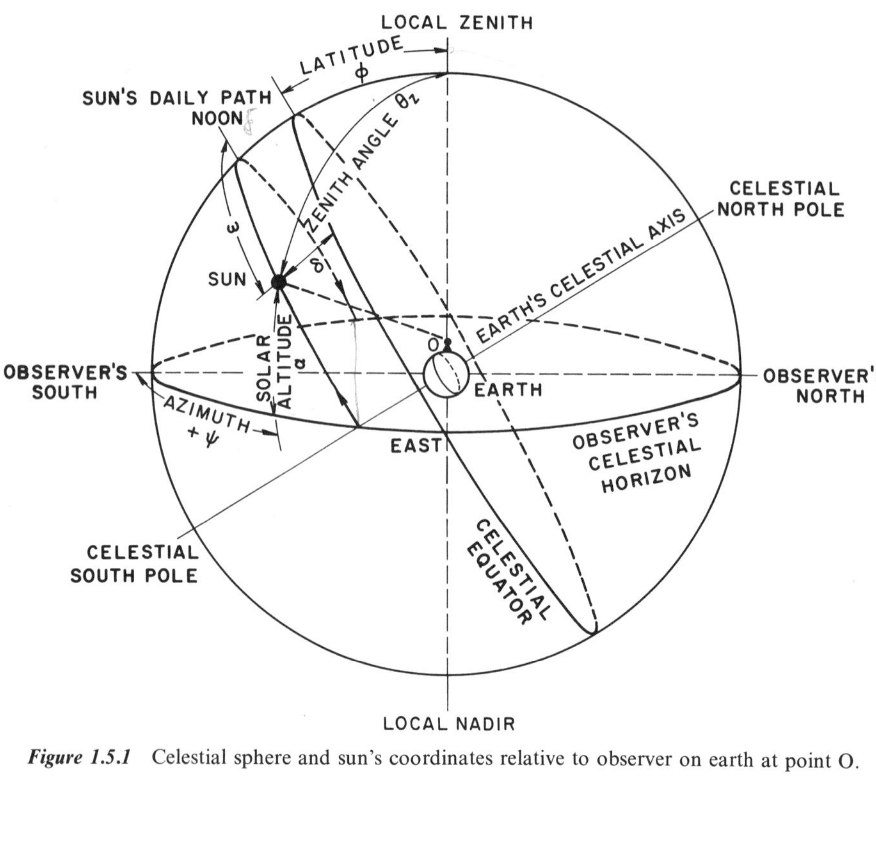

Figure 1.5.1 Celestial sphere and sun's coordinates relative to observer on earth at point O.

sun. It is an angle between 0° and 90°. The solar altitude α (also called the solar elevation) is the sun's angular height above the observer's celestial horizon. It is an angle between 0° and 90°. The solar altitude is the complement of the zenith angle.

The solar azimuth ψ is the angle at the local zenith between the plane of the observer's meridian and the plane of a great circle passing through the zenith and the sun. It is measured east positive, west negative (south zero) and thus varies between 0° and $\pm 180°$. The hour angle ω is the angle measured at the celestial pole between the observer's meridian and the solar meridian. Counting from midday, it changes 15° per hour.

For a given geographical position, in the absence of the earth's refractive atmosphere, the trigonometric relations between the sun (the center of the solar disk) and a horizontal surface are well known. These are as follows:

$$\cos\theta_z = \sin\delta\sin\phi + \cos\delta\cos\phi\cos\omega = \sin\alpha \tag{1.5.1}$$

and

$$\cos\psi = (\sin\alpha\sin\phi - \sin\delta)/\cos\alpha\cos\phi; \tag{1.5.2a}$$

$$0° \le \psi \le 90°, \qquad \cos\psi \ge 0,$$

$$90° \le \psi \le 180°, \qquad \cos\psi \le 0.$$

Sometimes the following equation is used to obtain ψ:

$$\sin\psi = \cos\delta\sin\omega/\cos\alpha. \tag{1.5.2b}$$

However, this equation gives improper values when $\psi > 90°$ and should be avoided. Definitions and sign conventions for the angles can be summarized:

θ_z is the zenith angle, also called the zenith distance, in degrees;

α is the solar altitude, also called the solar height, or solar elevation, in degrees; $\alpha = 90 - \theta_z$;

ω is the hour angle, noon zero and morning positive;

ϕ is the geographic latitude, in degrees, north positive;

ψ is the solar azimuth, in degrees, south zero, east positive (see Fig. 1.5.2 for a better description of this angle); and

δ is the declination, the angular position of the sun at solar noon with respect to the plane of the equator, north positive, in degrees.

For all practical purposes, the earth's refractive atmosphere will have negligible influence on the apparent solar altitude, which is slightly greater than the calculated altitude. The difference ranges from 0′ at zero zenith to about 34′ at the horizon. It is slightly dependent, however, on the atmospheric temperature and pressure, and is tabulated in the Almanac for θ_z between 80° and 90°, where the effect is maximum.

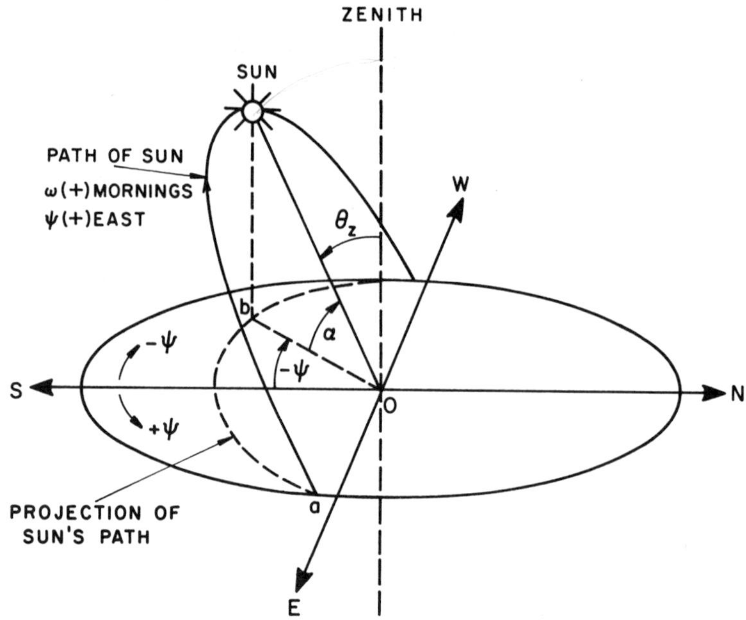

Figure 1.5.2 Definition of sun's zenith, altitude, and azimuth angles.

Which of the two equations (1.5.1) or (1.5.2) should be used depends on the particular application. Architects, for example, usually require the sun's azimuthal position at any hour during any month. For this purpose Eq. (1.5.2) can be employed.

Equation (1.5.1) can be solved for the sunrise hour angle ω_s. At sunrise, $\theta_z = 90°$. Therefore, Eq. (1.5.1) results in

$$\cos \omega_s = -\sin \phi \sin \delta / \cos \phi \cos \delta \tag{1.5.3}$$

or

$$\omega_s = \cos^{-1}(-\tan \phi \tan \delta). \tag{1.5.4}$$

Note that the sunrise hour angle is equal to the sunset hour angle except for the sign difference. From the preceding equation, day length N_d can easily be computed. The day length is $2\omega_s$, and when expressed in hours is as follows:

$$N_d - \tfrac{2}{15} \cos^{-1}(-\tan \phi \tan \delta). \tag{1.5.5}$$

A few interesting points with respect to Eq. (1.5.5) are worthy of note: (a) In the polar region, during the winter, the sun does not rise, and hence there is no day length when, from the right-hand side of Eq. (1.5.3), $\cos \omega_s > +1$. Similarly, during the summer there is a continuous day of about six months, and therefore there is no sunrise hour. For example, at the north pole the continuous duration of the polar day is 186 day, and of the polar night, 179 day (nights). (b) At the equator, $\phi = 0$; therefore, $\omega_s = \pi/2$, and day length is independent of solar declination (or season) and is always equal to 12 h. (c) At the equinoxes, $\delta = 0$; therefore, $\omega_s = \pi/2$, and day length is independent of latitude and is equal to 12 h.

□ EXAMPLE 1.5.1. Calculation of (a) the zenith angle and solar azimuth at 11:00 LAT, (b) the sunrise hour angle, and (c) the day length on 16 October at New York (40°7′ N).

Solution. At 11:00 solar time, $\omega = +15°$.
From Example 1.3.1., declination $\delta = -8.67°$.
(a) We obtain the zenith angle from Eq. (1.5.1):

$$\begin{aligned}\theta_z &= \cos^{-1}[(-0.1507)(0.6532) + (0.9886)(0.7572)(0.9659)]\\ &= 51.71°.\end{aligned}$$

Therefore the solar altitude $\alpha = 90 - \theta_z = 38.28\%$. The solar azimuth angle is obtained from Eq. (1.5.2a):

$$\psi = \cos^{-1}\left(\frac{\sin 38.28 \sin 40.16 - \sin(-8.67)}{\cos 38.28 \cos 40.16}\right) = 19.3°.$$

(b) The sunrise angle is given by Eq. (1.5.3):

$$\begin{aligned}\omega_s &= \cos^{-1}[-\tan 40.16 \tan(-8.67)]\\ &= 82.44°.\end{aligned}$$

In terms of the solar time, the sunrise is at $12 - 82.44/15 = 6.50$ h or 6:30:00 (LAT).
(c) The day length,

$$N_d = \tfrac{2}{15}\omega_s = 11 \text{ h.}$$ □

For accurate determination of the sunrise hour and day length, Eqs. (1.5.4) and (1.5.5) should be used. However, for many engineering purposes, the nomogram developed by Whillier [6], Fig. 1.5.3, may be readily employed. An illustrative example on the nomogram directly describes its use.

Before continuing, it is necessary to go back to Eqs. (1.5.1) and (1.5.2) and describe the sun path or polar diagrams. Consider the sun's trajectory shown in Fig. 1.5.2. A plane projection of this trajectory on a circle of unit radius describes the sun path. In this figure, point *a* is obtained from the solar azimuth at sunrise. The coordinates of point *b* at any hour of the day are given by the solar azimuth and length *Ob*, where *Ob* is equal to cos α. A diagram of this nature containing sun paths for various declinations (usually

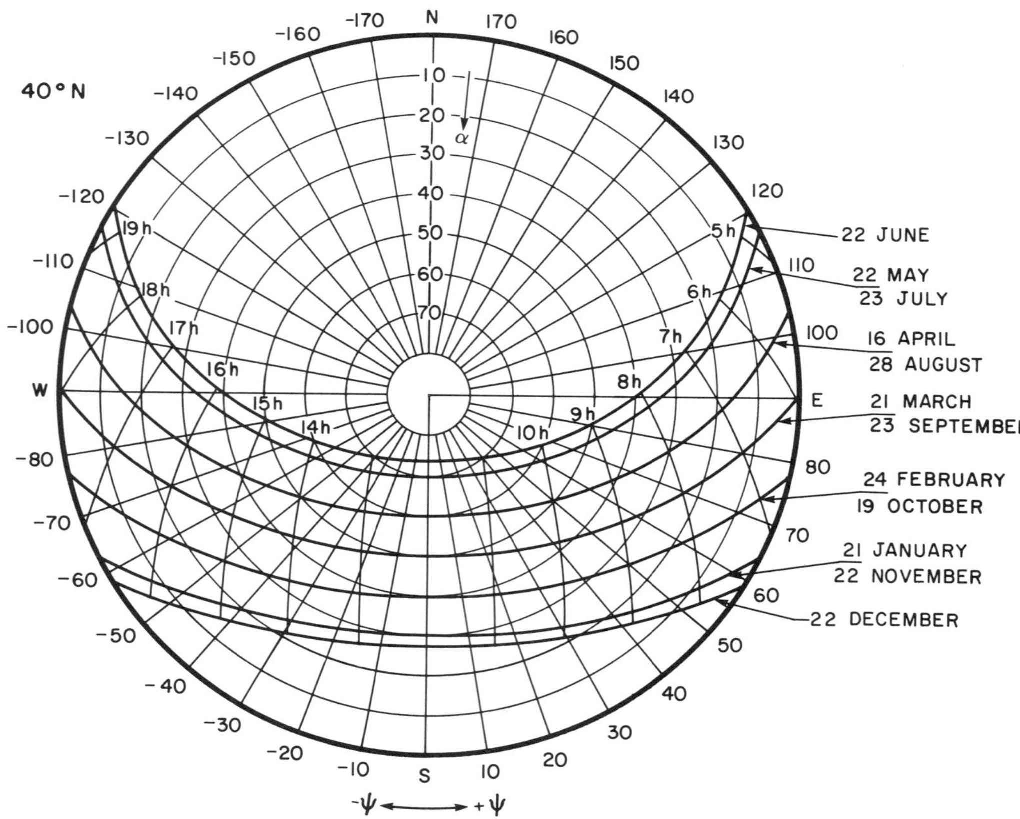

Figure 1.5.4 Polar diagram for 40° N latitude.

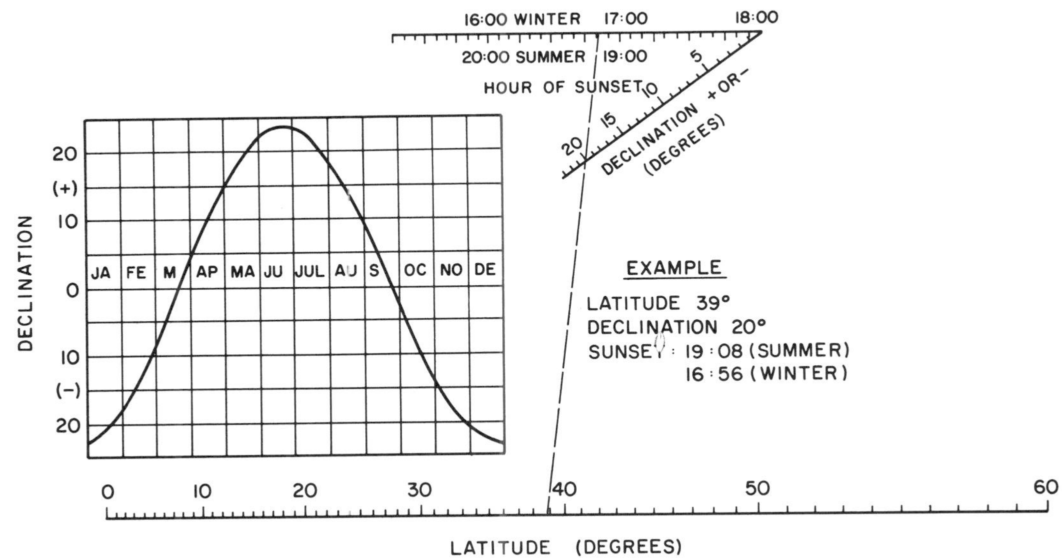

Figure 1.5.3 Nomogram to determine hour of sunset. Adapted with permission from Whillier [6]. Copyright 1965. Pergamon Press, Ltd.

one for each month) is often employed by architects to obtain graphically the solar altitude, azimuth, and eventually the shading effects of buildings. Naturally, a separate diagram is required for every individual latitude. As an example, a sun path diagram for 40° N latitude is shown in Fig. 1.5.4. In this diagram the outer circle is the horizon and the center represents the zenith. Inner circles represent solar altitudes. The altitude circles are equally spaced for better visual effect. Consequently the sun path lines are not simple projections on a horizontal surface of the diurnal motion of the sun. The radial lines indicate solar azimuths in 10° steps. The lines indicating various dates, the sun path lines, are crossed by hourly time lines which represent the local apparent time. Although this diagram is for 40° N latitude, when rotated by 180° it can be employed for the 40° S latitude.

1.6 Position of the Sun Relative to Inclined Surfaces

In order to determine the position of the sun with respect to an inclined surface, it is necessary to prescribe the slope of the surface with respect to the horizontal (or vertical) position and its orientation in relation to the local meridian. Fig. 1.6.1 describes such a surface where

β is the slope of the surface, measured from horizontal position, in degrees;
γ is the surface azimuth angle; that is, the deviation of the normal to the surface with respect to the local meridian, in degrees, east positive; and
θ is the angle of incidence for an arbitrarily oriented surface, the angle between normal to the surface and the sun–earth vector, in degrees.

Expressions for the incidence angle for surfaces facing the equator and those oriented toward the east or west now follow. These expressions will be written in a form suitable for solar energy applications where the radiation data are given in terms of the solar time. The trigonometric relations for a surface aligned with the local meridian will be treated first.

A. *Surfaces Facing the Equator*

Let θ_0 be the angle of incidence for a surface facing the equator in degrees. An expression for the incidence angle from Eq. (1.5.1) will now be developed. Consider Fig. 1.6.2 showing two surfaces located at latitudes ϕ and $\phi - \beta$. The surface at latitude ϕ is inclined at an angle β whereas the surface at latitude $\phi - \beta$ is horizontal. Following Liu and Jordan [7], a surface located at latitude ϕ and tilted toward the equator at an angle β from the horizontal position is parallel to a horizontal surface at latitude $\phi - \beta$.

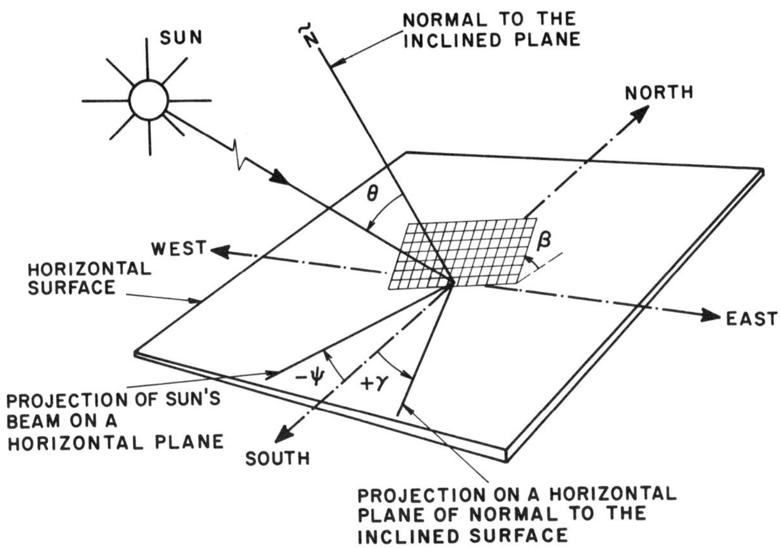

Figure 1.6.1 Position of sun relative to an inclined plane.

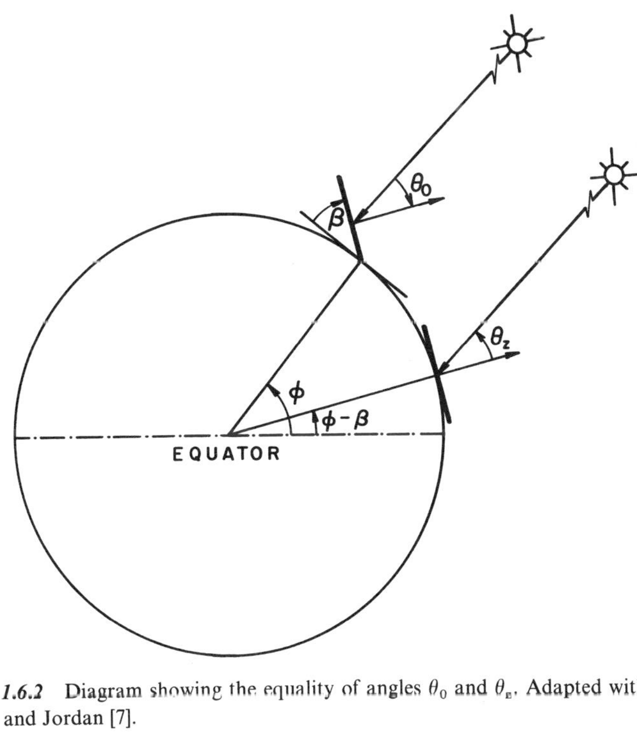

Figure 1.6.2 Diagram showing the equality of angles θ_0 and θ_z. Adapted with permission from Liu and Jordan [7].

Therefore, θ_0 at latitude ϕ is equal to θ_z at latitude $\phi - \beta$. Hence, from Eq. (1.5.1), the angle of incidence for a surface tilted toward the equator can be written as follows:

$$\cos\theta_0 = \sin\delta\sin(\phi - \beta) + \cos\delta\cos(\phi - \beta)\cos\omega. \qquad (1.6.1)$$

From the preceding, the angle at which direct solar radiation first strikes an inclined plane in the morning can be obtained. This angle is called the sunrise angle for an inclined plane. On a tilted surface the sun rises (or sets) when the incidence angle is 90°. Let ω_s' be the sunrise hour angle for a tilted surface facing the equator, in degrees. At sunrise, $\theta_0 = 90°$ and $\omega = \omega_s'$. Therefore

$$\cos\omega_s' = -\sin\delta\sin(\phi - \beta)/\cos\delta\cos(\phi - \beta) \qquad (1.6.2)$$

or

$$\omega'_s = \cos^{-1}[-\tan\delta\tan(\phi - \beta)]. \qquad (1.6.3)$$

For a surface tilted toward the equator, the sunrise hour angle is equal to the sunset hour angle except for the sign difference. Consider the following three particular cases:

(i) At the equinox, $\delta = 0$, Eq. (1.6.3) gives

$$\omega_s' = \pi/2.$$

That is, the sunrise hour angle is independent of the tilt or latitude.

(ii) During the summer (in the northern hemisphere), $\delta > 0$, and this results in $\omega_s > \omega_s'$. That is, the sun rises earlier on a horizontal than on an inclined surface. And

(iii) During the winter (in the northern hemisphere), $\delta < 0$. Mathematically, this results in the sunrise hour angle for a tilted surface being greater than that for a horizontal surface. Since this is not physically possible, we may write a general expression for the sunrise hour angle,

$$\omega_s' = \min\{\cos^{-1}(-\tan\delta\tan\phi), \cos^{-1}[-\tan\delta\tan(\phi - \beta)]\}, \qquad (1.6.4)$$

where min stands for minimum of the expression in braces, which can be applied to either (ii) or (iii) above.

□ EXAMPLE 1.6.1. Calculation of (a) the sunrise hour angle for the collector surface if the collector slope from the horizontal position is (i) 30°, (ii) 90°; (b) the angle of incidence at 11:00 (LAT) for each of the slopes in (a) above, for a flat-plate collector at London (51°31′ N) which is tilted toward the equator on 15 May.

Solution. From Table 1.3.1, $\delta = 18.77°$.

(a) The sunrise hour angle for a surface inclined toward the equator is given by Eq. (1.6.4), while that for a horizontal surface is given by Eq. (1.5.4).

For a horizontal surface,

$$\omega_s = \cos^{-1}[-\tan(51.5)\tan(18.77)]$$
$$= 115.29^\circ.$$

(i) For the collector sloped at $\beta = 30^\circ$,

$$\omega_s' = \min\{115.29^\circ, \cos^{-1}[-\tan(51.5 - 30)\tan(18.77)]\}$$
$$= \min\{115.29^\circ, 97.69^\circ\}$$
$$= 97.69^\circ.$$

(ii) For the collector sloped at $\beta = 90^\circ$,

$$\omega_s' = \min\{115.29^\circ, \cos^{-1}[-\tan(51.5 - 90)\tan(18.77)]\}$$
$$= \min\{115.29^\circ, 74.32^\circ\}$$
$$= 74.32^\circ.$$

(b) The angle of incidence for a surface tilted toward the equator is given by Eq. (1.6.1). At 11:00 (LAT), $\omega = 15^\circ$.

(i) For the collector sloped at $\beta = 30^\circ$,

$$\theta_0 = \cos^{-1}[\cos(21.5)\cos(18.77)\cos(15.0) + \sin(21.5)\sin(18.77)]$$
$$= 14.47^\circ.$$

(ii) For the collector sloped at $\beta = 90^\circ$,

$$\theta_0 = \cos^{-1}[\cos(-38.5)\cos(18.77)\cos(15.0) + \sin(-38.5)\sin(18.77)]$$
$$= 58.48^\circ.$$

□

B. *Arbitrarily Oriented Surfaces*

For a surface oriented in any direction with respect to the local meridian, the trigonometric relation for the incidence angle θ has been given by Benrod and Bock [8], Kondratyev [9], and in detail by Coffari [10]. This relation can be written in the following two forms:

$$\cos\theta = (\sin\phi\cos\beta - \cos\phi\sin\beta\cos\gamma)\sin\delta$$
$$+ (\cos\phi\cos\beta + \sin\phi\sin\beta\cos\gamma)\cos\delta\cos\omega$$
$$+ \cos\delta\sin\beta\sin\gamma\sin\omega \qquad (1.6.5a)$$

or

$$\cos\theta = \cos\beta\cos\theta_z + \sin\beta\sin\theta_z\cos(\psi - \gamma). \qquad (1.6.5b)$$

However, (1.6.5a) is preferred because it does not require a separate calculation of solar azimuth angle and solar zenith angle. For surfaces tilted toward the equator ($\gamma = 0$) it can be shown that the above equations reduce to Eq. (1.6.1).

It is also useful to write down the incidence angle equation for vertical walls, $\beta = 90°$;

$$\cos\theta = -\cos\phi\cos\gamma\sin\delta + \sin\phi\cos\gamma\cos\delta\cos\omega + \cos\delta\sin\gamma\sin\omega. \tag{1.6.6}$$

It is helpful (why it is so will be shown in Chapters 4 and 11) to determine the period during which the sun appears on an arbitrarily oriented surface. Let

ω_{sr} equal the sunrise hour angle for an arbitrarily oriented surface, in degrees, and

ω_{ss} equal the sunset hour angle for an arbitrarily oriented surface, in degrees.

Thus the period during which the sun is seen by the surface is

$$\omega_{sr} - \omega_{ss} \quad \text{in degrees.}$$

(Note: ω_{ss} in itself is negative.)

Naturally, the magnitudes of ω_{sr} and ω_{ss} are not identical. In addition, each one of these angles has to be evaluated separately for surfaces oriented toward the east and those oriented toward the west. Furthermore, it is necessary to watch for the two possible situations when the sunrise hour angle might be greater than the sunrise hour angle for horizontal surface, or the sunset hour angle greater than the corresponding angle for horizontal surfaces. ω_{sr} can be obtained numerically through iteration by setting $\theta = 90°$ in Eq. (1.6.5a). Here we shall develop[4] explicit expressions for each of the two surface orientations $\pm\gamma$. Keeping the symbol ω for the time being, we can write Eq. (1.6.5a) for sunrise or sunset as follows:

$$0 = A\sin\omega + B\cos\omega + C, \tag{1.6.7}$$

where

$$A = \cos\delta\sin\beta\sin\gamma, \tag{1.6.8}$$

$$B = \cos\delta\cos\phi\cos\beta + \cos\delta\sin\phi\sin\beta\cos\gamma, \tag{1.6.9}$$

$$C = \sin\delta\sin\phi\cos\beta - \sin\delta\cos\phi\sin\beta\cos\gamma. \tag{1.6.10}$$

To obtain an explicit expression for ω from Eq. (1.6.7), let us rewrite it as

[4] This development is courtesy of Cecilia Cameron.

follows:

$$\begin{aligned}0 &= A\sin[2(\omega/2)] + B\cos\omega + C\\ &= 2A\cos(\omega/2)\sin(\omega/2) + B\cos\omega + C\\ &= 2A\sqrt{(1+\cos\omega)/2}\sqrt{(1-\cos\omega)/2} + B\cos\omega + C.\end{aligned} \tag{1.6.11}$$

Rearranging and squaring both sides of the above, we obtain

$$(B^2 + A^2)\cos^2\omega + 2BC\cos\omega + (C^2 - A^2) = 0$$

or

$$\cos\omega = \frac{-BC \pm \sqrt{A^4 + A^2B^2 - A^2C^2}}{B^2 + A^2}. \tag{1.6.12}$$

It is possible to simplify the right-hand side of the above equation further. Divide through by A^2 and then let

$$x^2 = B^2/A^2, \qquad y^2 = C^2/A^2.$$

This results in

$$\cos\omega = \frac{-xy \pm \sqrt{x^2 - y^2 + 1}}{x^2 + 1}, \tag{1.6.13}$$

where

$$x = \frac{\cos\phi}{\sin\gamma\tan\beta} + \frac{\sin\phi}{\tan\gamma} \tag{1.6.14}$$

and

$$y = \tan\delta\left(\frac{\sin\phi}{\sin\gamma\tan\beta} - \frac{\cos\phi}{\tan\gamma}\right). \tag{1.6.15}$$

Keeping in mind that the sunrise or sunset hour angle for a tilted surface can never be greater than that for a horizontal surface, the two angles for surfaces oriented toward the east or west can be expressed as

$\gamma > 0$, surface oriented toward the east:

$$\omega_{sr} = \min\left[\omega_s, \cos^{-1}\left(\frac{-xy - \sqrt{x^2 - y^2 + 1}}{x^2 + 1}\right)\right], \tag{1.6.16}$$

$$\omega_{ss} = -\min\left[\omega_s, \cos^{-1}\left(\frac{-xy + \sqrt{x^2 - y^2 + 1}}{x^2 + 1}\right)\right]. \tag{1.6.17}$$

$\gamma < 0$, surface oriented toward the west:

$$\omega_{sr} = \min\left[\omega_s, \cos^{-1}\left(\frac{-xy + \sqrt{x^2 - y^2 + 1}}{x^2 + 1}\right)\right], \tag{1.6.18}$$

$$\omega_{ss} = -\min\left[\omega_s, \cos^{-1}\left(\frac{-xy - \sqrt{x^2 - y^2 + 1}}{x^2 + 1}\right)\right]. \tag{1.6.19}$$

The following example will illustrate the above points.

□ Example 1.6.2. Calculation of the sunrise and sunset hour angles on 22 November for (a) $\beta = 30°$, (b) $\beta = 90°$ for a flat-plate collector at Madrid (40°25′ N) oriented 30° east from its due south position.

Solution. $\phi = 40.42°$ N.

On 22 November, $\delta = -20.04°$ (from Table 1.3.1).

We calculate first the sunrise hour angle for a horizontal surface. From Eq. (1.5.4),

$$\omega_s = \cos^{-1}[-\tan(-20.04)\tan(40.42)] = 71.94°.$$

The collector is oriented 30° east of south. Therefore $\gamma = +30°$.

(a) The collector is inclined from the horizontal at an angle $\beta = 30°$. The sunrise hour angle is given by Eq. (1.6.16). In this equation

$$x = \frac{\cos(40.42)}{\sin(30)\tan(30)} + \frac{\sin(40.42)}{\tan(30)} = 3.7603$$

and

$$y = \tan(-20.04)\left(\frac{\sin(40.42)}{\sin(30)\tan(30)} - \frac{\cos(40.42)}{\tan(30)}\right)$$
$$= -0.3376.$$

The second term within the right-hand side of Eq. (1.6.16) is

$$\cos^{-1}\left(\frac{(3.7603)(0.3376) - \sqrt{(3.7603)^2 - (0.3376)^2 + 1}}{(3.7603)^2 + 1}\right) = 99.92°.$$

Therefore,

$$\omega_{sr} = \min(71.94°, 99.92°) = 71.94°.$$

The sunset hour angle is obtained from Eq. (1.6.17). We simply write down the result:

$$\omega_{ss} = -\min(71.94°, 70.13°) = -70.13°.$$

(b) The collector is inclined at an angle $\beta = 90°$. Repeating the foregoing, from Eq. (1.6.16) we have

$$\omega_{\text{sr}} = \min(71.94^\circ, 150.30^\circ) = 71.94^\circ,$$

and from Eq. (1.6.17), the sunset hour angle is

$$\omega_{\text{ss}} = -\min(71.94^\circ, 67.01^\circ) = -67.01^\circ. \quad \square$$

1.7 Further Reading

In this chapter a number of topics have been discussed. The material was gathered from different sources: calculation of the sun–earth distance, declination, and the equation of time are topics dating back almost to antiquity; Ref. [11] gives explanatory notes to the nautical almanac; Walraven [12] has presented a formulation for use in digital computers; Wilson and van Swaay [13, 14] have developed a series of ingenious charts for determining the solar azimuth and altitude angles; Hand [15] has presented sun path diagrams for obtaining the position of the sun for any minute of the day within the continental United States; Deris [16], Penrod [17], and Penrod and Prasanna [18] illustrate, with examples, the use of orthographic projections and spherical trigonometry to calculate angles of incidence for tilted surfaces.

Nomenclature

A, B, C	See Eqs. (1.6.7)–(1.6.10)
AU	Astronomical unit (1 AU = 1.496×10^8 km)
d_n	Day number of year, 1 on 1 January and 365 on 31 December; February is counted as having 28 days
E_0	Eccentricity correction factor of earth $(r_0/r)^2$
E_t	Equation of time (minutes)
L_e	Local longitude (degrees)
L_s	Standard longitude (degrees)
N_d	Day length (h)
r	Actual sun–earth distance (AU)
r_0	Mean sun–earth distance, 1 AU
α	Solar altitude, also called solar height, angular elevation of the sun above the true horizon (degrees)
β	Inclination of a surface from the horizontal position (degrees)
γ	Surface azimuth angle, that is, the deviation of the normal to the surface with respect to the local meridian, south zero, east positive (degrees)
δ	Declination, angular position of the sun at solar noon with respect to the plane of the equator, north positive (degrees)
θ	Angle of incidence for an arbitrarily oriented surface, the angle between normal to the surface and sun–earth vector (degrees)

θ_0	Angle of incidence for a surface sloped toward the equator (degrees)
θ_z	Zenith angle, also called zenith distance, the angular position of the sun with respect to the local vertical, $\theta_z = 90 - \alpha$ (degrees)
ϕ	Geographic latitude, north positive (degrees)
ψ	Solar azimuth, south zero, east positive (degrees)
ω	Hour angle, solar noon zero and morning positive (degrees); changes 15° every hour (e.g., $\omega = +15°$ at 11:00 and $\omega = -37.5°$ at 14:30)
ω_i	Hour angle at the middle of an hour (degrees)
ω_s	Sunrise hour angle for a horizontal surface (degrees)
ω_s'	Sunrise hour angle for a surface inclined toward the equator (degrees)
ω_{sr}	Sunrise hour angle for a surface oriented in any direction (degrees)
ω_{ss}	Sunset hour angle for a surface oriented in any direction (degrees)

References

1. J. W. Spencer, Fourier series representation of the position of the Sun. *Search* **2** (5), 172 (1971).
2. J. A. Duffie and W. A. Beckman, "Solar Engineering of Thermal Processes." Wiley, New York, 1980.
3. "American Ephemeris and Nautical Almanac." U.S. Government Printing Office, Washington, D.C., 1976.
4. Chr. Perrin de Brichambaut, "Cahiers A.F.E.D.E.S.," supplément au no 1. Editions Européennes Thermique et Industrie, Paris, 1975.
5. P. I. Cooper, The absorption of solar radiation in solar stills. *Sol. Energy* **12** (3), 333–346 (1969).
6. A. Whillier, Solar radiation graphs. *Sol. Energy* **9** (3), 164–165 (1965).
7. B. Y. H. Liu and R. C. Jordan, Daily insolation on surfaces tilted toward the equator, *ASHRAE J.* **3**(10) 53–59 (1961).
8. F. Benrod and J. E. Bock, A time analysis of sunshine. *Trans. Am. Illum. Eng. Soc.* **34,** 200–218 (1934).
9. K. Y. Kondratyev, "Radiation in the Atmosphere." Academic Press, New York, 1969.
10. E. Coffari, The sun and the celestial vault, "Solar Energy Engineering" (A. A. M. Sayigh, ed.), Chapter 2. Academic Press, New York, 1977.
11. Explanatory Supplement to the "Astronomical Ephemeris and the American Ephemeris and Nautical Almanac." U.S. Government Printing Office, Washington, D.C., 1961.
12. R. Walraven, Calculating the position of the sun. *Sol. Energy* **20** (5), 393–397 (1978).
13. M. J. Wilson and J. M. van Swaay, Where is the sun? *Heat. Vent.* **39,** 41–46 (1942).
14. M. J. Wilson and J. M. van Swaay, Where is the sun? *Heat. Vent.* **39,** 59–61 (1942).
15. I. F. Hand, Charts to obtain solar altitudes and azimuths. *Build. Syst. Design* **45** (10), 86–88 (1948).
16. N. Deris, Optimum angles of inclination, *Air Cond. Heat. Vent.* **58** (8), 57–60, 101 (1961).
17. E. B. Penrod, Solar load analysis by use of orthographic projections and spherical trigonometry, *Sol. Energy* **8** (4), 127–133 (1964).
18. E. B. Penrod and K. V. Prasanna, Design of a flat-plate collector for a solar earth pump. *Sol. Energy* **6** (1), 9–22 (1962).

Chapter 2

THERMAL RADIATION

2.1 Introduction

Matter can emit electromagnetic radiation by molecular and atomic agitation. The spectrum of electromagnetic radiation encompasses γ rays, x rays, ultraviolet radiation, light, heat, radio waves, and radar waves. Figure 2.1.1 shows the electromagnetic spectrum. In this text we are interested mainly in the thermal radiation portion of this spectrum. Thermal radiation, emitted by the agitation associated with the temperature of matter, is commonly called heat and light. Human eyes are sensitive detectors of light, but our bodies are rather poor detectors of heat. As will be shown later, most of the solar radiation reaching the surface of the earth is in the thermal radiation range.

Electromagnetic radiation is generally classified by wavelength, though frequency and wave number are also used. Frequency has the advantage over wavelength of not changing when radiation passes from one medium to another.

Frequency ν is defined as

$$\nu = c_0/\lambda_{\text{vac}} = c/\lambda_{\text{med}}, \tag{2.1.1}$$

where c_0 and c are the speeds of propagation for electromagnetic radiation in a vacuum and in a medium, respectively; λ is wavelength. In a vacuum, the speed of propagation is $c_0 = 2.998 \times 10^8$ m/sec. The speed c in a medium is given in terms of the index of refraction n, where

$$n = c_0/c. \tag{2.1.2}$$

For gases, n is very close to but greater than unity. The wave number η is defined as follows:

$$\eta = 1/\lambda = \text{number of waves per unit length.} \tag{2.1.3}$$

Wave number is sometimes used instead of wavelength in literature dealing with absorption of electromagnetic radiation by gases.

Common wavelength units are

micrometer (commonly called micron, μm); 1 μm $= 10^{-6}$ m $= 10^{-4}$ cm

millimeter: 1 mm $= 10^{-3}$ m

nanometer (also called millimicron, mμ): 1 nm $= 10^{-9}$ m

angstrom: 1 Å $= 10^{-10}$ m $= 10^{-4}$ μm

Thermal radiation encompasses the wavelength range 0.2–1000 μm. The visible spectrum may be considered within 0.39–0.77 μm, and the spectral division [1] of different colors (based on average judgment) as follows:

violet	0.390–0.455 μm
blue	0.455–0.492 μm
green	0.492–0.577 μm
yellow	0.577–0.597 μm
orange	0.597–0.622 μm
red	0.622–0.770 μm

The ultraviolet (uv) spectrum may be divided into three bands:

near uv	0.3–0.4 μm
far uv	0.2–0.3 μm
maximum uv	0.001–0.2 μm

The infrared (ir) portion of thermal radiation is generally divided into two parts: near ir, 0.77–25 μm, and far ir, 25–1000 μm.

Another subdivision of thermal radiation is short wave and long wave. The dividing line between the two is somewhat arbitrary: in solar energy terminology, the major portion of solar radiation is considered to be within the shortwave, the limit put variably from 3 to 4 μm, depending on one's own judgment. Therefore, radiation at wavelengths greater than 4 μm may be termed long-wave radiation. The long-wave radiation emitted by the earth and its atmosphere is often called terrestrial radiation.

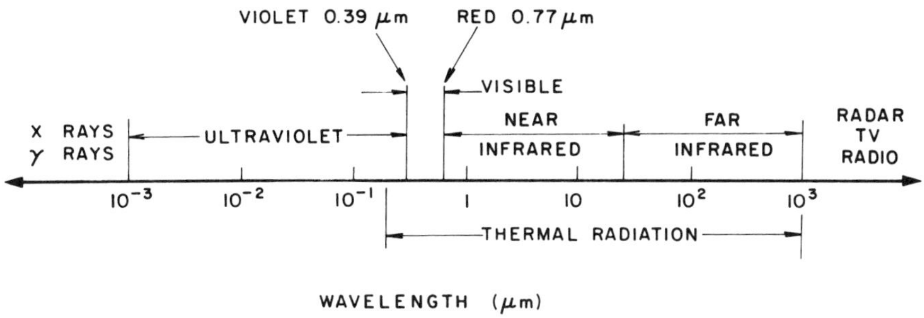

Figure 2.1.1 Spectrum of electromagnetic radiation.

2.2 Blackbody Radiation

A body (or a surface) emits energy at all wavelengths of the electromagnetic spectrum. At a given temperature, a blackbody is one that emits the maximum amount of energy at each wavelength and in all directions and also absorbs all incidental radiation at each wavelength and in all directions. A blackbody is an ideal surface with which the performance of real surfaces can be compared. In the next chapter, the radiation from the sun will be compared with that from a blackbody at an equivalent temperature. Therefore, it is useful to note down some fundamentals of blackbody emission.

A. *Planck's Law*

The emissive power of a blackbody at any wavelength and temperature, called its spectral[1] emissive power, is given by Planck's law. For radiation into a vacuum or a medium of refractive index of unity, this law in a simple form can be written

$$e_{b\lambda} = \frac{C_1}{\lambda^5[\exp(C_2/\lambda T) - 1]}, \tag{2.2.1}$$

where

$e_{b\lambda}$ is the hemispherical spectral emissive power of a blackbody (W m^{-2} μm^{-1}). (The word "hemispherical" indicates that power is streaming out radially in all directions above a surface. Henceforth, the prefix "hemispherical" shall be omitted in this text.)

[1] Throughout this book, the terms *spectral* and *monochromatic* will be employed synonymously and interchangeably. Both will represent quantities at a single wavelength.

C_1 is the first radiation constant (3.7427×10^8 W μm^4 m^{-2});
C_2 is the second radiation constant (1.4388×10^4 μm K);
λ is the wavelength (μm); and
T is the blackbody temperature (K).

A plot of Eq. (2.2.1) is shown in Fig. 2.2.1. Three observations may be made from this diagram. As temperature increases,

(i) the emissive power increases at each wavelength;
(ii) relatively more energy is emitted at shorter wavelengths; and
(iii) the position of maximum emissive power shifts toward shorter wavelengths.

As will be discussed later, the sun's equivalent surface temperature is 5777 K. Figure 2.2.1 shows that if the sun's surface were a blackbody, most of its energy would be in the short-wavelength range. Some flat-plate collectors are designed to operate at $\approx 100°C$. From Fig. 2.2.1 it can be concluded that the energy emitted by such collectors is mostly in the long-wave region.

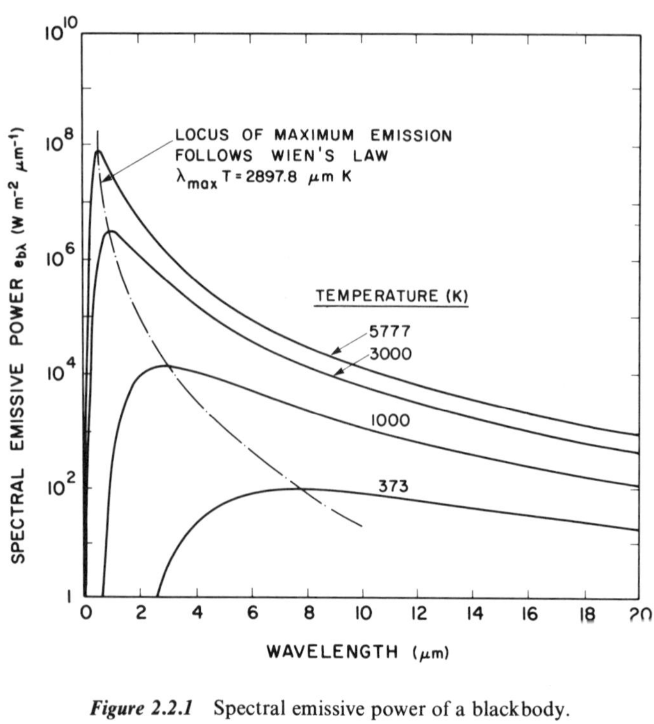

Figure 2.2.1 Spectral emissive power of a blackbody.

B. *Stefan–Boltzmann Law*

The emissive power of a blackbody within bandwidth $d\lambda$ is written $e_{b\lambda}\, d\lambda$. The radiation leaving a surface of unit area at all wavelengths is called the total emissive power e_b, where

$$e_b = \int_{\lambda=0}^{\lambda=\infty} e_{b\lambda}\, d\lambda = \int_{\lambda=0}^{\lambda=\infty} \frac{C_1}{\lambda^5[\exp(C_2/\lambda T) - 1]}\, d\lambda. \qquad (2.2.2)$$

Equation (2.2.2), when integrated, yields

$$\begin{aligned} e_b &= (C_1\pi^4/15C_2^4)T^4 \\ &= \sigma T^4, \end{aligned} \qquad (2.2.3)$$

where

$$\begin{aligned} \sigma &= \text{Stefan–Boltzmann constant} \\ &= 5.6697 \times 10^{-8} \quad \text{Wm}^{-2}\,\text{K}^{-4}. \end{aligned}$$

The above is the theoretical value. The measured value of this constant is 5.6866×10^{-8} W m^{-2} K^{-4} [2].

C. *Wien's Displacement Law*

Planck's law, Eq. (2.2.1), can be put into a more universal form. Dividing by T^5, we obtain the following:

$$\frac{e_{b\lambda}}{T^5} = \frac{C_1}{(\lambda T)^5[\exp(C_2/\lambda T) - 1]}. \qquad (2.2.4)$$

This equation expresses $e_{b\lambda}/T^5$ in terms of a single variable λT. A plot of this relation is given in Fig. 2.2.2, and replaces such diagrams as Fig. 2.2.1 containing multiple temperature curves. The locus of maximum λT called $\lambda_{max}T$ is 2897.8 μm K. The relation

$$\lambda_{max}T = 2897.8\ \mu\text{m K} \qquad (2.2.5)$$

is called Wien's displacement law. Maximum emissive power per unit wavelength is at

$$\lambda_{max} = 2897.8/T\ \mu\text{m}.$$

Assuming that the sun is a blackbody at 5777 K,

$$\lambda_{max} = 2897.8/5777 = 0.5016\ \mu\text{m},$$

which is in the green region. In the next chapter it will be shown that from the actual extraterrestrial spectral distribution of solar irradiation, λ_{max} is not exactly 0.5016 μm. For radiation from a flat-plate collector at 373 K,

$$\lambda_{max} = 2897.8/373 \approx 8\ \mu\text{m},$$

which is near infrared.

☐ EXAMPLE 2.2.1. Determination of the surface temperature of a blackbody radiating with a total emissive power of 7.250×10^4 W m^{-2} and of the wavelength of maximum spectral emissive power.

Solution. The blackbody temperature is $(e_b/\sigma)^{1/4}$, where

$$e_b = 7.250 \times 10^4\ \text{W m}^{-2},$$

$$\sigma = 5.6697 \times 10^{-8}\ \text{W m}^{-2}\ \text{K}^{-4},$$

$$T = (7.250/5.6697)^{1/4} \times 10^3 = 1063.4\ \text{K}.$$

The wavelength at which the maximum spectral emissive power occurs is

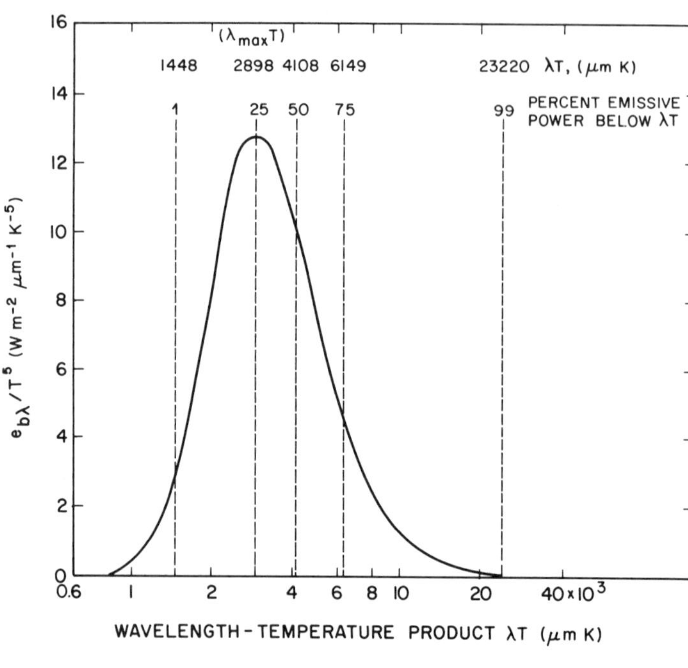

Figure 2.2.2 Spectral distribution of blackbody emissive power. Adapted from "Thermal Radiation Heat Transfer" by R. Siegel and J. R. Howell. Copyright © 1981 McGraw-Hill Book Company. Used with the permission of McGraw-Hill Book Company.

obtained from Wien's displacement law, Eq. (2.2.5):

$$\lambda_{max} = 2897.8/T = 2.73\ \mu m. \qquad \square$$

□ EXAMPLE 2.2.2. Plotting the spectral irradiance on a surface normal to the rays placed at the mean sun–earth distance, assuming that the sun is a blackbody at an effective temperature of 5777 K.

Solution. The mean sun–earth distance $r_0 = 149\,597\,890$ km.
The mean radius of the solar disk $r_s = 6.9598 \times 10^5$ km.
The spectral emissive power of the sun at the blackbody temperature of 5777 K is given by Planck's equation, Eq. (2.2.1):

$$e_{b\lambda} = \frac{C_1}{\lambda^5[\exp(C_2/\lambda 5777) - 1]} \quad \text{W m}^{-2}\ \mu\text{m}^{-1},$$

where

$$C_1 = 3.7427 \times 10^8\ \text{W}\ \mu\text{m}^4\ \text{m}^{-2},$$

$$C_2 = 1.4388 \times 10^4\ \mu\text{m K}.$$

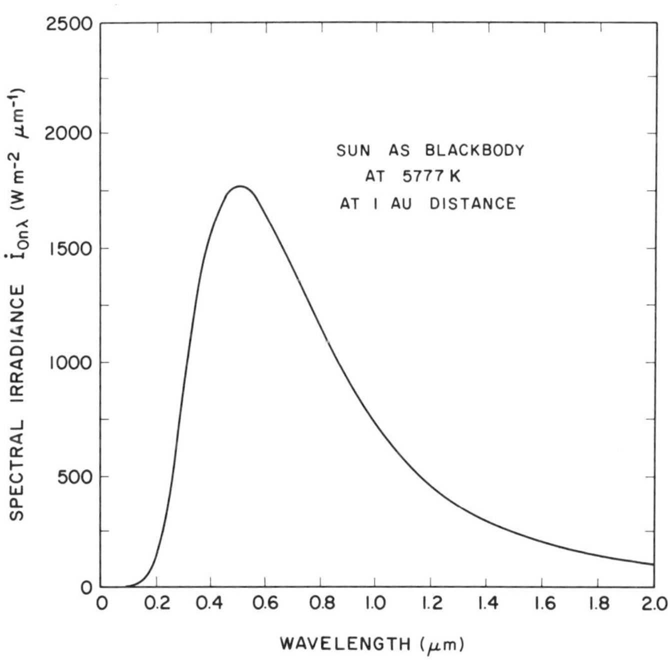

Figure 2.2.3 Spectral irradiance from the sun as a blackbody.

The spectral irradiance $\dot{I}_{0n\lambda}$ on a surface normal to the rays at the mean sun–earth distance is given by

$$\dot{I}_{0n\lambda} = e_{b\lambda}(r_s/r_0)^2.$$

This variation is plotted in Fig. 2.2.3. In the next chapter we compare this diagram with the measured values of the spectral irradiance. □

2.3 Blackbody Radiation in a Wavelength Interval

The fraction of the total emissive power emitted in a given wavelength band is written

$$F_{\lambda_1-\lambda_2} = \frac{1}{\sigma T^4}\int_{\lambda_1}^{\lambda_2} e_{b\lambda}\, d\lambda \tag{2.3.1}$$

The integral term in the above equation can be represented by the shaded area in Fig. 2.3.1; the denominator is the area under the entire curve. This equation can be divided into two parts:

$$\begin{aligned} F_{\lambda_1-\lambda_2} &= \frac{1}{\sigma T^4}\left(\int_0^{\lambda_2} e_{b\lambda}\, d\lambda - \int_0^{\lambda_1} e_{b\lambda}\, d\lambda\right) \\ &= F_{0-\lambda_2} - F_{0-\lambda_1}. \end{aligned} \tag{2.3.2}$$

As $e_{b\lambda}/T^5$ is a function of only λT, it can be shown that

$$\begin{aligned} F_{\lambda_1-\lambda_2} &= F_{\lambda_1 T-\lambda_2 T} \\ &= \frac{1}{\sigma}\left(\int_0^{\lambda_2 T} \frac{e_{b\lambda}}{T^5}\, d(\lambda T) - \int_0^{\lambda_1 T} \frac{e_{b\lambda}}{T^5}\, d(\lambda T)\right) \\ &= F_{0-\lambda_2 T} - F_{0-\lambda_1 T}. \end{aligned} \tag{2.3.3}$$

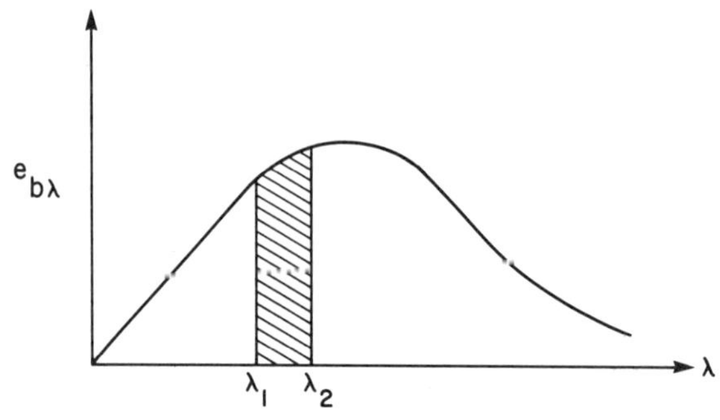

Figure 2.3.1 Representation of energy within a bandwidth λ_1 to λ_2.

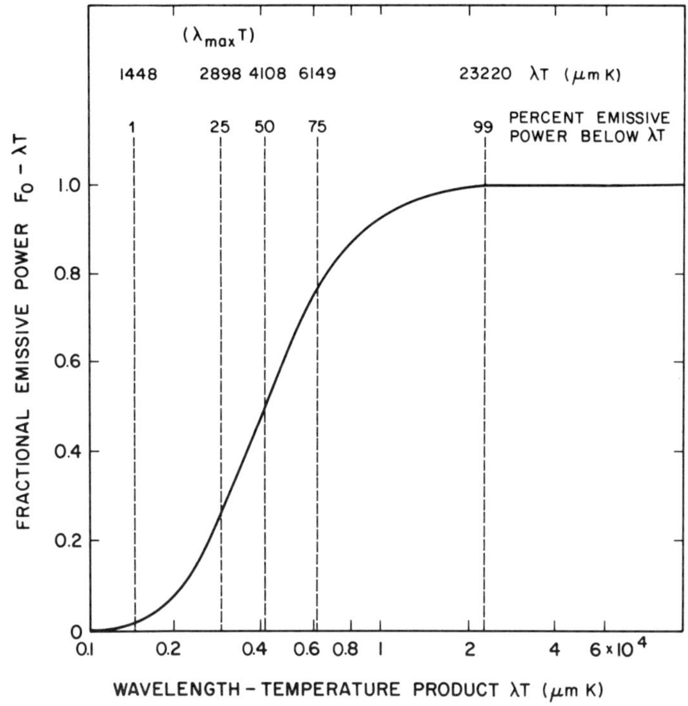

Figure 2.3.2 Fractional blackbody emissive power. Adapted from "Thermal Radiation Heat Transfer" by R. Siegel and J. R. Howell. Copyright © 1981 McGraw-Hill Book Company. Used with the permission of McGraw-Hill Book Company.

Tables of $F_{0-\lambda T}$ called "blackbody functions" are available in a number of texts treating thermal radiation heat transfer [3, 4]; a plot of the fractional emissive power $F_{0-\lambda T}$ is given in Fig. 2.3.2. Fifty percent of the energy lies on either side of $\lambda T = 4108$ μm K. In the next chapter we shall compare the blackbody fractional emissive power with that from the sun.

2.4 Angular Dependence of Radiation

In many practical applications, the rate of energy propagation in a given direction is required. This is described in terms of the *intensity of radiation.* In order to discuss intensity, it is necessary to explain the concept of solid angle. The solid angle is defined as the ratio of the area dS of a spherical surface to the square of its radius R, as shown in Fig. 2.4.1. It can be written as follows:

$$d\omega = dS/R^2. \tag{2.4.1}$$

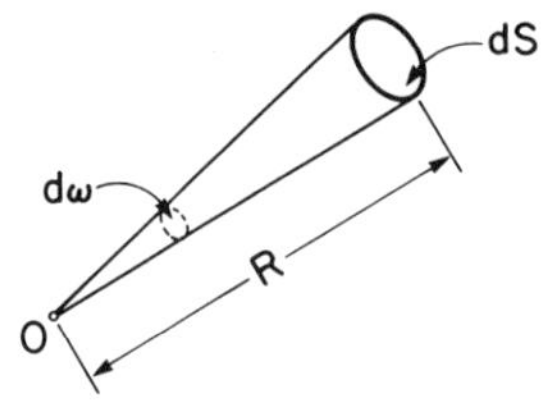

Figure 2.4.1 Description of solid angle.

Units of the solid angle are expressed in terms of the steradian (sr). From Eq. (2.4.1) it can be seen that the total solid angle subtended by a sphere is 4π sr.

By definition, the intensity of radiation of a surface is the rate of energy propagation *in a given direction per unit solid angle and per unit area perpendicular to the axis of the solid angle.* This definition applies to

(a) black as well as real surfaces;

(b) energy leaving a surface, whether emitted or both emitted and reflected;

(c) energy incident on a surface, coming from the sky or from any other source; and

(d) spectral as well as total energy.

Let us consider an elemental area dA radiating an amount of energy dF within a solid angle $d\omega$ (see Fig. 2.4.2). From the foregoing definition, the intensity of radiation, i, at an angle Φ from the surface normal is given by the following:

$$i = \frac{dF}{d\omega\, dA \cos \Phi}. \tag{2.4.2}$$

In many problems, a relation between F and i is required. In order to develop such a relation, let us assume the elemental area dA is situated at the base of a hemisphere of radius R (see Fig. 2.4.3). Let dF be the amount of radiation passing through an area dS on the surface of the hemisphere. From Eq. (2.4.2) we have

$$\frac{\mathrm{dF}}{dA} = i\, d\omega \cos \Phi. \tag{2.4.3}$$

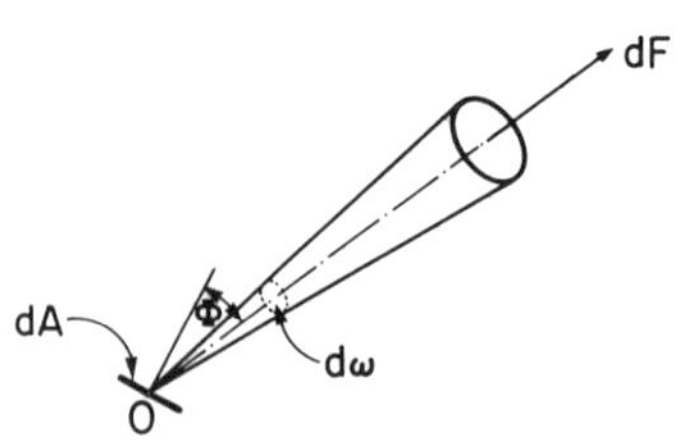

Figure 2.4.2 Radiation intensity.

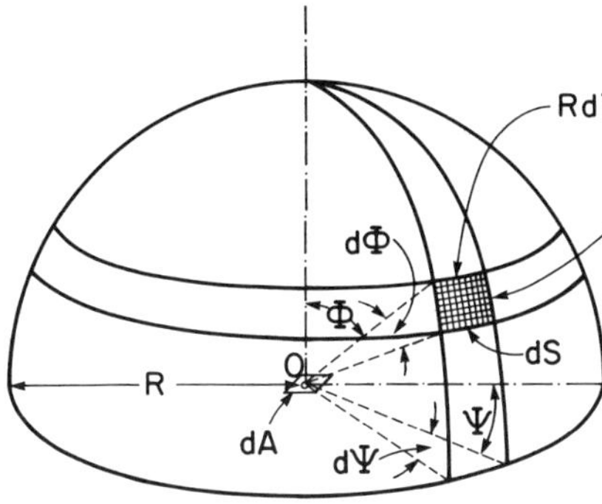

Figure 2.4.3 Radiation intensity over a hemisphere.

In order to evaluate the solid angle, we express the area dS in terms of its coordinates as follows:

$$dS = (R\,d\Phi)(R\,d\Psi\sin\Phi). \tag{2.4.4}$$

Combining Eqs. (2.4.1) and (2.4.4), we have

$$d\omega = \sin\Phi\,d\Phi\,d\Psi. \tag{2.4.5}$$

Equation (2.4.3) now reduces to

$$\frac{dF}{dA} = i\sin\Phi\cos\Phi\,d\Phi\,d\Psi. \tag{2.4.6}$$

The total rate of energy F passing through the entire hemispherical surface can now be written

$$\frac{F}{dA} = \int_0^{2\pi}\int_0^{\pi/2} i\sin\Phi\cos\Phi\,d\Phi\,d\Psi. \tag{2.4.7}$$

In most instances, the variation of i with Φ and Ψ is not known and the above integral cannot be evaluated. However, by definition, the intensity of radiation emitted by a blackbody, i_b, is independent of direction. In other words, a blackbody is perfectly *diffuse*, or *isotropic*. Consequently, for a blackbody, Eq. (2.4.7) reduces to

$$e_b = F/dA = \pi i_b. \tag{2.4.8}$$

Similarly, the energy emitted from any perfectly diffuse surface can also be written

$$e = \pi i. \tag{2.4.9}$$

2.5 Real-Body Properties

The term blackbody is used to describe an ideal surface or material that follows the laws of Planck, Stefan–Boltzmann, and Wien. An additional property of the blackbody is its absorption quality. By definition, a blackbody absorbs all radiation, at all wavelengths, incident upon it from any direction. This property leads to a further conclusion, that a blackbody neither reflects nor transmits any incident energy (or flux).

The blackbody concept serves as a useful standard for comparing the radiative properties of real surfaces with an ideal surface. A real surface will partly reflect and partly absorb incident radiation. Furthermore, if such a surface is not perfectly opaque, it will partly transmit the incident wave. Consider a unit quantity of monochromatic radiation incident on a real surface. We can write

$$1 = \alpha_\lambda + \rho_\lambda + \tau_\lambda, \tag{2.5.1}$$

where

α_λ is the monochromatic absorptance. It is the ratio of energy absorbed to incident energy.

ρ_λ is the monochromatic reflectance. It is the ratio of energy reflected to incident energy.

τ_λ is the monochromatic transmittance. It is the ratio of energy transmitted to incident energy.

The properties described apply to a substance as well as to the surface of a substance, whether the substance be a solid or a fluid.

In Chapters 6–9, these properties have been related to the earth's atmosphere, clouds, ground, and ground cover.

From Eq. (2.5.1), it follows directly that α, ρ, and τ are the spectrally integrated quantities of their corresponding monochromatic values. They are usually called total absorptance, total reflectance, and total transmittance, respectively. In general, there can be substantial differences between the total and its monochromatic counterpart. An imaginary surface whose monochromatic property is less than one, and which does not vary with wavelength, is called a *gray* body. Data on the monochromatic properties are generally not available. Consequently, the gray-body assumption is often utilized. For instance, the gray-body assumption is almost invariably made as far as reflectance of the ground and the ground cover is concerned.

Chapter 9 is devoted to the reflectance properties of the ground and the ground cover. When the incident radiation is from the sun, this reflectance is usually called *albedo*. In the meteorological literature, reflectance of the earth's atmosphere, as by clouds and air, is also called albedo. These terms are treated in detail later in this book.

2.6 Definitions

In this chapter, attention has been restricted to radiant energy emanating from a blackbody. However, the main aim of this text is to evaluate solar radiation arriving at a surface on the earth. In the literature, solar radiation arriving on a surface is variably termed *irradiation*, *insolation*,[2] *radiation*, *irradiance*, *radiance*, *intensity*, *radiant flux*, *radiant flux density*, etc. It is necessary to explain the meaning of these terms as used in the text. The main objective is to emphasize the distinction between terms expressing quantity of energy and those denoting rate of energy. Irradiation and insolation will be used interchangeably, both referring to quantity of solar energy arriving at a surface during a given period of time. Literally, irradiation simply refers to radiation arriving at a surface, whether or not the origin of radiation is the sun. Irradiance will indicate the rate of solar energy arriving at a surface per unit time and per unit area. Irradiance is the same as radiant flux density or flux. Since irradiance will mean the rate of incident energy, its units will be $W\ m^{-2}$, and the units of irradiation will be $kJ\ m^{-2}\ h^{-1}$ or $MJ\ m^{-2}\ day^{-1}$, etc. *Radiation* will be employed in a generic sense, and its meaning should be obvious from the context. *Intensity* will mean irradiance from a particular direction and contained within a unit solid angle. As defined earlier, intensity is expressed in $W\ m^{-2}\ sr^{-1}$ on an area normal to the direction of radiation. It is pertinent to point out that the term "intensity" is often loosely employed. For example, in meteorology, *intensity* is used for radiative flux as well as for quantity of radiation arriving from all over the sky dome.

2.7 Further Reading

The material in this chapter can be supplemented by Siegel and Howell [3, Chapters 1–3], and Sparrow and Cess [4, Chapter 1].

[2] This perfectly legitimate word seems to cause some confusion. *Insolation* derives from the Latin *insolatio*. Written records indicate that this word was introduced into the English language as far back as 1654. Insolation means the act of exposing or condition of being exposed to the rays of the sun, or the period of sunshine. In meteorology and engineering, it means the quantity of sunshine reaching a surface. Note, however, that in the often employed expression *solar insolation*, the word *solar* is redundant.

Nomenclature

Å	Angstrom (10^{-10} m)
c	Speed of electromagnetic radiation through a medium (m sec^{-1})
c_0	Speed of electromagnetic radiation through a vacuum (2.998×10^8 m sec^{-1})
C_1	First radiation constant (3.7427×10^8 W μm^4 m^{-2})
C_2	Second radiation constant (1.4388×10^4 μm K)
e_b	Total emissive power of a blackbody (W m^{-2})
$F_{0-\lambda T}$	Fractional emissive power of a blackbody from 0 to λT (dimensionless)
i	Intensity of radiation (W m^{-2} sr^{-1})
i_b	Intensity of radiation of a blackbody (W m^{-2} sr^{-1})
n	Refractive index (dimensionless)
T	Temperature (K)
α	Absorptance, the ratio of energy absorbed to the incident energy
η	$1/\lambda$, wave number, number of waves per unit length (μm^{-1})
λ	Wavelength (μm); as a subscript λ indicates monochromaticity
λ_{max}	Wavelength at which maximum emissive power occurs at a given temperature (μm)
ν	Frequency (sec^{-1} when λ is in m)
ρ	Reflectance, the ratio of energy reflected to the incident energy
σ	Stefan–Boltzmann constant (5.6697×10^{-8} W m^{-2} K^{-4}, theoretical value; 5.6866×10^{-8} W m^{-2} K^{-4}, experimental value)
τ	Transmittance, the ratio of energy transmitted to the incident energy
ω	Solid angle (sr)

References

1. "Van Nostrand's Scientific Encyclopedia," p. 615, 5th ed. Van Nostrand, Nostrand, New York, 1976.
2. J. M. Kendall and C. M. Berdahl, Two blackbody radiometers of high accuracy. *Appl. Opt.* **9** (5), 1082–1091 (1970).
3. R. Siegel and J. R. Howell, "Thermal Radiation Heat Transfer." McGraw-Hill, New York, 1981.
4. E. M. Sparrow and R. D. Cess, "Radiation Heat Transfer." McGraw-Hill, New York, 1978.

Chapter 3

THE SOLAR CONSTANT AND ITS SPECTRAL DISTRIBUTION

3.1 Introduction

This chapter discusses the physics of the sun, the nature of the radiant energy emanating from its surface, spectral distribution, and the total quantity of this energy arriving just outside the earth's atmosphere.

3.2 The Sun

The sun is the star closest to the earth, and its radiant energy is practically the only source of energy that influences atmospheric motions and our climate. The sun is a completely gaseous body composed mainly of hydrogen. Its physical structure is complex and may be considered to be composed of several regions: the core, the interior, the convecting zone, the photosphere, the reversing layer, the chromosphere, and the corona.

The innermost region, the core, is the sun's hottest and densest part. The core temperature is from 15×10^6 to 40×10^6 K and is composed of highly compressed gases at a density of 100–150 g/cm^3. Above the core is the interior, which contains practically all of the sun's mass: the core and the interior are thought of as a huge nuclear reactor and the source of almost all its energy. This energy is propagated mainly by radiation to the outer regions, which in turn transport this energy outward by convection and reradiation. The sun's "surface," called the photosphere, is the source of most visible radiation arriving at the earth's surface. This is the "crust," which can be seen through a blue glass by the naked eye. It is composed of very low-density inhomogeneous gases that form granulations and sunspots. The temperature in this region is 4000–6000 K.

Next to the photosphere is the reversing layer, which extends for a few hundred kilometers: this layer contains vapors of almost all the familiar elements of the earth's crust. Above it, extending over a distance of about 2500 km, is the chromosphere, which, with the reversing layer, forms the sun's atmosphere. Composed mainly of hydrogen and helium, it is visible to the naked eye during an eclipse. The temperature of the chromosphere is several times higher than that of the photosphere.

The outermost portion of the sun is called the corona and is composed of extremely rarefied gases called the solar winds, which are thought to extend into the solar system. The coronal gases are considered to be at temperatures several times those of the photosphere.

As can be seen from the preceding description, the sun does not have a fixed size. For the purpose of a number of calculations, however, its diameter is considered to be 1 391 960 km.

3.3 Extraterrestrial Solar Spectral Irradiance

The spectral distribution of radiation arriving outside the earth's atmosphere is very important in such applications as the photovoltaic power systems of satellites. A knowledge of this distribution provides a design input for the better thermal environment of a spacecraft and for the selection of suitable materials exposed to solar radiation. The spectral distribution of radiation arriving on the surface of the earth is indeed a function of its extraterrestrial distribution and the atmospheric constituents. This terrestrial distribution is also important in a number of applications, for example in earth-based photovoltaic systems, photosynthesis, the physics of the upper atmosphere, the albedo of the earth, and photochemical processes. As mentioned before, the sun has a highly variable temperature, which changes from region to region, and there is a continuous emission and absorption of radiation from one region to the other. Sunspot cycles, which have an average life of about 11 years, and other less predictable solar activities, such as faculae, prominences, and solar flares, produce spatial as well as temporal variations in the spectral distribution of radiant energy leaving the sun. Furthermore, there is the pronounced effect of the rotation of the sun, which takes about 27 earth days at its equator: irradiance decreases as a large number of sunspots pass across the point of observation. Therefore net spectral and total irradiance is a composite effect of all the elements mentioned earlier.

Ground-based measurements of solar spectral irradiance began around the beginning of this century, and from these measurements the extraterrestrial distribution was obtained by extrapolation. Such studies have been reported by Moon [1], Johnson [2], Allen [3], and Dunkelman and Scolnik [4], among others. Some measurements have also been made using rockets [5].

During 1968–1971, The National Aeronautics and Space Administration (NASA) carried out a series of measurements using the NASA 711 Galileo research aircraft [6–9], commercially known as the Convair 990A. A number of different instruments [10] (listed in Table 3.3.1) were used to cover the spectral range 0.3–15 μm. The spectral irradiance data obtained from these measurements did not yield identical results. Values were slightly modified in the visible and the near-infrared bands [11, 12]. The final results are available in a number of publications such as Refs. [10–12]. These data were accepted by the American Society for Testing and Materials (ASTM) [13] and are often referred to as the NASA/ASTM standard extraterrestrial solar

Table 3.3.1

Spectral Irradiance Instruments on Board a Research Aircraft, Used for Obtaining the Curve of Solar Spectral Irradiance[a]

Instrument	Energy detector	Type of instrument	Aircraft window material	Wavelength range (μm)
Perkin–Elmer	1P28 tube,	LiF prism	Sapphire	0.3–0.7
monochromator	thermocouple	LiF prism	Sapphire	0.7–4
Leiss	EMI 9558 QA,	Quartz double	Dynasil	0.3–0.7
monochromator	PbS tube	prism	Dynasil	0.7–1.6
Filter radiometer	Phototube	Dielectric thin films	Dynasil	0.3–1.2
P-4 interferometer	1P28 or R136,	Soleil prism	Infrasil	0.3–0.7
	PbS tube	Soleil prism	Infrasil	0.7–2.5
I-4 interferometer	Thermistor bolometer	Michelson mirror	Irtran 4	2.6–15

[a] Adapted with permission from Thekaekara [10], Copyright (1973) Pergamon Press, Ltd.

Table 3.3.2

Extraterrestrial Solar Spectral Irradiance at Mean Sun–Earth Distance (WRC spectrum)[a,b]

λ	$\dot{I}_{0n\lambda}$	$\sum_0^\lambda \dot{I}_{0n\lambda}$	$P_{0-\lambda}$	λ	$\dot{I}_{0n\lambda}$	$\sum_0^\lambda \dot{I}_{0n\lambda}$	$P_{0-\lambda}$
0.250	64.56	2.51	0.18	0.475	2016.25	247.45	18.10
0.255	91.25	2.84	0.21	0.480	2055.00	257.62	18.85
0.260	122.50	3.47	0.25	0.485	1901.26	267.64	19.58
0.265	253.75	4.29	0.31	0.490	1920.00	276.98	20.26
0.270	275.00	5.70	0.42	0.495	1965.00	286.59	20.97
0.275	212.50	6.87	0.50	0.500	1862.52	296.37	21.68
0.280	162.50	7.86	0.57	0.505	1943.75	305.80	22.37
0.285	286.25	9.12	0.67	0.510	1952.50	315.50	23.08
0.290	535.00	10.97	0.80	0.515	1835.01	325.05	23.78
0.295	560.00	13.91	1.02	0.520	1802.49	333.79	24.42
0.300	527.50	16.54	1.21	0.525	1894.99	343.33	25.12
0.305	557.50	19.26	1.41	0.530	1947.49	352.67	25.80
0.310	602.51	22.13	1.62	0.535	1926.24	362.34	26.51
0.315	705.00	25.51	1.87	0.540	1857.50	371.87	27.20
0.320	747.50	29.09	2.13	0.545	1895.01	381.22	27.89
0.325	782.50	32.70	2.39	0.550	1902.50	390.72	28.58
0.330	997.50	37.51	2.74	0.555	1885.00	400.17	29.27
0.335	906.25	42.34	3.10	0.560	1840.02	409.42	29.95
0.340	960.00	46.79	3.42	0.565	1850.00	418.71	30.63
0.345	877.50	51.45	3.76	0.570	1817.50	427.94	31.31
0.350	955.00	55.89	4.09	0.575	1848.76	437.11	31.98
0.355	1044.99	61.08	4.47	0.580	1840.00	446.22	32.64
0.360	940.00	65.72	4.81	0.585	1817.50	455.44	33.32
0.365	1125.01	71.01	5.20	0.590	1742.49	464.21	33.96
0.370	1165.00	76.92	5.63	0.595	1785.00	473.16	34.61
0.375	1081.25	82.15	6.01	0.600	1720.00	481.98	35.26
0.380	1210.00	88.32	6.46	0.605	1751.25	490.71	35.90
0.385	931.25	93.11	6.81	0.610	1715.00	499.35	36.53
0.390	1200.00	98.30	7.19	0.620	1715.00	516.51	37.79
0.395	1033.74	103.61	7.58	0.630	1637.50	533.22	39.01
0.400	1702.49	109.81	8.03	0.640	1622.50	549.73	40.22
0.405	1643.75	118.40	8.66	0.650	1597.50	565.79	41.39
0.410	1710.00	126.68	9.27	0.660	1555.00	581.10	42.51
0.415	1747.50	135.37	9.90	0.670	1505.00	596.65	43.65
0.420	1747.50	143.98	10.53	0.680	1472.50	611.50	44.73
0.425	1692.51	152.69	11.17	0.690	1415.02	625.86	45.78
0.430	1492.50	160.74	11.76	0.700	1427.50	640.28	46.84
0.435	1761.25	168.74	12.34	0.710	1402.50	654.28	47.86
0.440	1755.02	177.59	12.99	0.720	1355.00	668.10	48.87
0.445	1922.49	187.02	13.68	0.730	1355.00	681.84	49.88
0.450	2099.99	196.86	14.40	0.740	1300.00	695.28	50.86
0.455	2017.51	207.15	15.15	0.750	1272.52	708.17	51.81
0.460	2032.49	217.29	15.90	0.760	1222.50	720.62	52.72
0.465	2000.00	227.59	16.65	0.770	1187.50	732.70	53.60
0.470	1979.99	237.50	17.37	0.780	1195.00	744.52	54.47

Table 3.3.2 *(Continued)*

λ	$\dot{I}_{0n\lambda}$	$\sum_0^\lambda \dot{I}_{0n\lambda}$	$P_{0-\lambda}$	λ	$\dot{I}_{0n\lambda}$	$\sum_0^\lambda \dot{I}_{0n\lambda}$	$P_{0-\lambda}$
0.790	1142.50	756.23	55.32	1.900	136.01	1273.42	93.16
0.800	1144.70	767.69	56.16	1.950	126.00	1280.08	93.64
0.810	1113.00	779.02	56.99	2.000	118.50	1286.30	94.10
0.820	1070.00	789.93	57.79	2.100	93.00	1296.72	94.86
0.830	1041.00	800.50	58.56	2.200	74.75	1305.00	95.47
0.840	1019.99	810.77	59.31	2.300	63.25	1312.05	95.98
0.850	994.00	820.96	60.06	2.400	56.50	1317.96	96.41
0.860	1002.00	830.85	60.78	2.500	48.25	1323.16	96.80
0.870	972.00	840.61	61.49	2.600	42.00	1327.66	97.12
0.880	966.00	850.39	62.21	2.700	36.50	1331.57	97.41
0.890	945.00	859.94	62.91	2.800	32.00	1334.98	97.66
0.900	913.00	869.26	63.59	2.900	28.00	1337.97	97.88
0.910	876.00	878.16	64.24	3.000	24.75	1340.60	98.07
0.920	841.00	886.81	64.87	3.100	21.75	1342.93	98.24
0.930	830.00	895.10	65.48	3.200	19.75	1344.99	98.39
0.940	801.00	903.27	66.08	3.300	17.25	1346.84	98.53
0.950	778.00	911.18	66.66	3.400	15.75	1348.48	98.65
0.960	771.00	918.90	67.22	3.500	14.00	1349.96	98.76
0.970	764.00	926.58	67.78	3.600	12.75	1351.30	98.85
0.980	769.00	934.21	68.34	3.700	11.50	1352.51	98.94
0.990	762.00	941.88	68.90	3.800	10.50	1353.60	99.02
1.000	743.99	949.41	69.45	3.900	9.50	1354.59	99.09
1.050	665.98	984.76	72.04	4.000	8.50	1355.49	99.16
1.100	606.04	1016.27	74.35	4.100	7.75	1356.31	99.22
1.150	551.04	1045.16	76.46	4.200	7.00	1357.05	99.27
1.200	497.99	1071.43	78.38	4.300	6.50	1357.72	99.32
1.250	469.99	1095.66	80.15	4.400	6.00	1358.33	99.37
1.300	436.99	1117.96	81.78	4.500	5.50	1358.89	99.41
1.350	389.03	1138.51	83.29	4.600	5.00	1359.40	99.45
1.400	354.03	1156.97	84.64	4.700	4.50	1359.86	99.48
1.450	318.99	1173.91	85.88	4.800	4.00	1360.29	99.51
1.500	296.99	1189.28	87.00	4.900	3.75	1360.69	99.54
1.550	273.99	1203.52	88.04	5.000	3.47	1361.04	99.57
1.600	247.02	1216.48	88.99	6.000	1.75	1363.50	99.75
1.650	234.02	1228.52	89.87	7.000	0.95	1364.79	99.84
1.700	215.00	1239.74	90.69	8.000	0.55	1365.52	99.89
1.750	187.00	1249.69	91.42	9.000	0.35	1365.96	99.93
1.800	170.00	1258.55	92.07	10.000	0.20	1366.24	99.95
1.850	149.01	1266.42	92.64	25.000	0.12	1366.97	100.00

[a] Adapted from Frölich and Wehrli [15].

[b] λ is the wavelength (μm), $\dot{I}_{0n\lambda}$ the solar spectral irradiance averaged over small bandwidth centered at λ (W m^{-2} μm^{-1}), $\sum_0^\lambda \dot{I}_{0n\lambda}$ the integrated solar irradiance in the wavelength range 0–λ (W m^{-2}), and $P_{0-\lambda}$ the percentage of solar constant associated with wavelengths shorter than λ.

spectral irradiance or simply NASA standard. This standard has remained in vogue for almost a decade.

Since the adoption of the above standard, the old data have come under revised scrutiny [14]. Calibration problems of the earlier instruments have been reexamined. Fröhlich and colleagues [15, 16] at the World Radiation Center (WRC) in Davos, Switzerland, have proposed new values of the spectral distribution of solar irradiance. They have selected data from Arvesen *et al.*[1] [17], Smith and Gottlieb [18], Heath and Thekaekara [19], and Neckel and Labs [20]. The main part (0.330–1.247 μm) consists of Neckel and Labs' data. The data were slightly and uniformly adjusted to match the solar constant of 1367 W m^{-2} (to be discussed in the next section) also proposed by the WRC. The recommended spectrum with a resolution of 0.002 μm in the visible region is given in Table C-1, Appendix C. This table

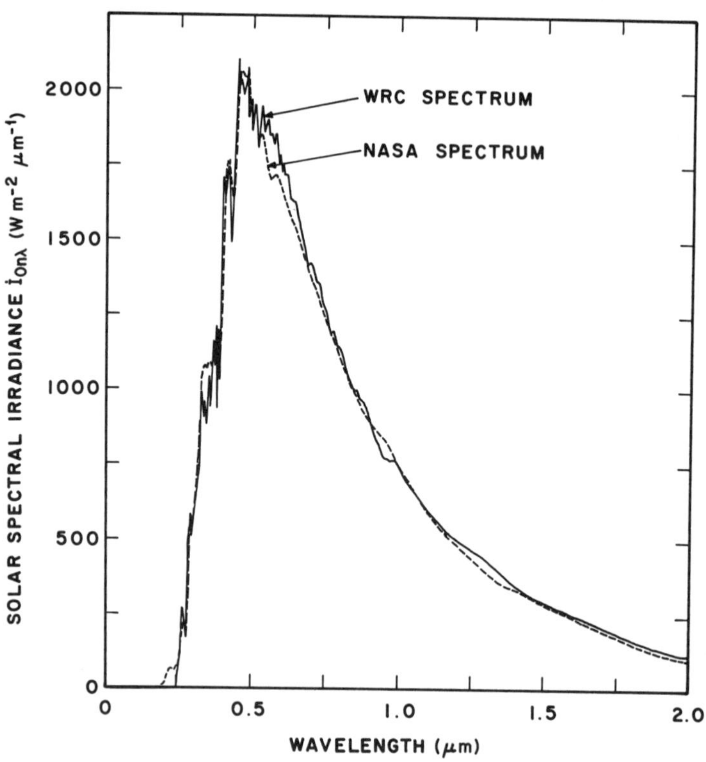

Figure 3.3.1 A comparison of the WRC spectrum with the NASA spectrum.

[1] *Et al.* is an abbreviation of the Latin *et alii*, which means *and others*. This abbreviation is commonly used when reference is made to more than two persons.

represents the state of the art up to the time of publication. The Commission for Instruments and Methods of Observation (CIMO) of the World Meteorological Organization (WMO), during its Eighth Session in October 1981 adopted the spectral distribution of Table C-1 recommended by Fröhlich and Wehrli [15]. In this text, this is called the WRC spectrum and will be utilized exclusively as far as possible.

An abridged version of Table C-1 is presented in Table 3.3.2. In this table the first column gives the wavelength λ in micrometer. The second gives the spectral irradiance $\dot{I}_{0n\lambda}$ at wavelength λ in W m^{-2} μm^{-1}. The spectral irradiance values were averaged over a narrow bandwidth for the shorter wavelengths. Wider bandwidths were used for longer wavelengths. The third column of this table gives total irradiance in W m^{-2} for the wavelength range 0–λ. At the end of this column, the irradiance totals 1367 W m^{-2}. The

Table 3.3.3

Approximate Division of Solar Spectrum in Various Color Bands and Energy Regions[a]

Color	λ (μm)	Band irradiance (W m^{-2})	Percentage of $\dot{I}_{SC}$
Violet	0.390–0.455	108.85	7.96
Blue	0.455–0.492	73.63	5.39
Green	0.492–0.577	160.00	11.70
Yellow	0.577–0.597	35.97	2.63
Orange	0.597–0.622	43.14	3.16
Red	0.622–0.770	212.82	15.57
Broader color divisions			
Ultraviolet	<0.4	109.81	8.03
Visible	0.390–0.770	634.40	46.41
Infrared	>0.77	634.40	46.40
Sun's energy (%)			
95	0.3–2.4		
1.2	<0.3		
3.6	>2.4		

[a] Solar constant $\dot{I}_{SC}$ = 1367 W m^{-2} (values obtained from Table 3.3.2).

fourth column gives the percentage of the solar constant associated with wavelengths shorter than λ. From this table it can be easily verified that 99% of solar energy is in the range 0.25–4.0 μm.

The NASA/ASTM standard and the WRC values are compared in Fig. 3.3.1. Substantial disagreements are apparent between the two curves. These differences are confined essentially to the ultraviolet and the visible portions of the spectrum. However, these differences can be quite significant when the spectral distribution of solar radiation arriving on the ground is considered. The cloudless-sky atmosphere is practically transparent to a substantial portion of the visible spectrum. Furthermore, as we shall see in Chapter 6, most of the diffuse irradiance is within the visible portion of the spectrum as well.

Table 3.3.3 demonstrates the distribution of extraterrestrial energy within different color bands. About half the solar energy lies in the visible region and almost the same amount in the infrared: little energy is found in the ultraviolet. Approximately 95% of the sun's energy lies within 0.3–2.4 μm, and approximately one percent lies in wavelengths longer than 4.0 μm.

3.4 The Solar Constant

The solar constant is the rate of total solar energy at all wavelengths incident on a unit area exposed normally to rays of the sun at one astronomical unit. It is not a true constant but seems to fluctuate slightly, a few tenths of a percent over periods of years. It should perhaps be called "solar factor."

The solar constant has been studied extensively since the beginning of this century, early measurements being made from ground-based observations. The ground-based observations are spectral measurements extrapolated to their predicted values at the top of the atmosphere by taking into account attenuation by the various atmospheric constituents. If such observations are made at the top of high mountains, attenuation by dust and water vapor is practically eliminated. Spectral integration of such values yields the solar constant. However, the ground-based spectral approach is subject to large errors because of uncertainties of the atmospheric constituents. Moreover, this method is quite tedious.

More recently, high-altitude measurements have been carried out with the use of high-altitude aircraft, balloons and space probes, whereby the corrections for atmospheric attenuation are very small. Pyrheliometers that integrate energy over all wavelengths are generally used. The absolute accuracy that can be achieved by such instruments is approximately $\pm 0.3\%$.

Table 3.4.1 from Thekaekara [10] lists the values of the solar constant reported by various authors. The lower part of the table contains values obtained from ground-based data and the upper part contains nine high-altitude measurements, one of them from the USSR. In each group, the values are listed in order of increasing magnitude. The ground-based values are consistently higher than those reported from the high-altitude measurements. Various sources of error in the ground-based values have been discussed by Thekaekara [21]. The value of the constant from the high-altitude measurements varies from 1338 to 1368 W m^{-2}. The NASA value of the solar constant [9, 13] is based on a weighted average of several values [11] and is

$$\dot{I}_{SC} = 1353 \text{ W m}^{-2}$$

(or in SI energy units, $I_{SC} = \dot{I}_{SC} \times 3.6 = 4871$ kJ m^{-2} h^{-1}). The estimated error in this value is ± 21 W m^{-2} or $\pm 1.5\%$. This value, adopted in 1971, is called the NASA design standard and is based on the international pyrheliometer scale[2] (IPS, 1956).

Under the auspices of the World Meteorological Organization (WMO), international pyrheliometer comparisons (IPCs) were carried out at Davos in 1970 and 1975. From a synthesis of many different measurements, a solar constant reference scale (SCRS) was established. Using this scale, Fröhlich [22] reexamined 12 previous measurements and some new ones. The selection of the measurements included those taken from aircraft, high-altitude balloons, satellites, and rockets. As a result of this reexamination Fröhlich recommended a solar constant value of 1373 W m^{-2} (sometimes quoted in the literature as 1377 W m^{-2}).

The preceding observations were a result of measurements up to 1970 and analysis of measurements up to 1975. Since 1975 a number of measurements of the solar constant have been made. For one thing, calibration of the instruments was reexamined when it was found that some instrument characteristics are different in space (vacuum) and in terrestrial measurements. Also, different instruments employ different window materials with nonidentical spectral transmission characteristics. Furthermore, in the recent measurements, modern cavity-type absolute instruments have been employed. Consequently, the World Meteorological Organization has now adopted a new scale, called the world radiometric reference (WRR), as a common base for all meteorological measurements. This reference is based on the results of over 25 000 measurements by a group of absolute radiometers of different type and design, the so-called world standard group, maintained at the World Radiation Center. Using the new reference,

[2] See Chapter 12 regarding pyrheliometer "scales."

Table 3.4.1[a]

Major Attempts to Evaluate the Solar Constant

Author	Source	Method	Value of the solar constant ($mW\ cm^{-2}$)	Estimated error ($\pm mW\ cm^{-2}$)	($cal\ cm^{-2}\ min^{-1}$)
High-altitude measurements					
Murcray	Univ. of Denver	Balloon Eppley NIP	133.8	0.6	1.919
McNutt and Riley	GSFC	NASA 711 Å 6618	134.3	2.6	1.926
Duncan and Webb	GSFC	NASA 711 Å 7635	134.9	4.0	1.935
McNutt and Riley	GSFC	NASA 711, Hy-Cal	135.2	2.2	1.939
Plamondon	JPL	Mariner–cone	135.3	2.0	1.940
Kondratyev *et al.*	Univ. of Leningrad	Balloon–Actinometer	135.3	1.4	1.940
Kruger	GSFC	NASA 711–cone	135.8	2.4	1.947
Laue and Drummond	JPL–Eppley	X-15, NASA 711, B-57B	136.0	1.3	1.950
Willson	JPL	Balloon, cavity	136.8	0.7	1.962
Ground-based measurements and revisions of earlier values					
Stair and Ellis	NBS	Mauna Loa 0.31–0.53 μm	135.9	6.8	1.949
Labs and Neckel	Germany	Jungfraujoch 0.33–1.25 μm	136.5	2.7	1.957
Gast	AFCRL	Revision	139.0		1.993
Johnson	NRL	Mt. Lemmon, Smithsonian	139.5	2.8	2.000
Makarova and Kharitonov	Univ. of Moscow	Revision	141.8	3.5	2.033

[a] Adapted with permission from Thekaekara [10], Copyright (1973) Pergamon Press, Ltd.

Fröhlich and colleagues [15, 16] examined eight solar constant measurements that were recorded from 1969 to 1980 (Table 3.4.2) and recommended the revised value[3]

$$\dot{I}_{SC} = 1367 \text{ W m}^{-2}$$

(or in SI energy units, $I_{SC} = \dot{I}_{SC} \times 3.6 = 4921$ kJ m^{-2} h^{-1}; in other common units, $I_{SC} = 433.3$ Btu ft^{-2} h^{-1} = 1.96 cal cm^{-2} min^{-1}), with a standard deviation of 1.6 W m^{-2} and largest deviation of ± 7 W m^{-2}.

As a result of the recent precise measurements reported above, the Commission for Instruments and Methods of Observation in October 1981 agreed to adopt for meteorological purposes the WRC solar constant. The difference between the NASA design standard and the WRC value is only 1%. In this text we use the WRC value.

Measurement of the solar constant and study of its variability will continue for many years. Although no significant changes are expected in the solar constant, the same cannot be said of its spectral distribution. There are larger uncertainties associated with the spectral data.

Table 3.4.2

Summary of Revised Solar Constant Determinations from **1969** *to* **1980**[a]

Date		Platform	Instrument[b]	Radiometric reference	Solar constant (W m^{-2})
August	1969	Balloon	ACR III	WRR	1369
June	1976	Rocket	ACR 402A	ACR	1369
November	1978	Rocket	ACR 402A	WRR	1365
June	1979	Balloon	PMO 6-9	WRR	1366
May	1980	Rocket	ACR 402A	WRR	1365
June	1980	Balloon	PMO 6-9	WRR	1367
February/July	1980	Solar-maximum satellite	ACRIM A	ACRIM	1367
				Mean	1367
				Standard deviation	1.6

[a] From Fröhlich and Brusa [16].

[b] See Chapter 12 regarding names of these modern electrical cavity-type absolute radiometers.

[3] This might seem to be a perfect example of a reason for the uninitiated to lose faith in the gurus. However, it is essential to emphasize that accurate measurement of the constant has been one of the biggest challenges of this century in radiometry. A final value of this "constant" remains elusive.

3.5 Blackbody Temperature of the Sun

As seen earlier, the temperature varies from one part of the sun to another. Radiation arriving outside the earth's atmosphere is a complicated function of the convection and reradiation processes in the outer layers of the sun and of a number of other factors mentioned before. The temperature of the sun is calculated from the two types of information we have: the solar constant and its spectral distribution. Using the solar constant in conjunction with the Stefan–Boltzmann equation [Eq. (2.2.3)], we can obtain the blackbody temperature of the sun from

$$T = (\dot{I}_{SC} r_0^2 / \sigma r_s^2)^{1/4}, \tag{3.5.1}$$

where

r_0 is the mean sun–earth distance, 149 597 890 km;
r_s is the mean radius of the solar disk, 695 980 km;
σ is the Stefan–Boltzmann constant, 5.6697×10^{-8} W m^{-2} K^{-4}, theoretical value; and
$\dot{I}_{SC}$ is the solar constant, 1367 W m^{-2}.

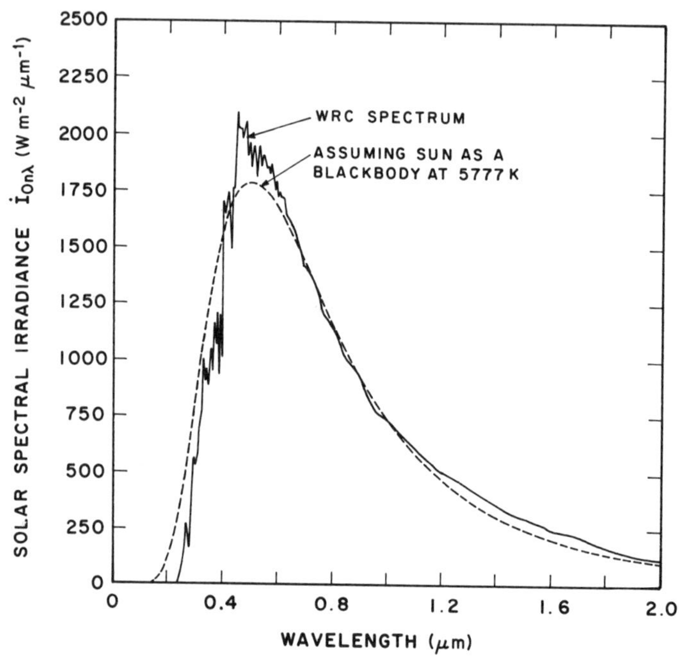

Figure 3.5.1 Comparison of the WRC spectrum with that from the sun assumed as a blackbody at 5777 K.

The above yields $T = 5777$ K.

We can compare actual spectral irradiance from the sun to the characteristics of a blackbody at 5777 K. Such a comparison is shown in Fig. 3.5.1 with the spectrum from the World Radiation Center. It is evident from this diagram that the measured spectrum does not strictly follow the blackbody curve. Furthermore, the maximum of the observed spectrum lies at a wavelength shorter than that of the blackbody at 5777 K.

Another useful examination of the spectrum can be carried out by summation of the fractional radiation following Fig. 2.3.2. Again, Fig. 3.5.2 shows that the spectrum does not strictly follow the blackbody law.

A further useful consideration is that the solar disk, as viewed from a point outside the earth's atmosphere, is not at a uniform temperature: the center appears to be hotter than the circumference. The term "limb darkening" comes from the fact that the edge of the solar disk appears darker than the center, where the layers are deeper.

The rate of energy radiated by the sun, 3.844×10^{23} kW, is calculated by multiplying the solar constant (1367 W m^{-2}) by the surface area of a sphere of radius 1 AU. If we assume that the earth's mean radius is 6370 km, the rate of solar energy incident on the earth is 1.743×10^{14} kW.

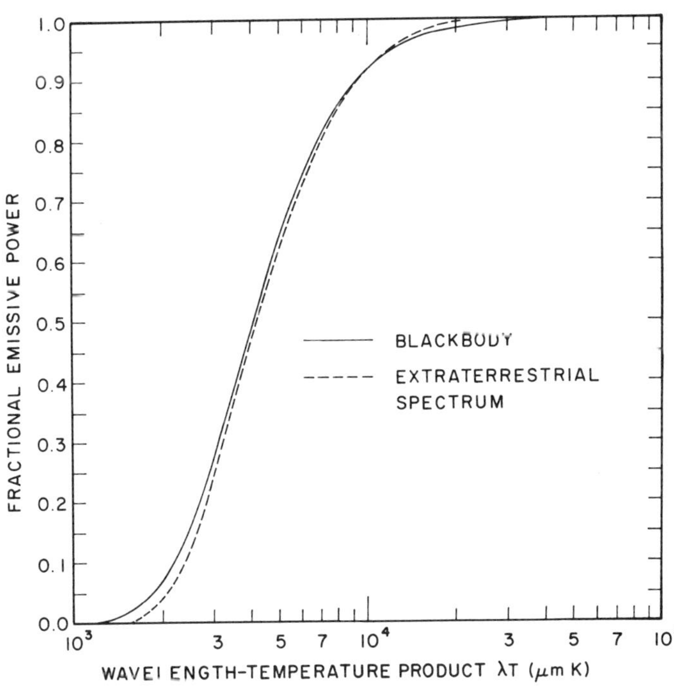

Figure 3.5.2 Fractional emissive power of a blackbody compared with that of the sun.

3.6 Further Reading

The physics of the sun is well explained in a number of books on astronomy and astrophysics. The material in this chapter was obtained largely from Elson [23]. Comprehensive studies regarding solar spectral irradiance and the solar constant are available in Drummond and Thekaekara [11] and White [14]. In this respect, a number of important points should be noted. The values in the NASA/ASTM standard curve and the solar constant are a result of a joint decision by the Committee on Solar Electromagnetic Radiation. This committee examined measurements made independently by different observers. A variety of instruments have been employed for each portion of the spectrum. These various aspects of radiometry and scales of radiation have been dealt with by Labs and Neckel [24], Kruger *et al.* [25], Thekaekara *et al.* [26], Drummond *et al.* [27], Laue [28], Drummond and Hickey [29], and Thekaekara [30].

The influence of sunspots and other solar activity on the solar constant has been investigated by Bossolasco *et al.* [31], Kondratyev and Nikolsky [32], and Hoyt [33]. The distribution of radiation over the sun's disk and the anisotropy of temperature has been studied by Ångström and Ångström [34].

In this chapter we have discussed more recent values of the solar constant and its spectral distribution. Naturally, this chapter would have been incomplete without mentioning Charles G. Abbott (1872–1973), who, from the turn of this century onward, devoted most of his working life at the Smithsonian Institution to studying the solar constant and its variation. Most of his work is recorded in the Annals of the Astrophysical Observatory of the Smithsonian Institution. One objective of the solar constant program of the observatory was to study a variation in the solar constant as a possible cause of climatic change, but a number of scientists have derived conflicting conclusions from the same data. A very interesting and detailed review of this program from 1902 to 1962 is given by Hoyt [35]. He reexamines the data and concludes that there is no evidence for cyclic variations or any long-term trend in the solar constant greater than a few tenths of a percent.

Nomenclature

$\dot{I}_{0n\lambda}$ Solar spectral irradiance at mean sun–earth distance averaged over small bandwidth centered at λ (W m^{-2} μm^{-1})

$\dot{I}_{SC}$ Solar constant (1367 W m^{-2}; I_{SC} without the overdot is in SI energy units, 4921 kJ m^{-2} h^{-1})

$P_{0-\lambda}$ Percentage of solar constant associated with wavelengths shorter than λ

r_0 Mean sun–earth distance (149 597 890 km)

r_s Mean radius of the solar disk (695 980 km)

T	Temperature (K)
λ	**Wavelength (μm)**
λ_{max}	Wavelength at which maximum **emissive** power occurs at a given temperature (μm)
σ	Stefan–Boltzmann constant (5.6697×10^{-8} W m^{-2} K^{-4}, theoretical value; 5.6866×10^{-8} W m^{-2} K^{-4}, experimental value)

References

1. P. Moon, Proposed standard solar radiation curves for engineering use, *J. Franklin Inst.* **230,** 583–617 (1940).
2. F. S. Johnson, The solar constant. *J. Meteorol.* **11**(6), 431–439 (1954).
3. C. W. Allen, Solar radiation. *Q. R. Meteorol. Soc.* **84**(362), 307–318 (1958).
4. L. Dunkelman and R. Scolnik, Solar spectral irradiance and vertical atmospheric attenuation in the visible and ultraviolet. *Opt. Soc. Am.* **49**(4), 356–367 (1959).
5. C. R. Detwiler, D. L. Garrett, J. D. Purcell, and R. Tousey, The intensity distribution in the ultraviolet solar spectrum. *Anna. Geophys.* **17**(3), 9–18 (1961).
6. M. P. Thekaekara, R. Kruger, and C. H. Duncan, Solar irradiance measurements from a research aircraft. *Appl. Opt.* **8**(8), 1717–1732 (1969).
7. M. P. Thekaekara, The solar constant and the solar spectrum measured from a research aircraft at 38,000 feet. NASA, Goddard Space Flight Center, Rep. X-322-68-304 (Greenbelt, Maryland) August 1968 (Also available as NASA TMX-63324).
8. M. P. Thekaekara, The solar constant and the solar spectrum measured from a research aircraft. NASA TR R-351 NASNA, National Aeronautics and Space Administration (October 1970).
9. Solar Electromagnetic Radiation, NASA SP-8005, National Aeronautics and Space Administration (May 1971).
10. M. P. Thekaekara, Solar energy outside the earth's atmosphere. *Sol. Energy* **14**(2), 109–127 (1973).
11. A. J. Drummond and M. P. Thekaekara (eds.), The Extraterrestrial Solar Spectrum. Institute of Environmental Science, Mount Prospect, Illinois (1973).
12. M. P. Thekaekara (ed), The Energy Crisis and Energy from the Sun. Institute of Environmental Science, Mount Prospect, Illinois (1974).
13. Standard solar constant and air mass zero solar spectral irradiance tables. ASTM Standard, E 490-73a. Annual Book of ASTM Standards. Part 41, ASTM, Philadelphia, Pennsylvania (1973).
14. O. R. White (ed.), "The Solar Output and Its Variation." Colorado Associated Univ. Press, Boulder, Colorado, 1977.
15. C. Fröhlich and C. Wehrli, Spectral distribution of solar irradiance from 25000 nm to 250 nm. World Radiation Center, Davos, Switzerland, private communication (1981).
16. C. Fröhlich and R. W. Brusa, Solar radiation and its variation in time. *Sol. Phys.* **74,** 209–215 (1981).
17. J. C. Arvesen, R. N. Griffin, and B. D. Pearson, Determination of extraterrestrial solar spectral irradiance from a research aircraft. *Appl. Opt.* **8**(11), 2215–2232 (1969).
18. E. V. P. Smith and D. M. Gottlieb, Solar flux and its variation. *Space Sci. Rev.* **16,** 771–802 (1974).
19. D. F. Heath and M. P. Thekaekara, The solar spectrum between 1200 and 3000 Å. "The Solar Output and its Variation" (O. R. White, ed.), pp. 193–212. Colorado Associated Univ. Press, Boulder, Colorado, 1977.

20. H. Neckel and D. Labs, Improved data of solar spectral irradiance from 0.33 to 1.25 μ. *Sol. Phys.* **74,** 231–249 (1981).
21. M. P. Thekaekara, The solar constant and spectral distribution of solar radiant flux. *Sol. Energy* **9**(1), 7–20 (1965).
22. C. Fröhlich, Contemporary measures of the solar constant. "The Solar Output and its Variation" (O. R. White, ed.), pp. 93–109. Colorado Associated Univ. Press, Boulder, Colorado, 1977.
23. B. M. Elson, Theoretical picture of the sun still evolving. *Aviat. Week Space Technol.*, **100**(2), 63–77 (1974).
24. D. Labs and H. Neckel, Transformation of the absolute solar radiation data into the international practical temperature scale of 1968. *Sol. Phys.* **15,** 79–87 (1970).
25. R. Kruger *et al.*, The solar irradiance experiments: instrumentation and data analysis. *Appl. Opt.* **8**(8), 1716–1732 (1969).
26. M. P. Thekaekara, R. H. Collingbourne, and A. J. Drummond, A comparison of working standard pyranometers. *Bull. Am. Meteorol. Soc.* **53**(1), 8–15 (1972).
27. A. J. Drummond, J. R. Hickey, W. J. Scholes, and E. G. Laue, Multichannel radiometer measurement of solar irradiance at different terrestrial elevations. *Sol. Energy* **4**(9), 1200–1206 (1967).
28. E. G. Laue, The measurement of solar spectral irradiance at different terrestrial elevations. *Sol. Energy* **13**(1), 43–57 (1970).
29. A. J. Drummond and J. R. Hickey, The Eppley–JPL solar constant measurement program. *Sol. Energy* **12**(2), 217–232 (1968).
30. M. P. Thekaekara, Solar radiation measurement: techniques and instrumentation. *Sol. Energy* **18**(4), 309–325 (1976).
31. M. Bossolasco, G. Cicconi, I. Dagnino, A. Elena, and G. Flocchini, Solar constant and sunspots. *Pure Appl. Geoph.* **62,** 207–214 (1965).
32. K. Ya Kondratyev and G. A. Nikolsky, Solar radiation and solar activity. *Q. R. Meteorol. Soc.* **96**(3), 509–522 (1970).
33. D. V. Hoyt, Variations in the solar constant caused by changes in the active features of the sun. *In* "Solar-Terrestrial Influences on Weather and Climate" (B. M. McCormac and T. A. Seliga, eds.), pp. 65–68. Reidel Publ., 1979.
34. A. K. Ångström and K. H. Ångström, The distribution of radiation over the sun's disk. *Sol. Energy* **13**(2), 243–250 (1971).
35. D. V. Hoyt, The Smithsonian Astrophysical Observatory solar constant program. *Rev. Geophys. Space Phys.* **17**(3), 427–458 (1979).

Chapter 4

EXTRATERRESTRIAL SOLAR IRRADIATION

4.1 Introduction

In this chapter the integrated value of the extraterrestrial solar radiation over all wavelengths is discussed. Mathematical formulations are developed to determine the irradiation at various surface orientations and for different time periods.

4.2 Extraterrestrial Irradiation on a Horizontal Surface

The expressions for radiation on horizontal surfaces will be formulated for different time periods: an hour, a day, a month, and so forth.

A. *Hourly Radiation on a Horizontal Surface*

On a given day, let $\dot{I}_{0n}$ be the extraterrestrial irradiance (*rate* of energy) on a surface normal to the rays from the sun, where

$$\dot{I}_{0n} = \dot{I}_{SC}(r_0/r)^2 = \dot{I}_{SC}E_0. \tag{4.2.1}$$

It is obvious from Fig. 4.2.1 that the irradiance on a horizontal surface can be written

$$\dot{I}_0 = \dot{I}_{0n} \cos \theta_z, \tag{4.2.2}$$

where $\cos \theta_z$ is given by Eq. (1.5.1) or

$$\dot{I}_0 = \dot{I}_{SC} E_0 (\sin \delta \sin \phi + \cos \delta \cos \phi \cos \omega). \tag{4.2.3}$$

The units of Eqs. (4.2.1)–(4.2.3) are W m^{-2}.

The irradiation (*amount* of energy) dI_0 during a short period of time dt will *be*

$$dI_0 = I_{SC} E_0 \cos \theta_z \, dt, \tag{4.2.4}$$

where dt is in hours and the solar constant (I_{SC} with no overdot) is in energy units, kJ m^{-2} h^{-1}. On the right-hand side of this equation, $\cos \theta_z$ contains the solar hour angle ω in degrees or radians. The time in hours can be converted to the hour angle as follows:
Let

$$\Omega = \text{rotational speed of the earth around its polar axis}$$
$$= \frac{2\pi \text{ rad}}{24\, h} = \frac{d\omega}{dt}.$$

From this

$$dt = (12/\pi)\, d\omega. \tag{4.2.5}$$

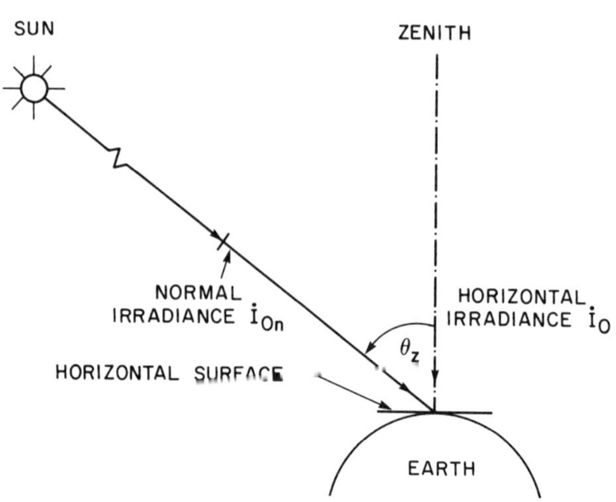

Figure 4.2.1 Relationship between the direct normal and the horizontal irradiance.

Equation (4.2.4) now reduces to the following:

$$dI_0 = (12/\pi)I_{SC}E_0(\sin\delta\sin\phi + \cos\delta\cos\phi\cos\omega)\,d\omega. \qquad (4.2.6)$$

One can now readily obtain I_0, the radiation over a period of 1 h. Let us consider the ith hour from solar noon, and ω_i the hour angle at the middle of this hour. The radiation over a period of 1 h will be [1]

$$I_0 = \frac{12}{\pi} I_{SC}E_0 \int_{\omega_i - \pi/24}^{\omega_i + \pi/24} (\sin\delta\sin\phi + \cos\delta\cos\phi\cos\omega)\,d\omega \qquad (4.2.7)$$

or

$$I_0 = I_{SC}E_0(\sin\delta\sin\phi + (24/\pi)\sin(\pi/24)\cos\delta\cos\phi\cos\omega_i). \qquad (4.2.8)$$

Since

$$(24/\pi)\sin(\pi/24) = 0.9972 \approx 1$$

we may write Eq. (4.2.8)

$$I_0 = I_{SC}E_0(\sin\delta\sin\phi + \cos\delta\cos\phi\cos\omega_i). \qquad (4.2.9)$$

The above yields extraterrestrial irradiation for 1 h centered around the hour angle ω_i.

□ EXAMPLE 4.2.1. Calculation of the extraterrestrial hourly solar irradiation on a horizontal surface over Vancouver (49°11′ N) for the hour ending at 11:00 (LAT) on 16 October. Compare the results obtained from Eqs. (4.2.8) and (4.2.9).

Solution. $\phi = 49.18°$ N.

On 16 October, $E_0 = 1.0064$ (from Table 1.2.1), $\delta = -8.67°$ (from Table 1.3.1). At the midhour between 10:00 and 11:00 (LAT), $\omega_i = 22.5°$. From Eq. (4.2.8)

$$\begin{aligned} I_0 &= 4921 \times 1.0064[\sin(-8.67)\sin(49.18) \\ &\quad + 0.9972\cos(-8.67)\cos(49.18)\cos(22.5)] \\ &= 2384 \text{ kJ m}^{-2}\text{ h}^{-1}. \end{aligned}$$

From Eq. (4.2.9)

$$\begin{aligned} I_0 &= 4921 \times 1.0064[\sin(-8.67)\sin(49.18) \\ &\quad + \cos(-8.67)\cos(49.18)\cos(22.5)] \\ &= 2392 \text{ kJ m}^{-2}\text{ h}^{-1}. \end{aligned}$$

The difference between the two results is very small, ~0.3%. □

Equation (4.2.9) can also be written in a slightly different form. From the expression for sunset hour angle on a horizontal surface, Eq. (1.5.3), the zenith angle can be written

$$\cos\theta_z = \cos\delta\cos\phi(\cos\omega - \cos\omega_s). \tag{4.2.10}$$

Consequently, Eq. (4.2.9) reduces to

$$I_0 = I_{SC}E_0\cos\delta\cos\phi(\cos\omega_i - \cos\omega_s). \tag{4.2.11}$$

In some instances, radiation for a period other than an exact hour may be required. Let us consider that the radiation on a horizontal surface between hours t_1 and t_2 is required. Counting the hours from midnight, and as long as t_1 and t_2 are during daylight, the radiation between any two hours will be as follows:

$$I_0\big|_{t_1}^{t_2} = I_{SC}E_0\{\sin\delta\sin\phi(t_2 - t_1) + (12/\pi)\cos\delta\cos\phi[\sin(15t_1) - \sin(15t_2)]\}. \tag{4.2.12}$$

Let the quantity $\bar{I}_0$ be defined as the extraterrestrial monthly average hourly radiation on a horizontal surface. In this text the terms "monthly average" and "monthly mean" are used interchangeably and synonymously.

Table 4.2.1

Characteristic Declinations,[a] δ, the Declinations on Which the Extraterrestrial Irradiation Is Identical to Its Monthly Average Value

Month	Date	δ (degrees)	Day number d_n
January	17	−20.84	17
February	14	−13.32	45
March	15	−2.40	74
April	15	+9.46	105
May	15	+18.78	135
June	10	+23.04	161
July	18	+21.11	199
August	18	+13.28	230
September	18	+1.97	261
October	19	−9.84	292
November	18	−19.02	322
December	13	−23.12	347

[a] Characteristic declinations are slightly but only slightly variable with latitude. This table is based on 35° N latitude. At a given latitude the extraterrestrial irradiation is a function of both δ and E_0. For solstice months (June and December) two values of δ_c are obtainable, at 20 days apart.

$\bar{I}_0$ is the average of all hourly irradiation received during the same local apparent time over a period of one month. This quantity will be needed in Chapter 8 to calculate solar radiation on the earth's surface in the presence of its atmosphere.

$\bar{I}_0$ can be expressed mathematically as

$$\bar{I}_0 = \frac{1}{n_2 - n_1} \sum_{n_1}^{n_2} I_0 \tag{4.2.13}$$

where n_1 and n_2 are the day numbers at the beginning and end of the month, respectively. It is possible to determine the particular day (that is, a particular

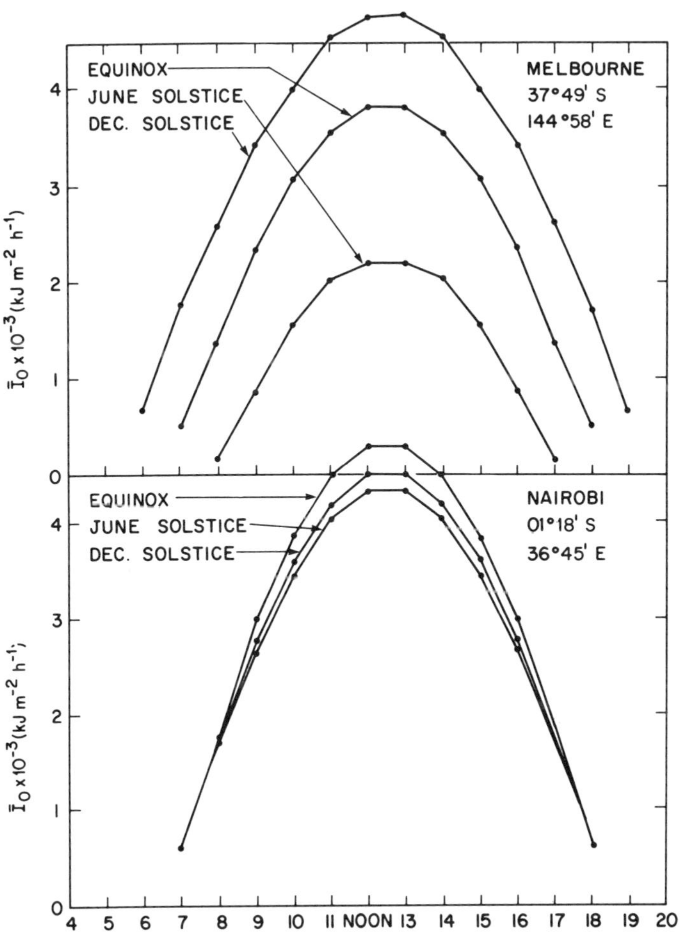

Figure 4.2.2 Diurnal variation of the extraterrestrial hourly irradiation on a horizontal surface (values at end of hour).

declination) which has irradiation equal to the monthly average hourly irradiation. Defining such a particular declination as the characteristic declination δ_c, we may now write $\bar{I}_0$ as follows:

$$\bar{I}_0 = I_0|_{\delta=\delta_c}. \tag{4.2.14}$$

Since declination does not change substantially over 24 h, one value of δ_c may be used for all hours during one month. Using Eq. (1.3.1), we compute values of δ_c for solar noon. Such values of the characteristic declination along with the corresponding day numbers are presented in Table 4.2.1.

Fig. 4.2.2 presents plots of the extraterrestrial hourly radiation on horizontal surfaces. These plots demonstrate the effect of latitude and declination (season) on the diurnal variation in insolation. Near the equator, effect of declination is minimal. At high altitudes, effect of declination (season) is greater.

B. *Daily Radiation on a Horizontal Surface*

Some examples of the diurnal variation of the extraterrestrial radiation on horizontal surfaces are presented in Fig. 4.2.2. The irradiation during a day, from sunrise (sr) to sunset (ss), is given by the area under the curves. Therefore

$$H_0 = \int_{sr}^{ss} I_0 \, dt \tag{4.2.15}$$

$$= 2\int_0^{ss} I_0 \, dt. \tag{4.2.16}$$

Assuming that E_0 remains constant during one day and after converting the time dt to the hour angle, we obtain

$$H_0 = \frac{24}{\pi} I_{SC} E_0 \int_0^{+\omega_s} (\sin\delta\sin\phi + \cos\delta\cos\phi\cos\omega)\, d\omega \tag{4.2.17}$$

or

$$H_0 = \frac{24}{\pi} I_{SC} E_0 [(\pi/180)\omega_s(\sin\delta\sin\phi) + (\cos\delta\cos\phi\sin\omega_s)], \tag{4.2.18}$$

where ω_s is in degrees.

It is possible to write the above expression for H_0 in two slightly different forms. Reconsider Eq. (1.5.3):

$$\cos\phi\cos\delta = -\sin\phi\sin\delta/\cos\omega_s.$$

Combining Eqs. (1.5.3) and (4.2.18) gives

$$H_0 = (24/\pi) I_{SC} E_0 \sin\phi \sin\delta[(\pi/180)\omega_s - \tan\omega_s] \tag{4.2.19}$$

or

$$H_0 = (24/\pi) I_{SC} E_0 \cos\phi \cos\delta[\sin\omega_s - (\pi/180)\omega_s \cos\omega_s]. \tag{4.2.20}$$

It should be noted that Eqs. (4.2.19) and (4.2.20) are not valid at either $\phi = 0°$ or $\phi = 90°$.

□ EXAMPLE 4.2.2. The extraterrestrial daily irradiation on 15 February of a horizontal surface at Melbourne (37°49′ S).

Solution. Latitude, $\phi = -37.82°$.
On 15 February $E_0 = 1.0256$, $\delta = -12.87°$.
From Eq. (1.5.3),

$$\omega_s = \cos^{-1}[-\tan(-37.82)\tan(-12.87)] = 100.22°.$$

To obtain H_0, we may employ any one of Eqs. (4.2.18)–(4.2.20). From Eq. (4.2.19), for instance,

$$\begin{aligned} H_0 &= (24/\pi) I_{SC} E_0 \sin(-37.82)\sin(-12.87)[100.22\pi/180 - \tan(100.22)] \\ &= 38.43 \text{ MJ m}^{-2}\text{ day}^{-1}. \end{aligned}$$

□

Two special cases of the extraterrestrial daily radiation on a horizontal surface are noteworthy:

(i) At the equator

$$\phi = 0,$$

and consequently

$$\omega_s = \pi/2.$$

Thus H_0 reduces to

$$H_0 = (24/\pi) I_{SC} E_0 \cos\delta. \tag{4.2.21}$$

(ii) In the polar regions, during the summer, there is no sunset or sunrise. In this situation, replace ω_s by π to obtain

$$H_0 = (24/\pi) I_{SC} E_0 \sin\phi \sin\delta(\pi^2/180). \tag{4.2.22}$$

In many engineering calculations, the monthly average extraterrestrial daily radiation on a horizontal surface, $\overline{H}_0$, is often required. This quantity is defined as follows:

$$\overline{H}_0 = \frac{1}{n_2 - n_1} \sum_{n_1}^{n_2} H_0. \tag{4.2.23}$$

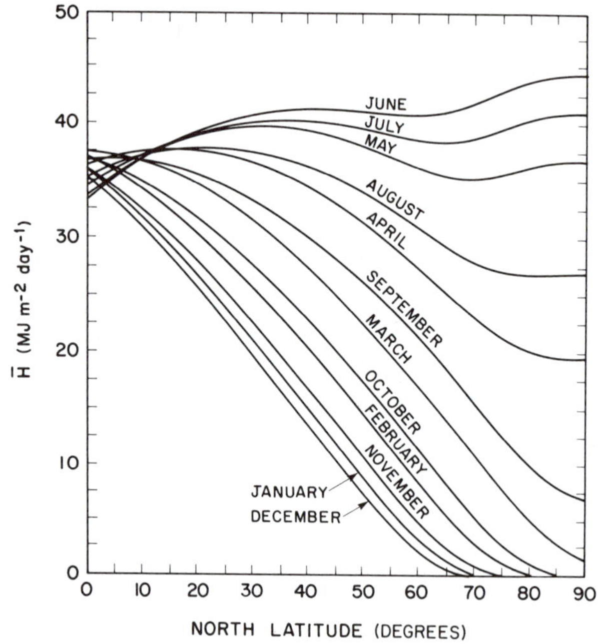

Figure 4.2.3 Monthly average extraterrestrial daily insolation on horizontal surfaces in the Northern Hemisphere. $\dot{I}_{SC} = 1367$ W m^{-2}.

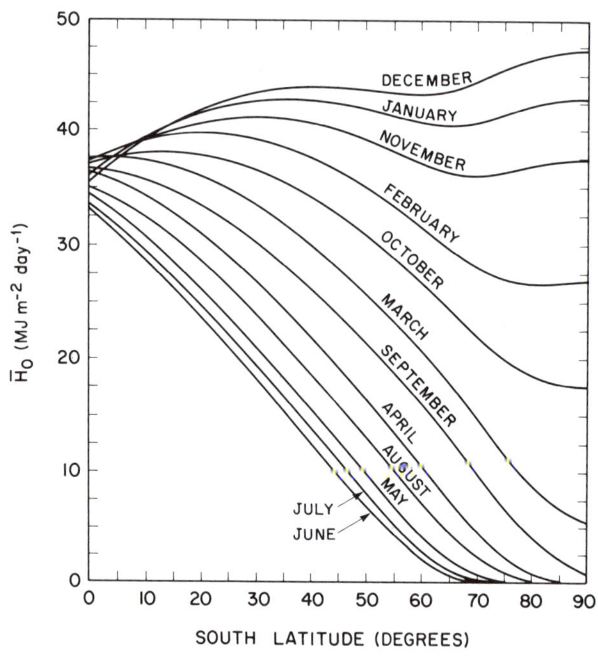

Figure 4.2.4 Monthly average extraterrestrial daily insolation on horizontal surfaces in the Southern Hemisphere. $\dot{I}_{SC} = 1367$ W m^{-2}.

Table 4.2.2

Yearly Variation of the Extraterrestrial Horizontal Daily Insolation for North Latitudes, $\overline{H}_0$ $(MJ\ m^{-2}\ day^{-1})$, $\dot{I}_{SC} = 1367\ W\ m^{-2}$

Month	North latitude (degrees)																		
	0	5	10	15	20	25	30	35	40	45	50	55	60	65	70	75	80	85	90
JAN	36.32	34.31	32.09	29.66	27.05	24.28	21.39	18.40	15.34	12.26	9.21	6.27	3.54	1.24	0.08	0.0	0.0	0.0	0.0
FEB	37.53	36.22	34.65	32.83	30.77	28.51	26.04	23.41	20.62	17.71	14.72	11.68	8.64	5.68	2.94	0.87	0.05	0.0	0.0
MAR	37.90	37.58	36.98	36.10	34.94	33.53	31.85	29.94	27.80	25.46	22.92	20.22	17.37	14.39	11.33	8.23	5.16	2.55	1.49
APR	36.75	37.47	37.92	38.08	37.97	37.58	36.92	35.98	34.80	33.36	31.70	29.84	27.80	25.62	23.39	21.29	19.96	19.61	19.68
MAY	34.78	36.28	37.54	38.54	39.29	39.77	40.00	39.97	39.70	39.19	38.48	37.61	36.65	35.74	35.32	35.90	36.60	37.02	37.17
JUN	33.50	35.35	36.97	38.37	39.53	40.44	41.11	41.54	41.74	41.73	41.56	41.28	41.00	41.00	42.15	43.33	44.18	44.69	44.86
JUL	33.89	35.59	37.05	38.27	39.24	39.97	40.44	40.67	40.66	40.44	40.03	39.48	38.90	38.47	38.90	39.96	40.74	41.21	41.37
AUG	35.56	36.62	37.43	37.96	38.23	38.22	37.95	37.41	36.63	35.60	34.35	32.91	31.31	29.64	28.04	27.00	26.92	27.22	27.32
SEP	37.07	37.19	37.03	36.60	35.88	34.90	33.65	32.15	30.40	28.43	26.25	23.87	21.33	18.63	15.82	12.93	10.08	7.75	6.74
OCT	37.34	36.43	35.24	33.80	32.10	30.17	28.02	25.67	23.14	20.45	17.63	14.70	11.70	8.68	5.71	2.95	0.94	0.07	0.0
NOV	36.47	34.69	32.68	30.46	28.03	25.44	22.69	19.81	16.85	13.83	10.79	7.81	4.96	2.39	0.54	0.00	0.0	0.0	0.0
DEC	35.74	33.55	31.16	28.58	25.84	22.96	19.97	16.92	13.82	10.74	7.73	4.88	2.34	0.40	0.0	0.0	0.0	0.0	0.0

Table 4.2.3

Yearly Variation of the Extraterrestrial Horizontal Daily Insolation for South Latitudes, $\overline{H}_0$ *(MJ m^{-2} day^{-1})*, $\dot{I}_{SC} = 1367\ W\ m^{-2}$

Month	South Latitude (degrees)																		
	0	5	10	15	20	25	30	35	40	45	50	55	60	65	70	75	80	85	90
JAN	36.32	38.09	39.61	40.87	41.86	42.59	43.04	43.24	43.18	42.89	42.39	41.75	41.05	40.50	40.78	41.84	42.66	43.15	43.32
FEB	37.53	38.58	39.34	39.83	40.03	39.94	39.57	38.92	38.01	36.84	35.45	33.84	32.07	30.20	28.36	27.00	26.69	26.95	27.06
MAR	37.90	37.93	37.68	37.14	36.31	35.22	33.85	32.23	30.37	28.28	25.98	23.49	20.83	18.02	15.09	12.09	9.10	6.53	5.49
APR	36.75	35.76	34.50	32.99	31.24	29.27	27.08	24.70	22.15	19.45	16.63	13.72	10.75	7.79	4.90	2.29	0.58	0.01	0.0
MAY	34.78	33.04	31.08	28.92	26.58	24.07	21.42	18.66	15.81	12.91	10.02	7.17	4.47	2.06	0.40	0.00	0.0	0.0	0.0
JUN	33.50	31.44	29.19	26.76	24.18	21.48	18.68	15.81	12.91	10.02	7.20	4.53	2.15	0.35	0.0	0.0	0.0	0.0	0.0
JUL	33.89	31.98	29.87	27.57	25.10	22.49	19.76	16.94	14.07	11.19	8.34	5.60	3.07	0.98	0.03	0.0	0.0	0.0	0.0
AUG	35.56	34.24	32.68	30.89	28.88	26.68	24.29	21.75	19.07	16.28	13.42	10.53	7.66	4.88	2.37	0.61	0.02	0.0	0.0
SEP	37.07	36.66	35.98	35.03	33.81	32.34	30.62	28.68	26.52	24.16	21.62	18.92	16.09	13.15	10.14	7.09	4.12	1.72	0.69
OCT	37.34	37.98	38.34	38.42	38.21	37.72	36.95	35.91	34.62	33.07	31.30	29.32	27.16	24.86	22.49	20.19	18.53	17.86	17.86
NOV	36.47	38.00	39.27	40.28	41.01	41.47	41.66	41.58	41.24	40.66	39.87	38.90	37.83	36.79	36.21	36.67	37.38	37.81	37.96
DEC	35.74	37.71	39.43	40.91	42.14	43.10	43.80	44.25	44.45	44.44	44.24	43.92	43.61	43.59	44.79	46.04	46.94	47.48	47.66

Like the monthly average hourly irradiation, the daily quantity $\overline{H}_0$ can also be calculated at a characteristic declination. Thus

$$\overline{H}_0 = H_0\big|_{\delta=\delta_c}. \tag{4.2.24}$$

The characteristic declinations for the hourly and daily irradiation remain the same as in Table 4.2.1. The variation of $\overline{H}_0$ with latitude and month is shown in Figs. 4.2.3 and 4.2.4. It is evident from these diagrams that at the equator the annual variation of the extraterrestrial radiation is minimal. However, as we move away from the equator, annual variations become quite pronounced. Tables 4.2.2 and 4.2.3 list the values of $\overline{H}_0$ for north and south latitudes, respectively.

Note: Before going further, it is useful to explain the term "monthly average daily" radiation. The word "daily" is used to emphasize that, since all months do not have the same number of days, computation is done daily for the sake of comparison. In this manner variations from month to month are brought out.

4.3 Radiation on Inclined Planes

In this section, we treat separately surfaces tilted toward the equator and those oriented arbitrarily. For each of these cases, expressions are developed for irradiance, hourly irradiation, and the daily irradiation.

A. *Radiation on Surfaces Tilted Toward the Equator*

(I) Hourly Irradiation

Consider Fig. 4.3.1, showing a surface tilted at an angle β toward the equator. Let $\dot{I}_{0\beta}$ be the extraterrestrial irradiance on a surface tilted toward the equator. Then

$$\dot{I}_{0\beta} = \dot{I}_{SC} E_0 \cos\theta_0, \tag{4.3.1}$$

where θ_0 is the incidence angle for a tilted surface facing the equator. The irradiation $I_{0\beta}$ between hour angles ω_1 and ω_2 is

$$I_{0\beta} = \frac{12}{\pi} I_{SC} E_0 \int_{\omega_1}^{\omega_2} [\sin\delta\sin(\phi-\beta) + \cos\delta\cos(\phi-\beta)\cos\omega]\,d\omega. \tag{4.3.2}$$

When integrating the above equation, one must pay attention to the signs of ω_1 and ω_2 and ensure that neither of them exceeds the sunrise (or sunset) hour angle for tilted surfaces.

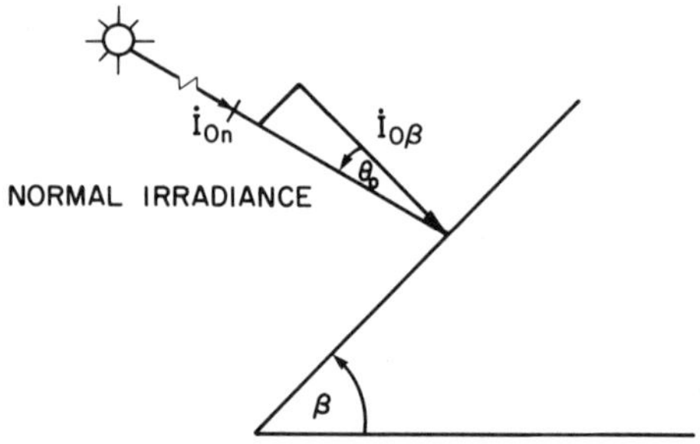

Figure 4.3.1 Irradiance on an inclined plane tilted toward the equator.

To obtain irradiation during a particular full hour, we follow the procedure employed to obtain the hourly irradiation on a horizontal surface. Thus, for one full hour, with ω_i the hour angle at midhour, we have the following:

$$I_{0\beta} = I_{SC}E_0[\sin\delta\sin(\phi-\beta) + 0.9972\cos\delta\cos(\phi-\beta)\cos\omega_i] \quad (4.3.3)$$

or

$$I_{0\beta} = I_{SC}E_0[\sin\delta\sin(\phi-\beta) + \cos\delta\cos(\phi-\beta)\cos\omega_i]. \quad (4.3.4)$$

When radiation is needed for a period shorter than an hour, for instance, between times t_1 and t_2, we have the following:

$$I_{0\beta}\big|_{t_1}^{t_2} = I_{SC}E_0\{\sin\delta\sin(\phi-\beta)(t_2-t_1) + (12/\pi)\cos\delta\cos(\phi-\beta)[\sin(15t_1) - \sin(15t_2)]\}. \quad (4.3.5)$$

In the above, the times t_1 and t_2 in hours are counted from midnight. Naturally these hours should be between sunset and sunrise.

Like $\bar{I}_0$, the monthly average extraterrestrial hourly irradiation $\bar{I}_{0\beta}$, on a surface tilted toward the equator can be computed from the characteristic declination. Thus

$$\bar{I}_{0\beta} = I_{0\beta}\big|_{\delta=\delta_c}. \quad (4.3.6)$$

(II) Daily Irradiation

The integration of Eq. (4.3.2) between the sunrise and sunset hour angle results in an expression for the daily irradiation $H_{0\beta}$. Consequently,

$$H_{0\beta} = \frac{24}{\pi} I_{SC}E_0 \int_0^{\omega=\omega_s,\,\omega_s'} [\sin\delta\sin(\phi-\beta) + \cos\delta\cos(\phi-\beta)\cos\omega]\,d\omega. \quad (4.3.7)$$

The limit of integration in the above should be such that only the minimum of ω_s or ω_s' is employed and that the final result is a positive quantity. Treating each of these two limits separately, we obtain

$$H_{0\beta} = (24/\pi)I_{SC}E_0\left[(\pi/180)\omega_s' \sin\delta \sin(\phi - \beta) + \cos\delta\cos(\phi - \beta)\sin\omega_s'\right], \qquad \omega_s' \leq \omega_s, \tag{4.3.8}$$

and

$$H_{0\beta} = (24/\pi)I_{SC}E_0\left[(\pi/180)\omega_S \sin\delta \sin(\phi - \beta) + \cos\delta\cos(\phi - \beta)\sin\omega_s\right], \qquad \omega_s \leq \omega_s'. \tag{4.3.9}$$

In the preceding two equations, ω_s and ω_s' are in degrees. These two equations can be combined:

$$\mathrm{H}_{0\beta} = (24/\pi)I_{SC}E_0\left[(\pi/180)\omega_s' \sin\delta \sin(\phi - \beta) + \cos\delta\cos(\phi - \beta)\sin\omega_s'\right], \tag{4.3.10}$$

where

$$\omega_s' = \min\{\omega_s, \cos^{-1}[-\tan\delta\tan(\phi - \beta)]\}. \tag{4.3.10$'$}$$

The monthly average extraterrestrial daily irradiation $\overline{H}_{0\beta}$ on a surface tilted toward the equator is calculated at the characteristic declination

$$\overline{H}_{0\beta} = H_{0\beta}\big|_{\delta=\delta_c}. \tag{4.3.11}$$

□ EXAMPLE 4.3.1. Monthly average daily insolation for the month of April on the roof of a house in Stockholm (59°20′ N) inclined at 60° from the horizontal position and facing due south in the absence of the earth's atmosphere.

Solution. From Table 4.2.1, for April $\delta_c = +9.46°$, $d_n = 105$ on 15 April. From Table 1.2.1, on 15 April $E_0 = 0.9932$. From Eq. (1.5.4),

$$\omega_s = \cos^{-1}[-\tan(59.33)\tan(9.46)] = 106.32.$$

From Eq. (1.6.3),

$$\omega_s' = \min\{\omega_s, \cos^{-1}[-\tan(59.33 - 60)\tan(9.46)]\} = 89.89°.$$

To compute the monthly average daily insolation during the month of April, we have from Eqs. (4.3.10) and (4.3.11)

$$\begin{aligned}\overline{H}_{0\beta} &= (24/\pi)(4921)(0.9932)[(\pi/180)(89.89)\sin(9.46)\sin(59.33 - 60) \\ &\quad + \cos(9.46)\cos(59.33 - 60)\sin(89.89)] \\ &= 36.71\ \mathrm{MJ\ m^{-2}\ day^{-1}}.\end{aligned}$$

□

B. *Extraterrestrial Radiation on an Arbitrarily Oriented Surface*

(I) Hourly Irradiation

The foregoing procedure for surfaces tilted toward the equator can be repeated in this case. The irradiance of a surface oriented at an azimuth γ is

$$\dot{I}_{0\beta\gamma} = \dot{I}_{SC} E_0 \cos\theta, \tag{4.3.12}$$

where $\cos\theta$ is obtained from Eq. (1.6.5).

The irradiation between hour angles ω_1 and ω_2 is written

$$I_{0\beta\gamma} = \frac{12}{\pi} I_{SC} E_0 \int_{\omega_1}^{\omega_2} \cos\theta \, d\omega. \tag{4.3.13}$$

Following the procedure employed to obtain I_0 and $I_{0\beta}$, we may write the above equation

$$I_{0\beta\gamma} = I_{SC} E_0 [(\sin\phi \cos\beta - \cos\phi \sin\beta \cos\gamma)\sin\delta + (\cos\phi \cos\beta + \sin\phi \sin\beta \cos\gamma)\cos\delta \cos\omega_i + \cos\delta \sin\beta \sin\gamma \sin\omega_i], \tag{4.3.14}$$

where ω_i at midhour is used.

The monthly average hourly value will be

$$\dot{I}_{0\beta\gamma} = I_{0\beta\gamma}|_{\delta=\delta_c}. \tag{4.3.15}$$

□ EXAMPLE 4.3.2. An exact expression for the extraterrestrial hourly irradiation on an arbitrarily oriented surface inclined at 30° and oriented 20° east situated over New York (40°27′ N); the irradiation is determined for the hour ending at 10:00 (LAT) on 4 July.

Solution. The exact expression is

$$\begin{aligned} I_{0\beta\gamma} &= \int^{1\,\mathrm{h}} I_{SC} E_0 \cos\theta \, dt \\ &= \frac{12}{\pi} I_{SC} E_0 \int_{\omega_i - \pi/24}^{\omega_i + \pi/24} \cos\theta \, d\omega \\ &= \frac{12}{\pi} I_{SC} E_0 \int_{\omega_i - \pi/24}^{\omega_i + \pi/24} (A \sin\omega + B \cos\omega + C) d\omega, \end{aligned}$$

where from Eqs. (1.6.8)–(1.6.10),

$$\begin{aligned} A &= \cos\delta \sin\beta \sin\gamma, \\ B &= \cos\delta \cos\phi \cos\beta + \cos\delta \sin\phi \sin\beta \cos\gamma, \\ C &= \sin\delta \sin\phi \cos\beta - \sin\delta \cos\phi \sin\beta \cos\gamma. \end{aligned}$$

Integration of the above yields

$$I_{0\beta\gamma} = (12/\pi)I_{SC}E_0(-A\cos\omega + B\sin\omega + C\omega)\big|_{\omega_i-\pi/24}^{\omega_i+\pi/24}$$
$$= (12/\pi)I_{SC}E_0[0.26A\sin\omega_i + 0.26B\cos\omega_i + (\pi/12)C].$$

In this particular example, $\gamma = 20°$, $\beta = 30°$, $\delta = 22.98°$, $\phi = 40.45°$, $\omega_i = 37.5°$, and $E_0 = 0.9666$. Employing these values, we obtain

$$A = 0.157, \qquad B = 0.887, \qquad C = 0.080.$$

The hourly radiation is

$$I_{0\beta\gamma} = (12/\pi)4921(0.9666)(0.025 + 0.184 + 0.021)$$
$$= 4173 \text{ kJ m}^{-2}\text{ h}^{-1}.$$ □

(II) Daily Irradiation

Continuing in a similar fashion, we may write the extraterrestrial daily irradiation

$$H_{0\beta\gamma} = \frac{12}{\pi} I_{SC}E_0 \int_{\omega_{sr}}^{\omega_{ss}} \cos\theta\, d\omega. \tag{4.3.16}$$

The expressions for ω_{ss} and ω_{sr}, the sunset and sunrise hour angles, respectively, have been treated in Chapter 1. In terms of these angles

$$\begin{aligned} H_{0\beta\gamma} = (12/\pi)I_{SC}E_0(&\cos\beta\sin\delta\sin\phi|\omega_{ss} - \omega_{sr}|\pi/180 \\ &- \sin\delta\cos\phi\sin\beta\cos\gamma|\omega_{ss} - \omega_{sr}|\pi/180 \\ &+ \cos\phi\cos\delta\cos\beta|\sin\omega_{ss} - \sin\omega_{sr}| \\ &+ \cos\delta\cos\gamma\sin\phi\sin\beta|\sin\omega_{ss} - \sin\omega_{sr}| \\ &+ \cos\delta\sin\beta\sin\gamma|\cos\omega_{ss} - \cos\omega_{sr}|). \end{aligned} \tag{4.3.17}$$

where $|\,|$ indicates the absolute value.

The monthly average daily value will be

$$\bar{H}_{0\beta\gamma} = H_{0\beta\gamma}|_{\delta=\delta_c}. \tag{4.3.18}$$

□ **Example** 4.3.3. (a) The hourly extraterrestrial irradiation between 12:00 and 13:00 (LAT), and (b) the daily extraterrestrial irradiation on a plane surface over Tokyo (35°40′ N) oriented at 30° west of south and inclined at an angle of 30° from the horizontal on 1 February.

Solution. (a) To obtain the hourly value, we can either employ the formula developed in Example 4.3.2. or Eq. (4.3.14). We shall employ the

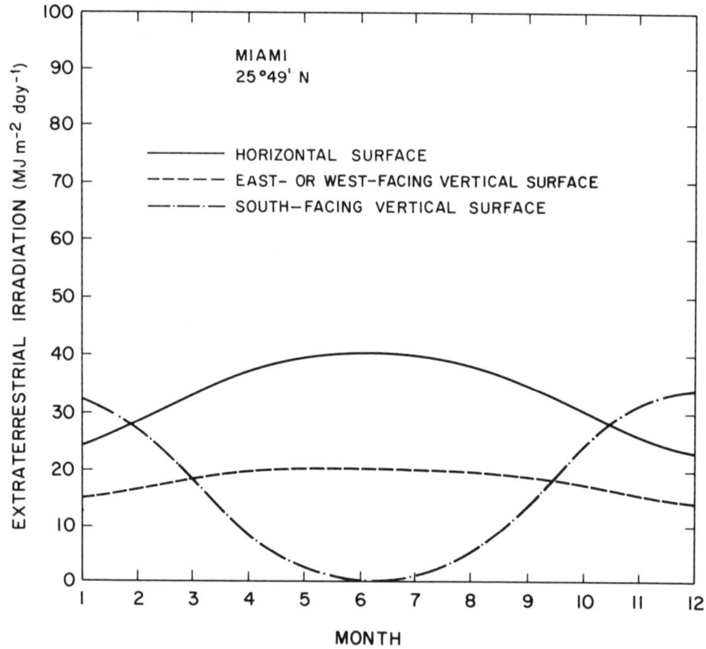

Figure 4.3.2 Annual variation of the extraterrestrial irradiation on vertical surfaces.

latter. For $\gamma = -30°$, $\beta = 30°$, $\phi = 35.67°$, $\delta = -17.28$, $\omega_i = -7.5°$, and $E_0 = 1.0306$,

$$
\begin{aligned}
I_{0\beta\gamma} &= 4921(1.0306)\{[\sin(35.67)\cos(30) \\
&\quad - \cos(35.67)\sin(30)\cos(-30)]\sin(-17.28) \\
&\quad + [\cos(35.67)\cos(30) \\
&\quad + \sin(35.67)\sin(30)\cos(-30)]\cos(-17.28)\cos(-7.5) \\
&\quad + [\cos(-17.28)\sin(30)\sin(-30)\sin(-7.5)]\} \\
&= 4518\ \text{kJ m}^{-2}\,\text{h}^{-1}.
\end{aligned}
$$

(b) $H_{0\beta\gamma}$ is evaluated from Eq. (4.3.17) where ω_{sr} and ω_{ss} are obtained through Eqs. (1.6.18) and (1.6.19). It can be shown that

$$\omega_{sr} = 72.10° = 1.26\ \text{rad}$$

and

$$\omega_{ss} = -76.90° = -1.34\ \text{rad};$$

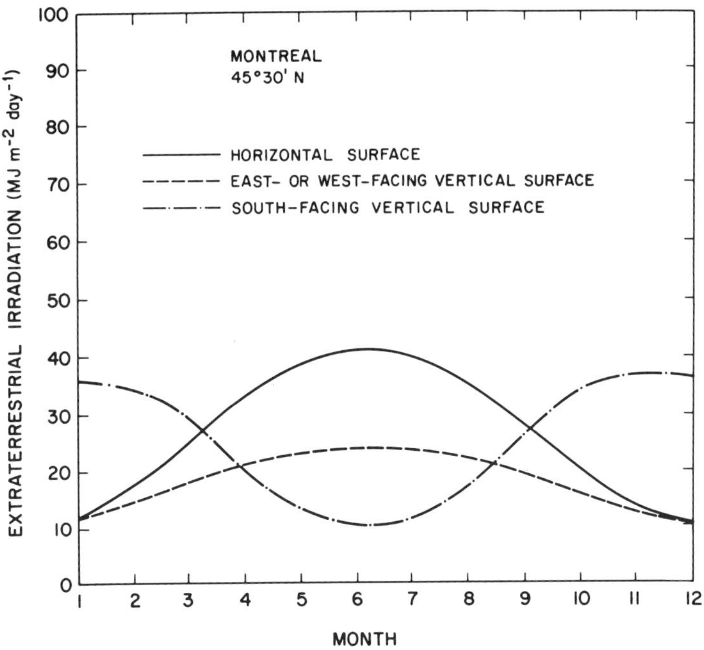

Figure 4.3.3 Annual variation of the extraterrestrial irradiation on vertical surfaces.

$H_{0\beta\gamma}$ can now be written

$$\begin{aligned} H_{0\beta\gamma} = {} & \frac{12}{\pi} 4921(1.0306)[\cos(30)\sin(-17.28)\sin(35.67)|1.26 + 1.34| \\ & - \sin(-17.28)\cos(35.67)\sin(30)\cos(-30)|-1.34 - 1.26| \\ & + \cos(35.67)\cos(-17.28)\cos(30)|0.96 + 0.97| \\ & + \cos(-17.28)\cos(-30)\sin(35.67)\sin(30)|0.96 + 0.97| \\ & + \cos(-17.28)\sin(30)\sin(-30)|0.30 - 0.23|] \\ = {} & 30.57 \text{ MJ m}^{-2}\text{ day}^{-1}. \end{aligned}$$

□

Figures 4.3.2 and 4.3.3 show the annual variation of radiation on vertical walls at different orientations. These plots are for two different locations, Miami and Montreal, separated by about 20° latitude. During the summer months, horizontal surfaces receive maximum insolation and the south-facing vertical surfaces minimum. During December and January, the south-facing surfaces receive maximum insolation. With decrease in latitude, the south-facing vertical walls exhibit large annual variations in radiation income.

4.4 Calculation of $\dot{r}_b$, r_b, $\bar{r}_b$, R_b, and $\bar{R}_b$

In Chapter 11 methods will be developed to compute insolation on inclined surfaces in the presence of the earth's atmosphere. In these methods, the ratio of the radiation on an inclined plane to that on a horizontal plane in the absence of the earth's atmosphere will be used. It seems best that the relevant mathematics now be developed. The above ratios pertaining to different time scales will be formulated one by one. It would facilitate matters to develop this material in two stages first for surfaces tilted toward the equator and second for arbitrarily oriented surfaces.

A. *Surface Tilted toward the Equator*

We begin by considering the ratios for instantaneous, hourly, and daily irradiation.

(i) The ratio of the irradiance on an inclined surface to that on a horizontal surface in the absence of the earth's atmosphere is termed $\dot{r}_b$. From Fig. 4.4.1 it is obvious that

$$\dot{I}_0 = \dot{I}_{0n} \cos \theta_z \tag{4.4.1}$$

and

$$\dot{I}_{0\beta} = \dot{I}_{0n} \cos \theta_0 . \tag{4.4.2}$$

Therefore

$$\dot{r}_b = \dot{I}_{0\beta}/\dot{I}_0 = \cos \theta_0 / \cos \theta_z . \tag{4.4.3}$$

Note: The above formulation appears trivial; however, it will be used to illustrate an important point in Chapter 11.

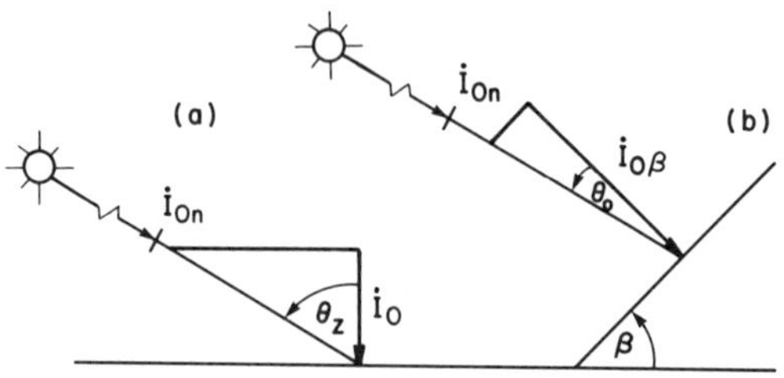

Figure 4.4.1 Irradiance on horizontal and inclined surfaces: (a) horizontal surface, (b) inclined surface tilted toward equator.

(ii) r_b is the ratio of the hourly radiation on an inclined plane to that on a horizontal plane in the absence of the earth's atmosphere. This ratio is written

$$r_b = \frac{I_{0\beta}}{I_0} = \frac{I_{SC}E_0 \int_{\omega_i-\pi/24}^{\omega_i+\pi/24} \cos\theta_0 \, d\omega}{I_{SC}E_0 \int_{\omega_i-\pi/24}^{\omega_i+\pi/24} \cos\theta_z \, d\omega}. \tag{4.4.4}$$

As illustrated previously, a very good approximation of the above integrals can be obtained by evaluating $\cos\theta_0$ and $\cos\theta_z$ at midhour. Consequently,

$$r_b = I_{0\beta}/I_0 \approx \cos\theta_0/\cos\theta_z. \tag{4.4.5}$$

It should be noted that at grazing angles (just at sunrise or at sunset) r_b can change rapidly and may approach infinity or zero because both the numerator and denominator are small numbers. This depends on slope, latitude, and date. Therefore caution is advised when calculating r_b (or $\dot{r}_b$) near the grazing angles.

(iii) The ratio of the monthly average hourly radiation on an inclined plane to that on a horizontal plane in the absence of the earth's atmosphere is termed $\bar{r}_b$. This is analogous to (ii) above, except that δ is evaluated at δ_c. Thus

$$\bar{r}_b = \frac{\bar{I}_{0\beta}}{\bar{I}_0} = \left.\frac{\cos\theta_0}{\cos\theta_z}\right|_{\delta=\delta_c}. \tag{4.4.6}$$

(iv) The ratio of the daily radiation on an inclined plane to that on a horizontal plane in the absence of the earth's atmosphere is termed R_b. Following Liu and Jordan [2], this ratio is as follows:

$$R_b = H_{0\beta}/H_0. \tag{4.4.7}$$

Referring to the earlier sections of this chapter, we have two principal expressions for $H_{0\beta}$, Eqs. (4.3.8) and (4.3.9). They are applicable to the two cases $\omega_s' \le \omega_s$ and $\omega_s \le \omega_s'$, respectively. For H_0 we can use Eq. (4.2.18), for instance. We treat the two cases separately.

Case 1, $\omega_s' \le \omega_s$. We have

$$R_b = \text{Eq. (4.3.8)/Eq. (4.2.18)}$$

or

$$R_b = \frac{(\pi/180)\omega_s' \sin\delta\sin(\phi-\beta) + \cos\delta\cos(\phi-\beta)\sin\omega_s'}{(\pi/180)\omega_s\sin\delta\sin\phi + \cos\delta\cos\phi\sin\omega_s} \tag{4.4.8}$$

$$= \frac{\cos(\phi-\beta)[(\pi/180)\omega_s'\sin\delta\tan(\phi-\beta) + \cos\delta\sin\omega_s']}{\cos\phi[(\pi/180)\omega_s\sin\delta\tan\phi + \cos\delta\sin\omega_s]}. \tag{4.4.9}$$

Using Eqs. (1.5.3) and (1.6.2) in the above expression, we reduce R_b to

$$R_b = \frac{\cos(\phi - \beta)[\sin\omega_s' - (\pi/180)\omega_s' \cos\omega_s']}{\cos\phi[\sin\omega_s - (\pi/180)\omega_s \cos\omega_s]}. \tag{4.4.10}$$

Case 2, $\omega_s \leq \omega_s'$. We have

$$R_b = \text{Eq. (4.3.9)/Eq. (4.2.18)}.$$

Following the procedure used previously, we can now write R_b

$$R_b = \frac{\cos(\phi - \beta)[\sin\omega_s - (\pi/180)\omega_s \cos\omega_s']}{\cos\phi[\sin\omega_s - (\pi/180)\omega_s \cos\omega_s]}. \tag{4.4.11}$$

It may be noted that R_b can also be written as a single expression through Eqs. (4.3.10) and (4.2.18), thus giving

$$R_b = \frac{(\pi/180)\omega_s' \sin\delta \sin(\phi - \beta) + \cos\delta \cos(\phi - \beta)\sin\omega_s'}{(\pi/180)\omega_s \sin\delta \sin\phi + \cos\delta \cos\phi \sin\omega_s}, \tag{4.4.12}$$

where ω_s' is given by Eq. (1.6.4).

(v) The ratio of the monthly average daily radiation on an inclined plane to that on a horizontal plane in the absence of the earth's atmosphere is termed $\bar{R}_b$. This ratio is written

$$\bar{R}_b = \frac{\bar{H}_{0\beta}}{\bar{H}_0} \tag{4.4.13}$$

$$= \left.\frac{H_{0\beta}}{H_0}\right|_{\delta = \delta_c}. \tag{4.4.14}$$

(vi) At equinoxes, $\delta = 0$ and $\omega_s = \omega_s' = \pi/2$. The following ratios, then, are identical:

$$\dot{r}_b = r_b = R_b = \cos(\phi - \beta)/\cos\phi. \tag{4.4.15}$$

This is an interesting result and will be elaborated upon in Chapter 11.

Because the ratio $\bar{R}_b$ is extensively employed in computing slope radiation, in Fig. 4.4.2 we produce a number of plots of this factor which can be employed for quick calculations in Chapter 11. The plots in this diagram are for the northern hemisphere. For the southern hemisphere, change the months as follows:

December → June, January → July, November → May,
February → August, October → April, March → September,
September → March, April → October, August → February,
May → November, July → January, June → December.

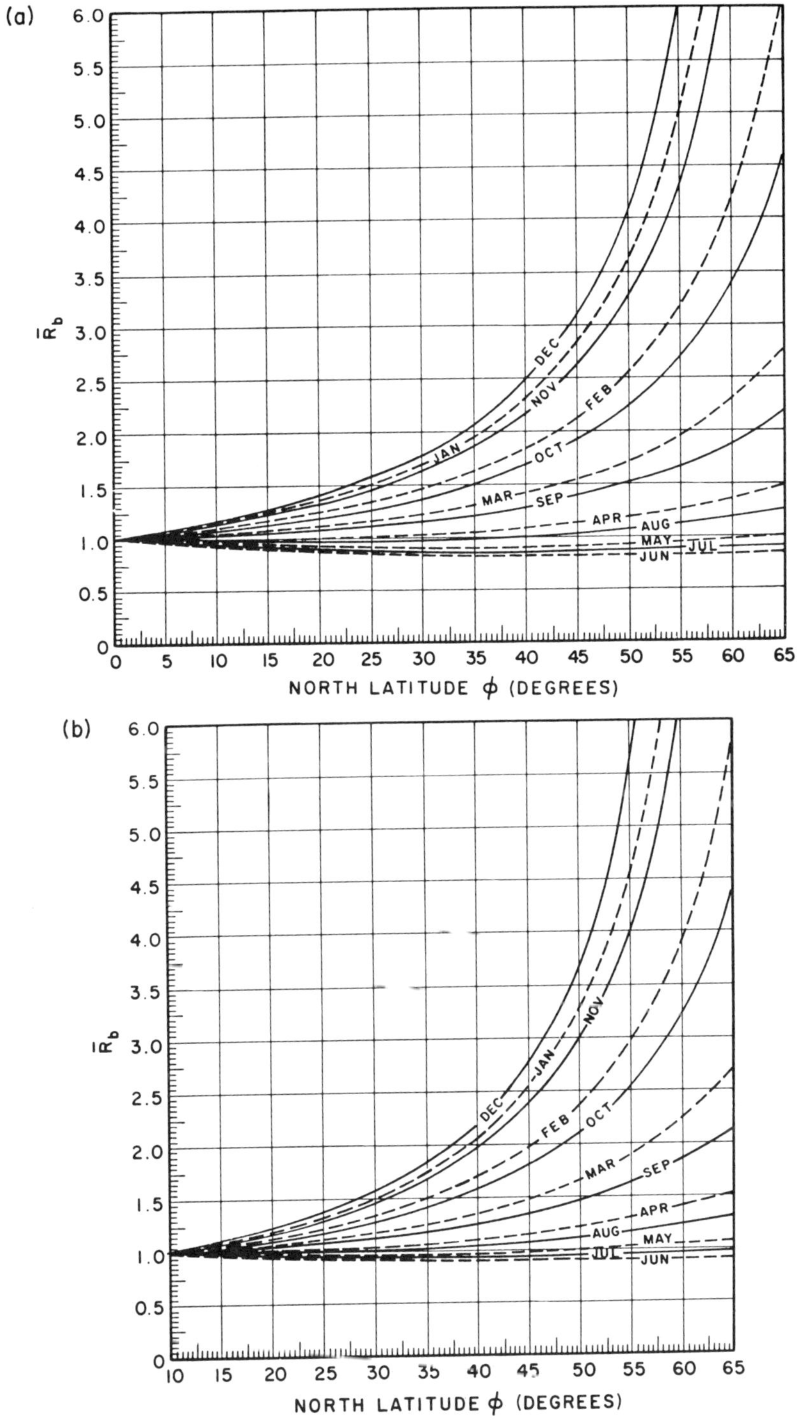

Figure 4.4.2 Variation of the parameter $\bar{R}_b$ as a function of the latitude for inclinations (a) $\phi - \beta = 0^\circ$ and (b) $\phi - \beta = 10^\circ$.

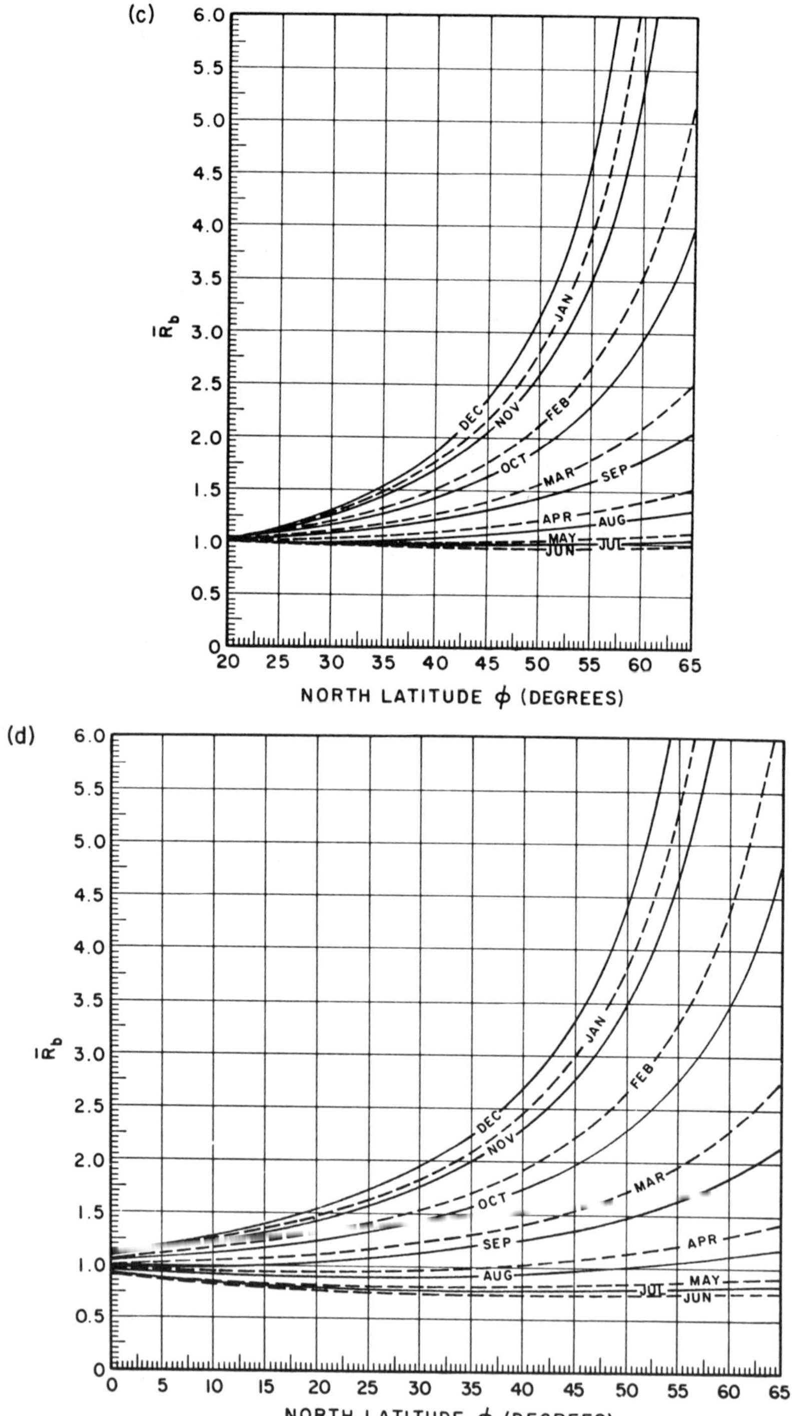

Figure 4.4.2 (continued) Variation of the parameter $\bar{R}_b$ as a function of the latitude for inclinations. (c) $\phi - \beta = 20^\circ$ and (d) $\phi - \beta = -10^\circ$.

(e)

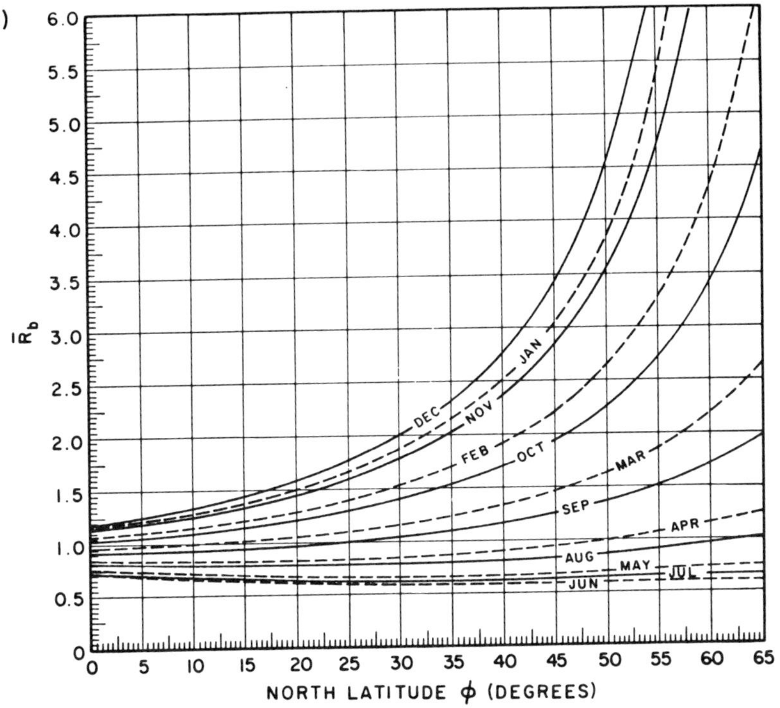

(f)

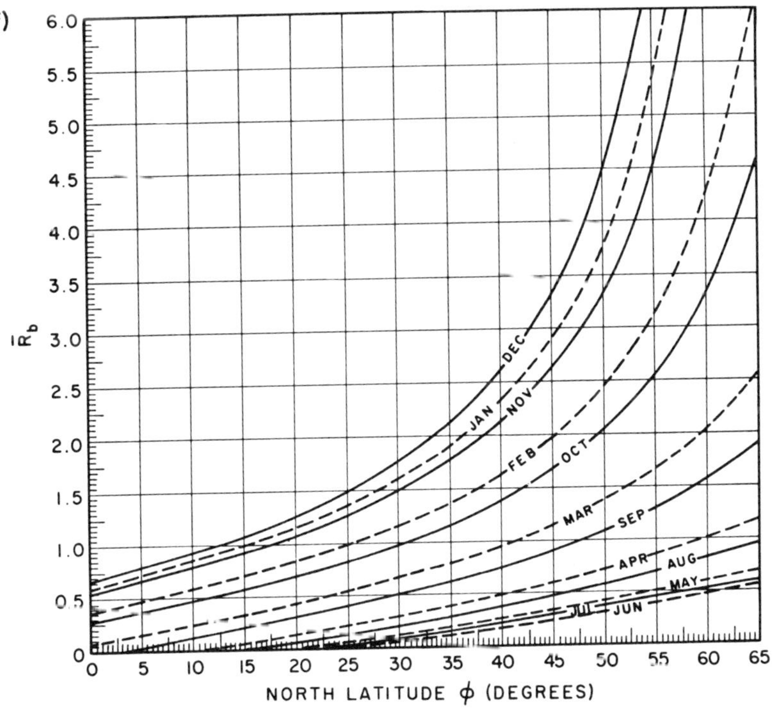

Figure 4.4.2 (continued) Variation of the parameter $\bar{R}_b$ as a function of the latitude for (e) inclination $\phi - \beta = -20^\circ$ and (f) vertical surfaces.

□ EXAMPLE 4.4.1. (a) r_b for the hour ending at 10:00 (LAT) and (b) R_b on a flat-plate collector at Los Angeles (34°0′ N) inclined toward the equator at an angle of 30° from the horizontal position for 15 May.

Solution. $\phi = 34°$, $\beta = 30°$, $\omega = 37.5°$. From Table 1.3.1, on 15 May, $\delta = +18.77°$.

(a) r_b is given by Eq. (4.4.5):

$$r_b = \frac{\sin(18.77)\sin(4) + \cos(18.77)\cos(4)\cos(37.5)}{\sin(18.77)\sin(34) + \cos(18.77)\cos(34)\cos(37.5)}$$

$$= 0.97.$$

(b) To obtain R_b, we should first calculate ω_s':

$$\omega_s = \cos^{-1}[-\tan(18.77)\tan(34)] = 103.25°,$$

$$\omega_s' = \min\{\omega_s, \cos^{-1}[-\tan(18.77)\tan(4)]\}$$

$$= \min(103.25°, 91.36°) = 91.36°.$$

R_b is obtained from Eq. (4.4.12):

$$R_b = \frac{[(\pi/180)91.36]\sin(18.77)\sin(4) + \cos(18.77)\cos(4)\sin(91.36)}{[(\pi/180)103.25]\sin(18.77)\sin(34) + \cos(18.77)\cos(34)\sin(103.25)}$$

$$= 0.90.$$ □

B. *Arbitrarily Oriented Surfaces*

In this case, the expressions for the ratios are rather complicated. However, each one shall be treated separately.

(i) The ratio of the hourly extraterrestrial radiation on an arbitrarily oriented plane to that on a horizontal plane, r_b. This ratio is

$$r_b = \cos\theta/\cos\theta_z, \tag{4.4.16}$$

where the numerator and denominator are evaluated at midhour of the hour concerned.

(ii) R_b is the ratio of the daily extraterrestrial radiation on an arbitrarily oriented plane to that on a horizontal plane:

$$R_b = H_{0\beta\gamma}/H_0 = \text{Eq. (4.3.17)/Eq. (4.2.18)}. \tag{4.4.17}$$

Equation (4.4.17) can be reduced to

$$R_b = \{\cos\beta \sin\delta \sin\phi\,|\omega_{ss} - \omega_{sr}|\,\pi/180 - \sin\delta\cos\phi\sin\beta\cos\gamma\,|\omega_{ss} - \omega_{sr}|\,\pi/180 + \cos\phi\cos\delta\cos\beta\,|\sin\omega_{ss} - \sin\omega_{sr}| + \cos\delta\cos\gamma\sin\phi\sin\beta\,|\sin\omega_{ss} - \sin\omega_{sr}| + \cos\delta\sin\beta\sin\gamma\,|\cos\omega_{ss} - \cos\omega_{sr}|\} \times \{2[\cos\phi\cos\delta\sin\omega_s + (\pi/180)\omega_s\sin\phi\sin\delta]\}^{-1}. \quad (4.4.18)$$

4.5 Further Reading

Liu and Jordan [2] laid down the foundation for calculating insolation on inclined planes tilted toward the equator. Klein [3] developed an algorithm to compute extraterrestrial insolation on arbitrarily oriented surfaces.

Nomenclature

E_0	Eccentricity correction factor (dimensionless), Eq. (1.2.1)
H_0	Extraterrestrial daily radiation incident on a horizontal surface (MJ m^{-2} day^{-1})
$H_{0\beta}$	Extraterrestrial daily radiation incident on a surface inclined toward the equator (MJ m^{-2} day^{-1})
$H_{0\beta\gamma}$	Extraterrestrial daily radiation incident on an inclined surface oriented in any direction (MJ m^{-2} day^{-1})
$I_{0\beta}$	Extraterrestrial hourly radiation incident on a surface inclined toward the equator (kJ m^{-2} h^{-1})
$I_{0\beta\gamma}$	Extraterrestrial hourly radiation incident on an inclined surface oriented in any direction (kJ m^{-2} h^{-1})
$\dot{I}_{SC}$	Solar constant (1367 W m^{-2})
I_{SC}	Solar constant in energy units, $I_{SC} = 3.6\,\dot{I}_{SC}$ (kJ m^{-2} h^{-1})
n_1, n_2	Day numbers at the beginning and end of a month
R_b	$H_{0\beta\gamma}/H_0$ or $H_{0\beta}/H_0$
r_b	$I_{0\beta\gamma}/I_0$ or $I_{0\beta}/I_0$
t	Time, counting from midnight in the standard 24-hr day system (h)
β	Inclination of a surface from the horizontal position (degrees)
γ	Surface azimuth angle, east positive, west negative (degrees)
δ	Declination, north positive, south negative (degrees)
δ_c	Declination on characteristic days (degrees); see Table 4.2.1
θ	Angle between beam radiation and surface normal, for a surface inclined in any arbitrary direction (degrees)
θ_0	Angle between beam radiation and surface normal for a surface inclined toward the equator (degrees)
θ_z	Zenith angle, the angle between the beam from the sun and the vertical (degrees)
Σ	Summation

ϕ Latitude, north positive (degrees)
ω Hour angle, solar noon zero, morning positive (degrees)
ω_i Hour angle at the middle of an hour
ω_s Sunset hour angle for a horizontal surface (degrees)
ω_s' Sunrise hour angle for a surface inclined toward the equator (degrees)
ω_{sr} Sunrise hour angle for an inclined surface oriented in any direction (degrees)
ω_{ss} Sunset hour angle for an inclined surface oriented in any direction (degrees)

An overdot denotes an instantaneous value, an overbar a monthly average value.

References

1. A. Whillier, The determination of hourly values of total solar radiation from daily summations. *Arch. Meteorol. Geophys. Bioklimatol. Ser. B* **7**(2), 197–204 (1956).
2. B. Y. H. Liu and R. C. Jordan, Daily insolation on surfaces tilted toward the equator. *ASHRAE J.* **3**(10), 53–59 (1961).
3. S. A. Klein, Calculation of monthly average insolation on tilted surfaces. *Sol. Energy* **19**(4), 325–329 (1977). See also *ibid.* **20**(5), 441 (1978) and Letters to the editor, *ibid.* **25**(3), 287 (1981).

Chapter 5

A CLOUDLESS-SKY ATMOSPHERE AND ITS OPTICS

5.1 Introduction

Solar radiation emanating from the sun is attenuated, before reaching the ground, by the earth's atmosphere, which can be classified into two broad types: (1) atmosphere without clouds and (2) atmosphere with clouds. In this chapter, constituents of a cloudless atmosphere are described.

Maximum radiation on the earth is received under cloudless and clear skies. Most solar devices operate when radiation is at a maximum or at least above a certain threshold level. However, high levels of radiation can create serious problems in, for instance, agriculture and architecture. Consequently, the cloudless condition is important from both the utilization and the control point of view.

5.2 The Earth's Atmosphere

Since about the beginning of this century, the vertical structure of the earth's atmosphere has been generally described by a succession of *standard*

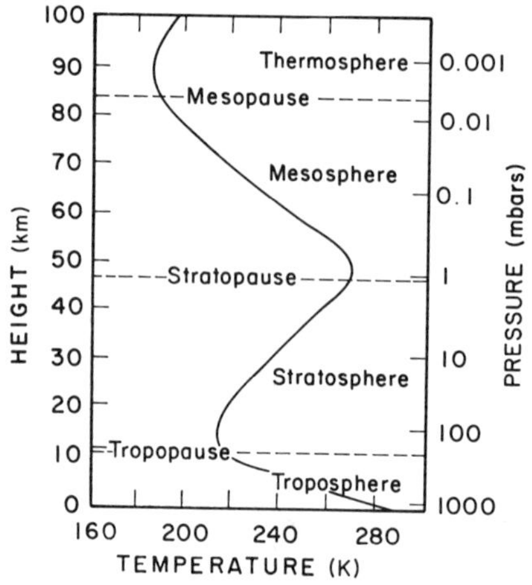

Figure 5.2.1 Variation of the atmospheric temperature and pressure with geographic altitude. After the U.S. Standard Atmosphere [1].

atmospheres. The model used today in the United States was adopted in 1976 [1] and supersedes those of 1956, 1959, and 1962. Figure 5.2.1 shows vertical temperature profiles and variation of pressure to a height of 100 km [1].

The earth's atmosphere is divided into a number of concentric spheres: troposphere, stratosphere, mesosphere, thermosphere, etc., and important temperature variations exist in each of these. However, density and pressure decrease continuously. The U.S. standard atmosphere (U.S.S.A.) 1976 sea-level values of pressure, temperature, and density are as follows:

Pressure: 1013.25 mbars, 760 mm Hg or 101.325 kPa
Temperature: 288 K or 15°C
Density: 1.225 kg m^{-3}

Up to a height of about 90 km, molecular weights of gases remain fairly constant; beyond this height, molecular weights slowly decrease. Fig. 5.2.2 shows the vertical variation of molecular weights up to 700 km from the earth [2].

The earth's atmosphere consists mainly of molecular nitrogen and molecular oxygen. Clean dry air contains about 78% nitrogen, 21% oxygen, 1% argon, and 0.33% carbon dioxide by volume. In addition, the earth's atmosphere contains water vapor and particulate matter (aerosols) such as dust, soot, water drops, and ice crystals, which are highly variable in time and space. Consequently, in order to determine the transmittance of the atmosphere to solar radiation, the total atmosphere of the earth is usually divided into three groups: dry air molecules, water vapor, and aerosols. Solar radiation entering the earth's atmosphere is attenuated by each of these groups. The composition and distribution of these three groups will now be discussed.

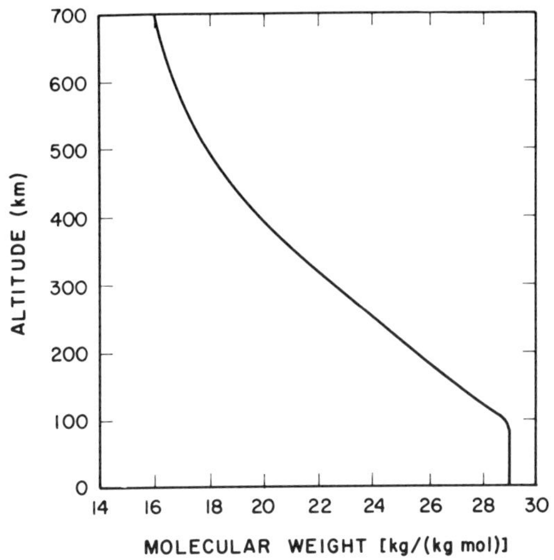

Figure 5.2.2 Variation of molecular weight of the atmospheric gases with geographic altitude. Adapted from "Handbook of Geophysics and Space Environments," edited by S. L. Valley. Copyright © 1965 McGraw-Hill Book Company. Used with the permission of McGraw-Hill Book Company.

5.3 Clean Dry Air

The actual composition and concentration of the constituents of clean air vary with geographic location, elevation, and season. The "normal" composition of the U.S. Standard Atmosphere 1976 is given in Table 5.3.1. The concentration of some gases such as carbon dioxide, ozone, carbon monoxide, and methane can be highly variable. These gases are not homogeneously distributed, in space or in time, throughout the atmosphere. These variations are a function of the industrial and agricultural activity of the place, its surroundings, and the general dynamic nature of the atmosphere.

The dissociation of molecular oxygen (O_2) by solar ultraviolet radiation begins beyond about 90 km in the vertical direction. Consequently as altitude increases, concentration of O_2 decreases and concentration of atomic oxygen (O) increases. Because molecular nitrogen is much more difficult to dissociate, concentration of atomic nitrogen (N) remains very small, even at high altitudes. Fig. 5.3.1 shows relative proportions of the three principal constituents up to a 500-km altitude. Beyond this elevation, however, there are further changes in the atmosphere: above 600 km, helium becomes a major constituent, and at about 2000 km, the principal constituents are ionized helium, ionized hydrogen, and electrons.

Table 5.3.1

Normal Composition of Clean Atmospheric Air[a]

Constituent gas	Content (% by volume)
Nitrogen (N_2)	78.084
Oxygen (O_2)	20.948
Argon (Ar)	0.934
Carbon dioxide (CO_2)	0.333
Neon (Ne)	18.18×10^{-4}
Helium (He)	5.24×10^{-4}
Krypton (Kr)	1.14×10^{-4}
Xenon (Xe)	0.089×10^{-4}
Hydrogen (H_2)	0.5×10^{-4}
Methane (CH_4)	1.5×10^{-4}
Nitrous oxide[b] (N_2O)	0.27×10^{-4}
Ozone (O_3)	$0–12 \times 10^{-4}$
Sulfur dioxide[b] (SO_2)	0.001×10^{-4}
Nitrogen dioxide[b] (NO_2)	0.001×10^{-4}
Ammonia[b] (NH_3)	0.004×10^{-4}
Carbon monoxide[b] (CO)	0.19×10^{-4}
Water vapor[b] (H_2O)	$0–0.04 \times 10^{-4}$
Nitric Oxide[b] (NO)	0.0005×10^{-4}
Hydrogen sulfide[b] (H_2S)	0.00005×10^{-4}
Nitric acid vapor	Traces

[a] From U.S. Standard Atmosphere [1].
[b] Concentration near the earth's surface.

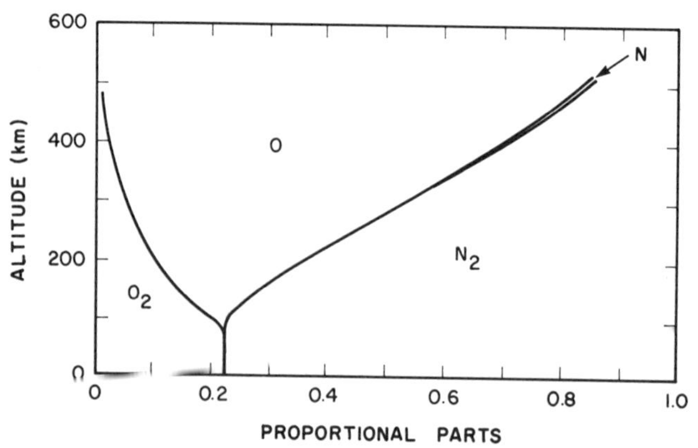

Figure 5.3.1 Variation of the atmospheric oxygen and nitrogen with geographic altitude. Adapted from "Handbook of Geophysics and Space Environments," edited by S. L. Valley. Copyright © 1965 McGraw-Hill Book Company. Used with the permission of McGraw-Hill Book Company.

All molecules of air deplete solar energy by scattering, which takes place at all wavelengths, and which therefore is called a continuum process. Air molecules, though, absorb solar radiation only at selective wavelengths. The most important of the dry air absorbers are ozone, carbon dioxide, oxygen, oxides of nitrogen, nitrogen, and hydrocarbon combinations. The basics of selective absorption and the particular wavelengths at which some of these gases absorb solar radiation will be discussed in the following chapter. Here, however, the atmospheric distribution of ozone will be described.

In the upper atmosphere, ozone is created mainly by ultraviolet solar radiation. On the ground, it is formed through decomposition of nitrogen oxide that enters the atmosphere from factory smoke and forest fires, for example. The amount of total ozone l in a vertical column of air is given in units of atmosphere centimeters (atm cm). This is the height of gaseous ozone if all the ozone in a vertical column of unit area were brought to normal temperature and surface pressure (NTP). In engineering literature, this thickness is usually expressed in cm(NTP). Around the equator, total ozone averages about 0.24 cm(NTP), and the amount increases with latitude: in

Table 5.3.2

Seasonal Variation of Atmospheric Ozone [in cm (NTP)][a]

	Month											
Latitude	Jan	Feb	Mar	Apr	May	June	July	Aug	Sept	Oct	Nov	Dec
90° N	0.33	0.39	0.46	0.42	0.39	0.34	0.32	0.30	0.27	0.26	0.28	0.30
80° N	0.34	0.40	0.46	0.43	0.40	0.36	0.33	0.30	0.28	0.27	0.29	0.31
70° N	0.34	0.40	0.45	0.42	0.40	0.36	0.34	0.31	0.29	0.28	0.29	0.31
60° N	0.33	0.39	0.42	0.40	0.39	0.36	0.34	0.32	0.30	0.38	0.30	0.31
50° N	0.32	0.36	0.38	0.38	0.37	0.35	0.33	0.31	0.30	0.28	0.29	0.30
40° N	0.30	0.32	0.33	0.34	0.34	0.33	0.31	0.30	0.28	0.27	0.28	0.29
30° N	0.27	0.28	0.29	0.30	0.30	0.30	0.29	0.28	0.27	0.26	0.26	0.27
20° N	0.24	0.26	0.26	0.27	0.28	0.27	0.26	0.26	0.26	0.25	0.25	0.25
10° N	0.23	0.24	0.24	0.25	0.26	0.25	0.25	0.24	0.24	0.23	0.23	0.23
0°	0.22	0.22	0.23	0.23	0.24	0.24	0.24	0.23	0.23	0.22	0.22	0.22
10° S	0.23	0.24	0.24	0.24	0.24	0.24	0.24	0.24	0.24	0.24	0.24	0.23
20° S	0.24	0.25	0.24	0.25	0.25	0.25	0.25	0.26	0.26	0.26	0.26	0.25
30° S	0.27	0.28	0.26	0.27	0.28	0.28	0.29	0.31	0.32	0.32	0.29	0.29
40° S	0.30	0.29	0.28	0.29	0.31	0.33	0.35	0.37	0.38	0.37	0.34	0.32
50° S	0.31	0.30	0.29	0.30	0.32	0.36	0.39	0.40	0.40	0.39	0.37	0.35
60° S	0.32	0.31	0.30	0.30	0.33	0.38	0.41	0.42	0.42	0.40	0.39	0.35
70° S	0.32	0.31	0.31	0.29	0.34	0.39	0.43	0.45	0.43	0.40	0.38	0.34
80° S	0.31	0.31	0.31	0.28	0.35	0.40	0.44	0.46	0.42	0.38	0.36	0.32
90° S	0.31	0.30	0.30	0.27	0.34	0.38	0.43	0.45	0.41	0.37	0.34	0.31

[a] From Robinson [3].

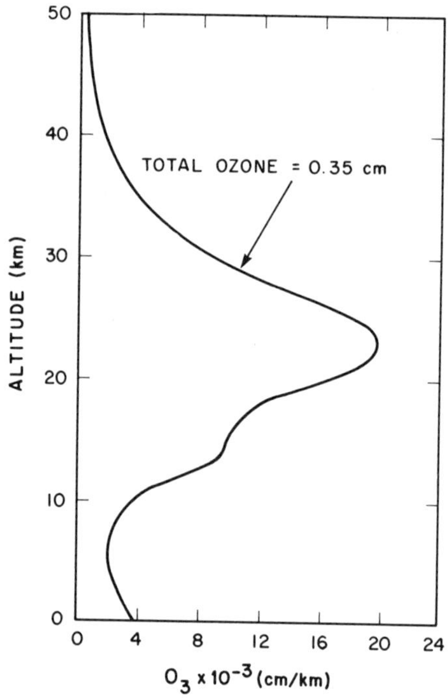

Figure 5.3.2 Variation of ozone concentration with geographic altitude. Adapted from Elterman [4]. This figure represents average conditions at midlatitudes.

the polar regions, total ozone may be as much as 0.46 cm(NTP). At high latitudes, there are distinct seasonal variations: in each hemisphere, it is maximum in the spring and minimum in the fall (in the sense of seasons in each hemisphere). Table 5.3.2 gives the seasonal variation of ozone content for the northern and the southern hemispheres [3].

The vertical distribution of ozone varies with latitude and season. It is mainly concentrated between a 10- and 35-km altitude. The profile given in Fig. 5.3.2 has been considered to represent average conditions [2, 4] at midlatitudes. This profile results in 0.35 cm(NTP) ozone. However the U.S. standard average is assumed to be 0.34 cm(NTP), and this value has been employed in establishing standard spectral irradiance curves [5].

5.4 Water Vapor

Water can exist in the atmosphere in three states, as gas, liquid, and ice. Water in gaseous state is called water vapor. The amount of water vapor present in the atmosphere can be defined in a number of ways. For the present context, we give two definitions: *mixing ratio* M_r and *precipitable water* w'. The *mixing ratio* is the ratio of the mass of water vapor present to the mass of dry air present in a unit volume. Precipitable water is the total amount of water vapor in the zenith direction, between the surface of the earth (or a surface at a certain elevation) and the top of the atmosphere. Therefore, precipitable water can be written

$$w' = \frac{1}{g}\int_0^\infty M_r \, dz, \qquad (5.4.1)$$

where z is the vertical height and g is the acceleration due to gravity. From this equation, the units of w' are mass per unit area. However, precipitable water is often described as the thickness of the liquid water that would be formed if all the vapor in the zenith direction were condensed at the surface of

Table 5.4.1

Mean Annual Mixing Ratio M_r (ppm) and Precipitable Water (μm) in the 2-km Layers[a]

Altitude (km)	Mean mixing ratio M_r (ppm)	Precipitable water w' (μm)
0	6300	11 224
2	3800	5 206
4	1900	2 057
6	900	710
8	270	150
10	37	22
12	17	8
14	10	4
16	9	3
18	13	3
20	18	4
22	27	4
24	38	4
26	58	4
28	86	5
30	128	2
31	150	–
	Total	19.41 mm

[a] From "Handbook of Geophysics and Space Environments," edited by S. L. Valley. Copyright © 1965 McGraw-Hill Book Company. Used with the permission of McGraw-Hill Book Company.

a unit area. Equation (5.4.1) can still be employed since a height of 1 cm corresponds to 1 g cm^{-2} of precipitable water. Few data on the vertical distribution of water vapor are available. Table 5.4.1 contains the mean annual mixing ratios and precipitable water in 2 km altitude intervals [2]. This table is based on a particular model of the mixing ratio distribution combined with the U.S.S.A. 1962. It is evident from this table that about one-half the precipitable water is concentrated in the first 2 km above sea level; beyond 12 km, precipitable water is practically nonexistent. The total amount of precipitable water is about 2 cm. The mean mixing ratio is highest in the first 2-km altitude. It decreases rapidly to its minimum value at a height of 16 km, and then slowly increases again.

The amount of precipitable water can be highly variable. It varies with season. An extremely dry atmosphere may contain as little as 1 mm of precipitable water and a humid atmosphere may contain more that 40 mm. Table 5.4.2 indicates the mean monthly distribution of precipitable water for a number of locations in the United States [6]. Within the coterminous United States, the average precipitable water varies from about 0.5 to 4.3 cm;

Table 5.4.2

Mean Monthly Distribution of Precipitable Water (cm)

	Jan	Feb	Mar	Apr	May	June	July	Aug	Sept	Oct	Nov	Dec
Albuquerque, NM	0.50	0.48	0.50	0.56	0.77	1.11	1.86	1.86	1.40	0.90	0.63	0.52
Bismarck, ND	0.52	0.56	0.63	0.91	1.36	1.97	2.27	2.16	1.59	1.13	0.75	0.60
Brownsville, TX	2.21	2.22	2.32	2.75	3.74	3.65	3.85	3.94	4.03	3.24	2.64	2.38
Cape Hatteras, NC	1.45	1.27	1.38	1.72	2.36	2.96	3.85	3.85	3.08	2.38	1.64	1.54
Caribou, ME	0.55	0.52	0.64	0.87	1.33	1.93	2.32	2.21	1.79	1.33	0.97	0.63
Charleston, SC	1.59	1.54	1.67	2.03	2.74	3.50	4.08	4.08	3.52	2.52	1.94	1.62
Columbia, MO	0.89	0.78	1.04	1.52	1.93	2.71	3.08	2.96	2.39	1.78	1.27	1.03
Dodge City, KA	0.66	0.65	0.73	1.04	1.49	2.10	2.64	2.52	1.98	1.29	0.89	0.72
El Paso, TX	0.69	0.67	0.73	0.79	1.05	1.62	2.36	2.39	2.01	1.25	0.89	0.76
Ely, NE	0.50	0.48	0.47	0.53	0.73	1.00	1.24	1.34	0.90	0.69	0.63	0.48
Ft. Worth, TX	1.18	1.26	1.41	1.96	2.68	3.25	3.63	3.59	3.15	2.22	1.51	1.30
Great Falls, MN	0.55	0.53	0.54	0.66	0.95	1.30	1.40	1.39	1.10	0.81	0.67	0.55
Lake Charles, LA	1.83	1.76	1.88	2.35	2.89	3.58	4.17	4.09	3.69	2.59	2.04	1.91
Medford, OR	1.13	1.02	0.97	1.00	1.30	1.59	1.64	1.65	1.43	1.27	1.30	1.06
Miami, FL	2.38	2.34	2.48	2.70	3.24	4.02	4.34	4.25	4.33	3.59	2.86	2.49
Nashville, TN	1.10	0.98	1.15	1.72	2.09	2.77	3.28	3.24	2.92	1.89	1.34	1.24
New York, NY	0.82	0.81	0.97	1.34	1.87	2.52	2.90	2.85	2.45	1.69	1.33	1.02
N. Omaha, NB	0.67	0.72	0.87	1.21	1.89	2.52	2.69	2.78	2.12	1.53	0.96	0.79
Phoenix, AZ	1.01	0.92	0.91	1.11	1.23	1.62	3.12	3.18	2.24	1.54	1.05	0.97
Santa Maria, CA	1.17	1.18	1.18	1.24	1.65	1.65	2.00	1.95	1.81	1.53	1.35	1.21
Washington, DC	0.99	0.94	1.08	1.52	2.13	2.79	3.24	3.17	2.65	1.83	1.33	1.04

[a] From [6].

the overall mean monthly values are given in Table 5.4.3 [7]. It can be seen from this table that the precipitable water during the summer months in the United States is about three times as high as during the winter months.

Many countries publish national contour maps of the monthly mean values of precipitable water. U.S. and Canadian data are available in [7] and [8], respectively. Bannon and Steele [9] have published world maps of precipitable water. For locations without such maps or tables, some simple correlation between precipitable water and some other atmospheric measurement is necessary. Numerous authors have proven the existence of correlations between precipitable water and partial pressure of water vapor, or dew-point temperature, or relative humidity. Some examples of such correlations are given below.

Using Bannon and Steele's maps, Cole [10] has developed the following correlation:

$$w' = 0.125 \exp(0.295 p_v^{1/2} - 0.803), \tag{5.4.2}$$

where w' is in centimeters and p_v (the vapor pressure at ground level) in millibars. This equation is valid for mean values and should be treated as an approximation of global values.

Smith [11] developed a correlation between precipitable water and dew-point temperature. His correlation is valid for the United States. The coefficients in this correlation vary with latitude and season. Atwater and Ball [12] have simplified Smith's correlation to

$$w' = \exp(0.07074 t_d + y), \tag{5.4.3}$$

where w' is in centimeters, t_d is the station dew-point temperature in degrees Celsius, and $y = -0.02290$ from April to June and 0.02023 for the remaining months.

For all seasons, Won [13] has developed a simple correlation based on Canadian data as follows:

$$w' = 0.1 \exp(2.2572 + 0.05454 t_d), \tag{5.4.4}$$

Table 5.4.3

Mean Monthly Values of Precipitable Water (cm) for the United States, Based on 1946–1956 Data[a]

Jan	Feb	Mar	Apr	May	June	July	Aug	Sept	Oct	Nov	Dec
0.98	0.94	1.05	1.37	1.89	2.49	2.99	2.90	2.28	1.72	1.20	1.11

[a] From Reitan [7].

where t_d is the station dew-point temperature in degrees Celsius and w' is in centimeters.

The precipitable water as obtained from Eqs. (5.4.2)–(5.4.4) applies to prevailing station pressure and temperature. However, the attenuation equations (discussed in the following two chapters) often require this quantity to be reduced to a datum of 1013.25 mbars pressure and 273 K temperature. Paltridge and Platt [14] suggest the following formula for reduction of w' to the datum conditions:

$$w = w'\left(\frac{p}{1013.25}\right)^{3/4}\left(\frac{273}{T}\right)^{1/2}, \tag{5.4.5}$$

where w is the reduced precipitable water in centimeters, p is station pressure in millibars, and T is the surface (dry-bulb) temperature in degrees Kelvin.

Leckner [15] has presented the following formula, which expresses precipitable water in terms of relative humidity:

$$w = 0.493\phi_r p_s/T, \tag{5.4.6}$$

where ϕ_r is relative humidity in fractions of one, T is ambient temperature in degrees Kelvin, and p_s is the partial pressure of water vapor in saturated air and is given by the following semiempirical equation:

$$p_s = \exp(26.23 - 5416/T). \tag{5.4.7}$$

The pressure and temperature correction is not necessary in Eq. (5.4.6) since it is already included in its numerical constant.

In this chapter, although we are interested in conditions under cloudless skies, data on precipitable water are usually given as the mean of all days. According to Hoyt [16], precipitable water on a cloudless day is 81% of the mean.

Among the various correlations given above, Eqs. (5.4.3) and (5.4.6) are believed to be most accurate. Through the following example we show a comparison of some of these correlations.

□ EXAMPLE 5.4.1. The pressure-corrected precipitable water vapor at a station where the ambient pressure is 990 mbars, the temperature 20°C, and the relative humidity 100%.

Solution. As the humidity is 100%, the dew-point temperature is equal to the ambient temperature. We can obtain the actual precipitable water either from Smith's, Won's, or Leckner's correlation.

(a) Smith's correlation, Eq. (5.4.3),

$$\begin{aligned} w' &= \exp(0.07074 \times 20 - 0.02290) = 4.02 \text{ cm from April to June} \\ &= \exp(0.07074 \times 20 + 0.02023) = 4.20 \text{ cm from July to March.} \end{aligned}$$

To obtain the pressure- and temperature-corrected precipitable water, we calculate the factor

$$\left(\frac{990}{1013.25}\right)^{3/4}\left(\frac{273}{293}\right)^{1/2} = 0.95.$$

Therefore,

$$\begin{aligned} w &= 4.02 \times 0.95 = 3.82 \text{ cm from April to June} \\ &= 4.20 \times 0.95 = 3.99 \text{ cm from July to March.} \end{aligned}$$

(b) Won's correlation, Eq. (5.4.4),

$$\begin{aligned} w' &= 0.1 \exp(2.2572 + 0.05454 \times 20) \\ &= 2.84 \text{ cm.} \end{aligned}$$

The pressure- and temperature-corrected precipitable water is

$$w = 2.84 \times 0.95 = 2.70 \text{ cm.}$$

(c) Leckner's method, Eq. (5.4.6),

$$w = 0.493[\exp(26.23 - 5416/293)(293)^{-1} = 3.89 \text{ cm},$$

and does not need further correction.

It is interesting to note that there is a wide difference between results obtained from the different correlations. However, Leckner's correlation is recommended to evaluate w, and when relative humidity is not known, Smith's correlation is recommended. □

5.5 Aerosols

An *aerosol* is a small solid or liquid particle that remains suspended in the air and follows the motion of the air within certain broad limits. Obviously, rain, snow, and hail are not aerosol particles. However, coagulated water-vapor molecules that follow the motion of the air are considered aerosols. In contrast to molecules of the permanent atmospheric gases, suspended particles within the atmosphere display considerable diversity in volume, size, distribution, form, and material composition. These particles are either of terrestrial origin (deriving, for example, from industrial smoke, pollen, volcanic eruptions, meteoric dust, sandstorms, forest fires, and agricultural and slash burning), or of marine origin (from salt crystals, ocean spray, and nuclei of hygroscopic salts on which water is condensed, for instance). Suspended water and ice particles in fog and clouds lie on the borderline of the definition of aerosol particles.

Natural aerosol particles range in radius from 10^{-3} to 10^2 μm; very small particles (called *Aitken particles*) from 10^{-3} to 10^{-1} μm; and large particles from 0.1 to 1 μm. Particles in the 1–100 μm range are called *giant particles.* The amount of aerosols in the atmosphere is sometimes specified in terms of the number of particles per cubic centimeter. The number of dust particles is usually greater over land than over water and greater in the drier seasons. Also, the number of dust particles is lower in cold polar or arctic air and higher in tropical air. To this number must be added the large amount of particulate matter injected into the atmosphere from domestic heating plants during winter.

Fig. 5.5.1 demonstrates an approximate distribution of the number density with altitude of the three particle size ranges [2]. The number density decreases exponentially with altitude, up to about 5 km, and remains constant at 10–15-km altitude. As a result of the thermal and dynamic nature of the atmosphere, the number density increases slightly in a layer centered around 20 km (*Junge layer*). Below 20 km, condensation and precipitation are the chief cleaners of "dry" dust from the atmosphere. Rain reduces the number of aerosol particles but increases the size of those that remain. Consequently, turbidity (in the optical sense) remains unchanged immediately after rain [17].

An atmosphere containing aerosols is also called turbid or hazy. A property of an aerosol-laden atmosphere that depletes direct solar radiation is called atmospheric turbidity. Turbidity is an optical parameter of the

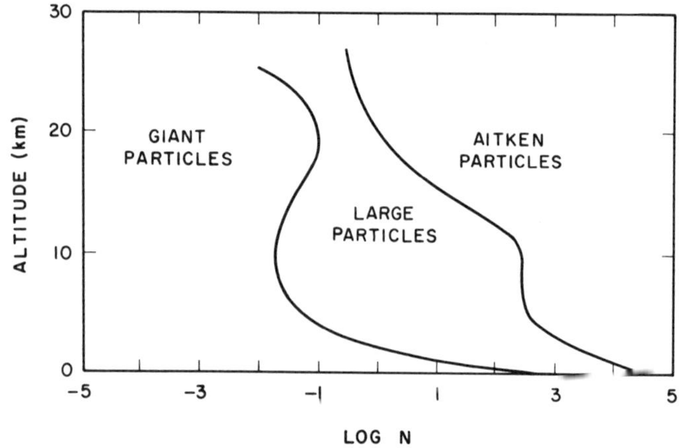

Figure 5.5.1 Variation of aerosol concentration with geographic altitude (N is number of particles per cubic centimeter). Adapted from "Handbook of Geophysics and Space Environments," edited by S. L. Valley. Copyright © 1965 McGraw-Hill Book Company. Used with the permission of McGraw-Hill Book Company.

atmosphere and can be roughly related to horizontal visibility, which in itself is a subjective measurement. Although we shall discuss the implications of the turbidity parameter in greater detail in the following two chapters, it is useful to state here that the presence of aerosols in the atmosphere can be quantified by any one of the following three parameters:

(1) number of dust particles per cubic centimeter,
(2) atmospheric turbidity, and
(3) visibility.

The number of dust particles per unit volume is not routinely measured. However, data on the remaining two climatic quantities are more common. In this book, we employ the turbidity parameter in the computation of direct and diffuse radiation. Values of this parameter for a number of U.S. locations are given in the next chapter (Table 6.6.2).

5.6 Relative Optical Path Length,[1] Relative Optical Mass *m*

When monochromatic radiation traverses a medium, each molecule (or particle, in the case of aerosols) attenuates energy. Attenuation is a function of the type and the number of molecules in the path of a solar ray. The number of molecules a solar ray strikes before reaching the ground is related to the distance traversed by the ray. Calculation of this distance, called the path length or slant path, is the subject of this section.

The density multiplied by the path length represents the mass of a substance in a column of unit cross section; this is also called the optical mass. The actual optical mass can be written

$$m_{\text{act}} = \int_0^\infty \rho\, ds, \tag{5.6.1}$$

where ds is the geometrical path length of the light ray from the sun and ρ is the density of the substance at ds. The integration is along a path s (called the slant path) traversed by the beam of radiation from the upper limits of the atmosphere to the ground (or to a surface at a certain height). Since refraction is wavelength dependent, the slant path varies with wavelength, and consequently Eq. (5.6.1) applies to monochromatic radiation. When the sun is at its zenith, the light path goes straight downward and ds equals the height of

[1] *Relative optical path length* accurately describes the subject of this section. However, in place of this lengthy expression, *optical mass* or simply *air mass* is often used.

an element dz, where z is the distance along the vertical direction. Thus the actual optical mass in the vertical direction is as follows:

$$m_{\text{act},v} = \int_0^\infty \rho \, dz. \tag{5.6.2}$$

This is the mass of a substance in a vertical column of unit cross section.

The relative optical mass m_r is defined as the ratio of the optical path along the oblique trajectory to the vertical path in the zenith direction. Thus

$$m_r = \int_0^\infty \rho \, ds \bigg/ \int_0^\infty \rho \, dz. \tag{5.6.3}$$

In the foregoing, the word "air" has been deliberately avoided since attenuation of a solar beam takes place not only by dry air molecules, but also by water vapor and aerosols, etc. Therefore, Eq. (5.6.3) should be solved separately for each one of the attenuating components of the atmosphere. Optical masses for the various components are discussed below.

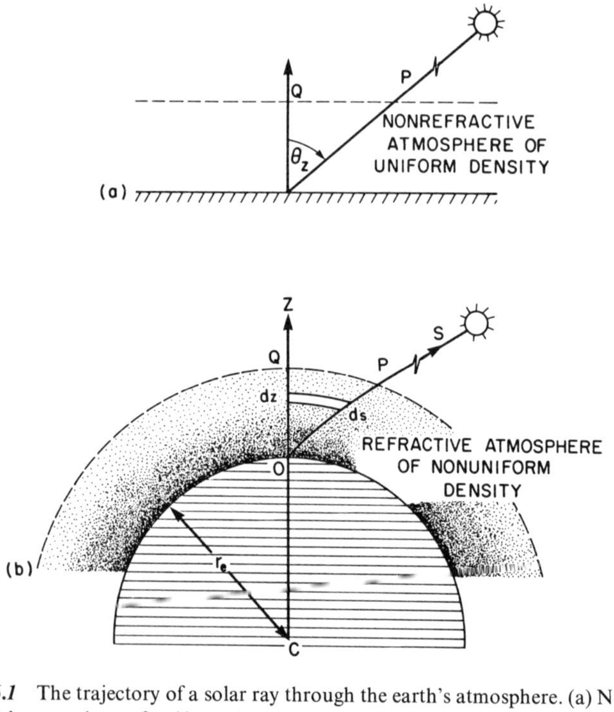

Figure 5.6.1 The trajectory of a solar ray through the earth's atmosphere. (a) Nonrefractive plane parallel atmosphere of uniform density. (b) Refractive spherical atmosphere of variable density.

Ignoring the earth's curvature and assuming that the atmosphere is nonrefractive and completely homogeneous (Fig. 5.6.1a), it can be seen that the relative optical mass applied to all the atmospheric constituents is

$$m'_r = \sec \theta_z. \tag{5.6.4}$$

The error in this equation, because the earth's curvature and the refraction of the real atmosphere have been neglected, is 0.25% at $\theta_z = 60°$, and increases to 10% at $\theta_z = 85°$.

However, density is actually variable with height. Furthermore, because of the curvature of the earth and refraction of the atmosphere, the slant path of the beam radiation will follow the path OP (Fig. 5.6.1b). Therefore Eq. (5.6.3) has to be evaluated through integration along the slant path and the zenith direction. The relative optical mass is obviously a function of the distribution, with height, of atmospheric density and refractive index: the relative mass from a mountain location will be different from that recorded at sea level.

Among the individual optical masses, we begin with that for clean dry air. Because ozone has a distinct concentration profile different from the rest of the air molecules, its relative optical mass will be treated separately from that of clean dry air.

5.7 Relative Optical Air Mass m_a

To solve Eq. (5.6.3), the density variation of an actual atmosphere is required. However, this information is generally available only for a Standard Atmosphere. From Kasten [18], the above equation can be written as

$$m_r = \frac{1}{\rho_0 z_0} \int_0^\infty \left[1 - \left(\frac{r_e}{r_e + z_0} \right)^2 \left(\frac{n_0}{n} \right)^2 \sin^2 \theta_{ob} \right]^{-1/2} \rho \, dz, \tag{5.7.1}$$

where

ρ_0 is the density on the ground,
z_0 is the height of a homogeneous atmosphere of density ρ_0 (8.43 km at NTP),
r_e is the mean earth radius (6370 km),
n_0 is the refractive index at ground level,
n is the refractive index at height z, and
θ_{ob} is the observed zenith angle.

Kasten solved this equation using the air density profile of the ARDC Model Atmosphere 1959, and the refractive index at wavelength 0.7 μm. Kasten presented tables of the relative optical air mass and the following formula, which approximates his tables:

$$m_r = [\cos\theta_z + 0.15(93.885 - \theta_z)^{-1.253}]^{-1}. \tag{5.7.2}$$

For ground-level measurements, this formula is accurate to better than 0.1% for zenith angles up to 86°. The greatest deviation, 1.25%, occurs at $\theta_z =$ 89.5°. The error is slightly greater for measurements at very high altitudes.

Equation (5.7.2) is applicable to a standard pressure of 1013.25 mbars at sea level; for other pressures, it should be modified. Although this task is very complicated, it is common to employ the following approximation to obtain the relative optical air mass for local conditions:

$$m_a = m_r(p/1013.25), \tag{5.7.3}$$

where p is local pressure in millibars.

In general, pressure correction must be applied to stations at 2000 m or higher. Even at stations not far above sea level, a pressure correction is necessary if the difference between the standard and local pressure is more than 20 mbars. The pressure above sea level may be obtained from [19]

$$p/p_0 = \exp(-0.0001184z), \tag{5.7.4}$$

where z is the station altitude in meters above sea level.

In the preceding discussion, we have developed expressions for the optical air mass; these will be used in the study of attenuation of solar radiation by dry air molecules. The corresponding expressions for relative optical water-vapor mass and relative optical ozone mass (since ozone is concentrated within a relatively thin layer) are now presented.

5.8 Relative Optical Water-Vapor Mass m_w

Equation (5.6.3) holds here as well; however, the density term in this equation now refers to the density of water vapor. The vertical water-vapor density profile is very different from that of dry air. Furthermore, since water vapor is concentrated mainly in the lower layers of the atmosphere, the limits of integration in this equation need not go to upper limits of the atmosphere. Kasten [18], on the basis of a study by Schnaidt [20], has developed the following formula:

$$m_w = [\cos\theta_z + 0.0548(92.650 - \theta_z)^{-1.452}]^{-1}. \tag{5.8.1}$$

Table 5.8.1

Values of Optical Masses for Air, Water Vapor, and Ozone

Zenith angle θ_z (deg)	Optical mass			
		Air	Water vapor	Ozone
	$m'_r = \sec\theta_z$ Eq. (5.6.4)	m_r (Kasten) Eq. (5.7.2)	m_w (Schnaidt) Eq. (5.8.1)	m_0 (Robinson) Eq. (5.9.1)
0	1.00	1.00	1.00	1.00
30	1.15	1.15	1.15	1.15
60	2.00	1.99	2.00	1.99
65	2.37	2.35	2.36	2.34
70	2.92	2.90	2.92	2.86
75	3.86	3.81	3.85	3.70
80	5.76	5.58	5.71	5.23
85	11.47	10.32	11.11	8.36
86	14.34	12.34	13.65	9.28
87	19.11	15.22	17.61	10.25
88	28.65	19.54	24.52	11.17
89	57.30	26.31	38.74	11.86
90	∞	36.51	75.12	12.12

In the attenuation equations to be studied in the following two chapters,

$$(w)(\text{relative optical water-vapor mass})$$

appear as one combined factor. The uncertainty involved in accurate determination of w is much greater than that in determination of its relative optical mass. Therefore, it is a common practice in the literature (and followed in this text) to use m_r, Eq. (5.7.2), instead of Eqs. (5.8.1) or (5.6.4). Use of Eq. (5.6.4) is not encouraged at $\theta_z > 70°$ (see Table 5.8.1). Furthermore, because w is a pressure-corrected quantity, it should not further be multiplied by $p/1013.25$ even if local pressure is different from the standard pressure.

5.9 Relative Optical Ozone Mass m_o

Figure 5.3.2 gives the concentration profile of ozone in the atmosphere. It is very different from the profiles of dry air or water vapor. Let us assume that all ozone is concentrated in a thin layer centered at a height z_3. Robinson [3], through simple geometry, has shown the following:

$$m_o = \frac{1 + z_3/r_e}{[\cos^2\theta_z + 2(z_3/r_e)]^{1/2}}. \qquad (5.9.1)$$

Table 5.8.1 also lists the optical mass for ozone. At $\theta_z = 80°$, m_o is substantially different from the other optical masses. In this table, ozone is assumed to be concentrated at a height of 22 km. Like water vapor, ozone also appears as

$$(l)(\text{relative optical ozone mass})$$

in the attenuation equations to be studied in the following two chapters. Again, the uncertainty involved in the accurate determination of ozone layer thickness is much greater than that in determination of its relative optical mass. Therefore, use of Eq. (5.9.1) is rare. In this text we shall replace m_o by m_r and use Eq. (5.7.2). Thereby a consistency can be established in machine calculations. Note, however, that this optical mass should not be further corrected for any variation of local pressure from the standard pressure.

□ EXAMPLE 5.9.1. Optical masses (a) m_a, (b) m_w, (c) m_o calculated when the sun is 15° above the horizon at a station located 2000 m above sea level, assuming that ozone is concentrated at 20 km above sea level. The values obtained are compared with $m = \sec\theta_z$.

Solution. Zenith angle $\theta_z = 75°$.
The station pressure is obtained from Eq. (5.7.4):

$$p = 1013.25\exp(-0.0001184 \times 2000) = 799.6 \text{ mbars.}$$

(a) Relative optical air mass m_a. m_a is obtained from Eq. (5.7.3) and m_r, the optical air mass at standard pressure is obtained from Eq. (5.7.2):

$$m_r = [\cos(75°) + 0.15(93.885 - 75)^{-1.253}]^{-1} = 3.81,$$

$$m_a = (3.81)(799.6/1013.25) = 3.01.$$

(b) Relative optical water-vapor mass m_w. m_w is obtained from Eq. (5.8.1):

$$m_w = [\cos(75°) + 0.0548(92.650 - 75)^{-1.452}]^{-1} = 3.85.$$

No pressure correction is to be applied to m_w.

(c) Relative optical ozone mass m_o. To evaluate m_o, we employ Eq. (5.9.1) with $z_3 = 20$ km in air and $r_e = 6370$ km:

$$m_o = \frac{1 + 20/6370}{[\cos^2(75°) + 2(20/6370)]^{1/2}} = 3.71.$$

The above optical masses can now be compared with that obtained from

$$m_r' = \sec(75°) = 3.86.$$ □

5.10 Relative Optical Aerosol Mass m_d

Aerosols are the most uncertain parameters in calculating solar radiation on the ground. They are highly variable in size, distribution, composition, and optical properties. For lack of more information, the optical air mass m_a [Eq. (5.7.3)] will be employed to compute optical aerosol mass.

Before ending this chapter, it is necessary to draw attention again to the meaning of two terms often used in the literature: *relative optical mass* and *absolute optical mass.* In this text, *relative* means relative to the zenith direction, whether or not the local conditions are at standard temperature and pressure. *Absolute optical mass* is defined by Eq. (5.6.1).

In the rest of the text, quite often we shall use the symbol m in a generic sense to represent either one of the relative optical path lengths.

5.11 Further Reading

In a single volume, Ref. [2] covers a large amount of material on atmospheric composition. Robinson [3], Kondratyev [21], and McCartney [22] give in detail the development of equations for the relative optical masses. Rodgers [23] has presented a simple formula for the relative optical air mass:

$$m_r = 35/(1224 \cos \theta_z + 1)^{1/2}. \tag{5.11.1}$$

This equation replaces Kasten's formula [Eq. (5.7.2)]. However, in the current literature it is common to use either of the two equations.

Further data on ozone distribution in the atmosphere are available in [24]. Van Heuken [25] presents a method of estimating atmospheric ozone which can be used in the absence of real data. Monthly means of the precipitable water over the United States are available in [6], and the atmospheric turbidity for the world in [26]. Shettle and Fenn [27] describe a number of atmospheric aerosol models and their optical properties.

Nomenclature

g	Earth's gravitational constant
l	Ozone layer thickness [cm(NTP)]
M_r	Mean mixing ratio (particles per million, ppm)
m	Optical mass (dimensionless)
m_a	Optical mass for air, under actual conditions (dimensionless)
m_d	Optical mass for dust (aerosol) particles (dimensionless)
m_o	Optical mass for ozone (dimensionless)

m'_r Optical mass for a homogeneous plane-parallel layer (dimensionless)
m_r Optical mass for air at standard conditions (dimensionless)
m_w Optical mass for water vapor (dimensionless)
N Number of particles per cubic centimeter
n Refractive index at height h (dimensionless)
n_0 Refractive index at ground level (dimensionless)
p Atmospheric pressure (mbars)
p_s Partial pressure of water vapor in saturated air (mbars)
p_v Partial pressure of water vapor (mbars)
r_e Mean earth radius, 6370 km
s Slant path (km)
T Ambient temperature (K)
t_d Dew-point temperature (°C)
w Precipitable water reduced to 1013.25-mbars pressure and 273 K temperature (cm)
w' Precipitable water under actual conditions (cm)
z Vertical path (km)
z_0 Height of a homogeneous atmosphere of density ρ_0, 8.43 km at NTP
z_3 Height of ozone layer concentration (km)
θ_z Zenith angle (degrees)
θ_{ob} Observed zenith angle (degrees)
ρ Density of air (kg m^{-3})
ρ_0 Density of air on ground (kg m^{-3})

References

1. U.S. Standard Atmosphere. U.S. Government Printing Office, Washington, D.C., 1976.
2. S. L. Valley (ed.), "Handbook of Geophysics and Space Environments," U.S.A.F. Cambridge Research Labs. McGraw-Hill, New York, 1965.
3. N. Robinson (ed.), "Solar Radiation." American Elsevier, New York, 1966.
4. L. Elterman, U.S., Visible and I.R. attenuation for altitudes to 50 km. AFCRL-68-0153, Environmental Research Paper No. 285 (1968).
5. M. P. Thekaekara (ed.), "The Energy Crisis and the Energy from the sun," pp. 21–49. Institute of Environmental Sciences, Mount Prospect, Illinois, 1974.
6. Precipitable Water over the United States, Vol. 1, Monthly means. NOAA Technical Report NWS 20 (1976).
7. C. H. Reitan, Distribution of precipitable water over the continental United States, *Bull. Am. Meteorol. Soc.* **41**(2), 79–87 (1960).
8. J. E. Hay, Precipitable water over Canada: II Distribution. *Atmosphere* **9**(14), 101–111 (1971).
9. J. K. Bannon and L. P. Steele, Average water vapor content of the air. Air Ministry, Meteorological Office, Geophysical Memoirs No. 102, London (1960).
10. R. J. Cole, Direct solar radiation data as input into mathematical models describing the thermal performance of buildings—II. Development of relationships, *Build. Environ.* **11**, 181–186 (1976)
11. W. Smith, Note on the relationship between total precipitable water and surface dew point. *J. Appl. Meteorol.* **5**, 726–727 (1966).
12. M. A. Atwater and J. T. Ball, Comparisons of radiation computations using observed and estimated precipitable water. *Appl. Meteorol.* **15**, 1319–1320 (1976).

13. T. Won, The simulation of hourly global radiation from hourly reported meteorological parameters—Canadian Prairie Area. *Conference, 3rd*, Canadian Solar Energy Society Inc., Edmonton, Alberta (1977).
14. G. W. Paltridge and C. M. R. Platt, "Radiative Processes in Meteorology and Climatology." American Elsevier, New York, 1976.
15. B. Leckner, The spectral distribution of solar radiation at the earth's surface—elements of a model. *Sol. Energy* **20**(2), 143–150 (1978).
16. D. V. Hoyt, A model for the calculation of solar global insolation. *Sol. Energy* **21**(1), 27–35 (1978).
17. E. C. Flowers, R. A. McCormick, and K. R. Kurfis, Atmospheric turbidity over the United States. *J. Appl. Meteorol.* **8,** 955–962 (1969).
18. F. Kasten, A new table and approximate formula for relative optical air mass. *Arch. Meteorol. Geophys. Bioklimatol. Ser. B* **14,** 206–223 (1966).
19. P. J. Lunde, "Solar Thermal Engineering." Wiley, New York, 1980.
20. F. Schnaidt, Berechnung der relativen Schichtdicken des Wasserdampfes in der Atmosphäre. *Meteorol. Z.* **55,** 296–299 (1938).
21. K. Y. Kondratyev, "Radiation in the Atmosphere." Academic Press, New York, (1969).
22. E. J. McCartney, "Optics of the Atmosphere." Wiley, New York, 1976.
23. C. D. Rodgers, The radiative heat budget of the troposphere and lower stratosphere. Res. Rep. No. A2, Planetary Circulations Project, Dept. of Meteorology, M.I.T., Cambridge, Massachusetts, (1967).
24. Ozone Data for the World. Meteorological Service for Canada. Issued Monthly.
25. T. K. Van Heuken, Estimating atmospheric ozone for solar radiation models. *Sol. Energy* **22**(1), 63–68 (1979).
26. Atmospheric Turbidity and Precipitable Chemistry Data for the World, Environmental Data Information Services, Asheville, North Carolina (1975).
27. E. P. Skettle and R. W. Fenn, Models of the atmospheric aerosols and their optical properties. *Proc. AGARD Conf.* No. 183, Optical Properties in the Atmosphere, pp. 2.1–2.16 (1975).

Chapter 6

SOLAR SPECTRAL RADIATION UNDER CLOUDLESS SKIES

6.1 Introduction

This chapter studies the basics of scattering and absorption in the atmosphere. Simple formulations are given describing attenuation of direct radiation by the various atmospheric constituents. An algorithm is presented by which direct and diffuse spectral radiation arriving on the ground can be easily computed. The equations are designed in a manner that permits the atmospheric parameters such as ozone layer thickness, precipitable water vapor, turbidity, and ground albedo[1] to be varied independently. Throughout this chapter we shall use for calculation purposes the extraterrestrial spectrum listed in Table 3.3.2.

6.2 Attenuation of Direct Solar Radiation: Basic Formulation

When solar radiation enters the earth's atmosphere, a part of the incident energy is removed by scattering and a part by absorption. Both influence the

[1] See Chapter 9 for full details on ground albedo, reflectance of the earth.

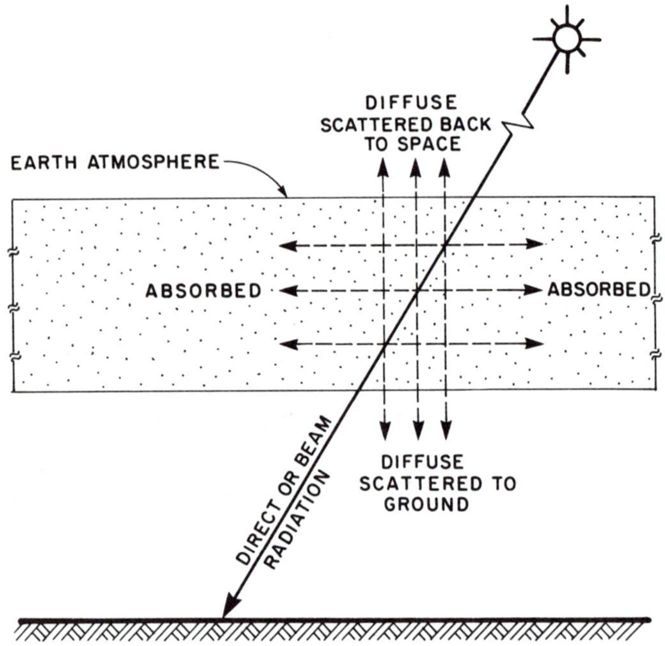

Figure 6.2.1 Distribution of direct, diffuse, and absorbed solar radiation.

extraterrestrial spectrum by considerably modifying the spectral energy passing through the atmosphere. The scattered radiation is called *diffuse radiation.* A portion of this diffuse radiation goes back to space and a portion reaches the ground. The radiation arriving on the ground directly in line from the solar disk is called *direct or beam radiation* (Fig. 6.2.1). A knowledge of spectral irradiance (direct and diffuse) arriving at the earth's surface is important for the design of certain solar energy applications such as photovoltaics. The quantity of total (integrated over all wavelengths, also called "broadband") direct and diffuse radiation is needed in the calculations of heating and cooling loads in architecture and in the design of flat-plate collectors, etc. However, only a knowledge of direct radiation is needed in designing many concentrating systems. This chapter develops a general procedure describing the attenuation of monochromatic radiation.

Consider a beam of parallel monochromatic radiation $\dot{I}_{0n\lambda}$ entering a homogeneous medium, as in Fig. 6.2.2. $\dot{I}_{n\lambda}$ is the flux emerging from the medium after traversing a distance m. (The subscript n indicates that the flux is measured normal to the rays.) The amount attenuated in passing through the medium can be evaluated by Bouguer's[2] law, also called Lambert's law

[2] See Section 6.23 for historical notes on Bouguer, Lambert, and Beer.

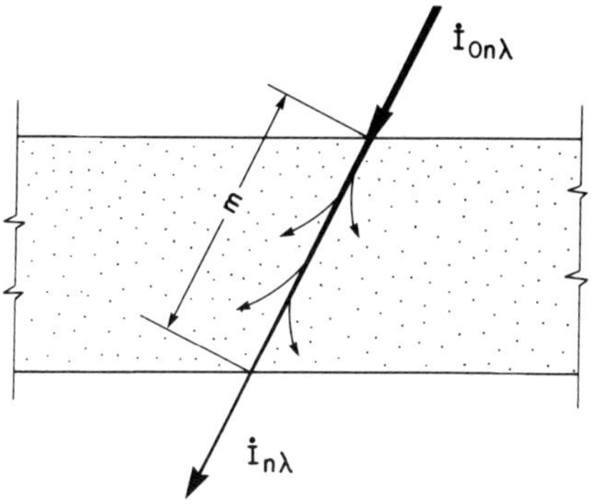

Figure 6.2.2 Attenuation of direct radiation through a homogeneous atmosphere.

or Beer's law. According to Bouguer's law, the attenuation of light through a medium is proportional to the distance traversed in the medium and to the local flux of radiation. This simple law applies in most instances (with the exception of absorption by gases), whether the energy is absorbed or scattered. Moreover, this law applies to monochromatic radiation only. In terms of this law,

$$\dot{I}_{n\lambda} = \dot{I}_{0n\lambda} \exp(-k_\lambda m), \tag{6.2.1}$$

where

k_λ is the monochromatic extinction or attenuation coefficient,
m is the optical path length, and
$k_\lambda m$ is the monochromatic extinction optical thickness[3] (dimensionless).

(Sometimes k_λ is called the extinction optical thickness.) In atmospheric applications, k_λ is usually measured relative to the zenith direction and should be dimensionless when m (in this case the relative optical mass) is dimensionless; otherwise, its unit is $(\text{length})^{-1}$.

Now, imagine that the medium in Fig. 6.2.2 is the earth's atmosphere. $\dot{I}_{0n\lambda}$ is the monochromatic extraterrestrial irradiance at the mean sun–earth distance. As this flux enters the earth's atmosphere, it will be depleted by various processes with extinction coefficients $k_{i\lambda}$. Since all extinction processes occur independently of each other (except for some small considerations as to which one comes first), the overall extinction optical thickness due to all processes can be written as the sum of individual thicknesses

$$k_\lambda m = \sum_{i=1}^{i=j} k_{i\lambda} m_i, \tag{6.2.2}$$

where

$k_{i\lambda}$ is the monochromatic attenuation coefficient for a single process, such

[3] *Optical thickness* should not be confused with *optical mass*.

as absorption by ozone, i varies from 1 to j, and j is the total number of processes, and

m_i is the optical path length for the process under consideration.

Equation (6.2.2) holds as long as a process obeys Bouguer's law.

The transmittance or transmission coefficient of the atmosphere is also useful. It is the ratio of radiation emerging from a medium to incident radiation. It is particularly useful when one or more of the attenuation processes does not obey the simple Bouguer's law. The monochromatic transmittance due to direct radiation can be written as

$$\tau_\lambda = \frac{\dot{I}_{n\lambda}}{\dot{I}_{0n\lambda}} = \prod_{i=1}^{i=j} \tau_{i\lambda}, \tag{6.2.3}$$

where $\tau_{i\lambda}$ is the transmittance due to a single process i and τ_λ is due to all the processes combined. The transmittance due to a single process is given by the following:

$$\tau_{i\lambda} = \exp(-k_{i\lambda} m_i). \tag{6.2.4}$$

Bouguer's law accurately represents attenuation processes of scattering by gases and scattering and absorption by aerosols. Absorption of radiation by gases is highly complex. Therefore, strictly speaking, this law is not valid as far as absorption by gases is concerned. Nevertheless, the application of Bouguer's law is usually extended to gas absorption [1]. In any case, once the spectral transmittance is known, either through Bouguer's law or otherwise, direct spectral irradiance on the ground can be evaluated through Eq. (6.2.3).

It is necessary now to emphasize that the attenuation coefficients $k_{i\lambda}$ as presented in this text are overall values applied to the total path length from the top of the atmosphere to the earth. A more accurate approach would be to divide the atmosphere into several layers, determine the coefficients $k_{i\lambda}$ and path lengths separately for each process, and then to calculate the transmittance. The latter approach has been used by a number of researchers.

The total broadband (summed over all wavelengths) direct normal irradiance on the ground can now be written as follows:

$$\dot{I}_n = \left(\frac{r_0}{r}\right)^2 \sum_{\lambda=0}^{\infty} \dot{I}_{n\lambda} \Delta\lambda, \tag{6.2.5}$$

or

$$\dot{I}_n = \left(\frac{r_0}{r}\right)^2 \sum_{\lambda=0}^{\infty} \left(\dot{I}_{0n\lambda} \prod_{i=1}^{i=j} \tau_{i\lambda}\right) \Delta\lambda. \tag{6.2.6}$$

Once the direct normal irradiance on the ground is known, beam[4] irradiance on a horizontal surface $\dot{I}_b$ can be calculated from

$$\dot{I}_b = \dot{I}_n \cos \theta_z, \tag{6.2.7}$$

where θ_z is the zenith angle.

It can now be said that the problem of computing beam radiation on earth reduces to the problem of determining the atmospheric transmission coefficients $\tau_{i\lambda}$.

In the following sections, we shall discuss the scattering and absorption processes and present the required transmission coefficients.

Before proceeding further, it seems appropriate to point out that in dealing with attenuation of direct solar radiation, it is quite common in the literature to express Bouguer's law also in decadal form,

$$\dot{I}_{n\lambda}/\dot{I}_{0n\lambda} = 10^{-k'_\lambda m}, \tag{6.2.8}$$

where, naturally,

$$k_\lambda = k'_\lambda \log 10 = 2.3k'_\lambda. \tag{6.2.9}$$

6.3 Scattering of Direct Solar Radiation

When an electromagnetic wave strikes a particle, a part of the incident energy is scattered in all directions. This scattered energy is called diffuse radiation. The energy scattered by spherical particles can be theoretically obtained by solution of Maxwell's electromagnetic wave equation in spherical polar coordinates. All particles in nature, whether the size of an electron or a planet, scatter radiation. A particularly simple solution is obtained when the particle is spherical and is much smaller than the wavelength of incident radiation. This solution was derived late in the nineteenth century by Lord Rayleigh and in his honor is called *Rayleigh's theory*. This theory is particularly useful in studying scattering of solar radiation by air molecules. Rayleigh, through his theory, explained the blue of the sky under extremely clear conditions.

When the particle size is of the order of wavelength of incident radiation, the solution of the wave equation becomes formidable. The solution was first successfully attempted at the beginning of this century by Gustav Mie and in his honor is named *Mie's theory*. Rayleigh's solution forms one individual case of Mie's theory.

[4] The expressions "direct" radiation and "beam" radiation will be used synonymously and interchangeably. The subscript n is reserved for direct normal incidence only.

For mathematical treatment, a convenient parameter to express the size of a scattering particle is $\pi D/\lambda$, where D is the particle diameter. Let n be the index of refraction and λ the wavelength in micrometers. It is considered that [2]

(1) when $\pi D/\lambda < 0.6/n$, scattering is governed by Rayleigh's theory, and in a cloudless atmosphere applies to air molecules, most of which have a size ≈ 1 Å (note that radio wave scattering by clouds is also Rayleigh scattering);

(2) when $\pi D/\lambda > 5$, scattering is chiefly a diffuse reflection process seldom occuring in the earth's atmosphere; and

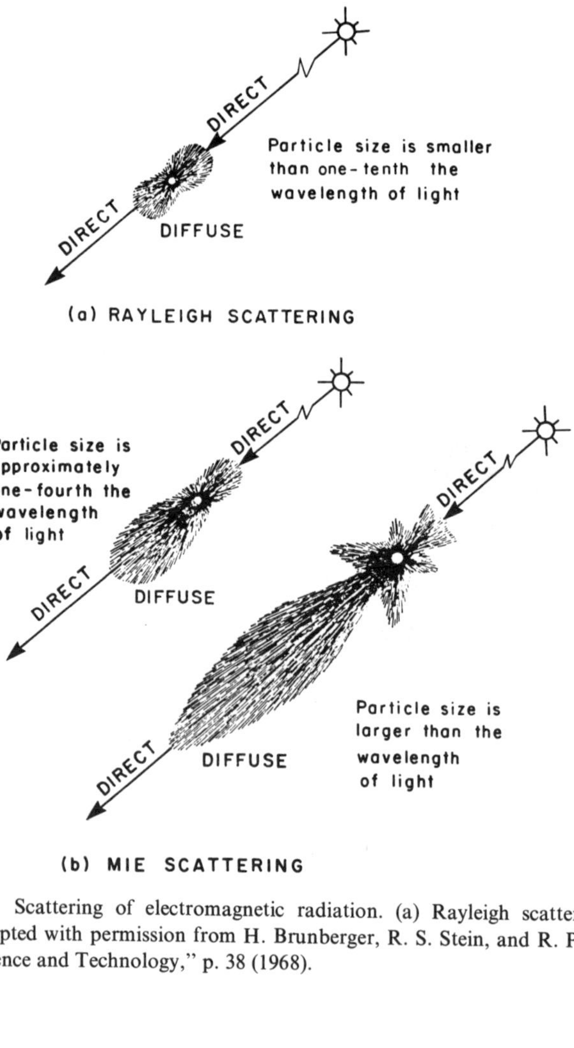

Figure 6.3.1 Scattering of electromagnetic radiation. (a) Rayleigh scattering. (b) Mie scattering. Adapted with permission from H. Brunberger, R. S. Stein, and R. Powell, "Light Scattering: Science and Technology," p. 38 (1968).

(3) when $0.6/n < \pi D/\lambda < 5$, scattering is governed by Mie's theory, and applies to scattering by particles of size greater than 10 Å, such as aerosols.

Figure 6.3.1 shows the difference between the Rayleigh and Mie modes of scattering. In the Rayleigh mode (Fig. 6.3.1a) the scattering process is identical in forward and backward directions. In addition, scattering is maximum in forward and backward directions. It is minimum at 90° to the line of incidence. Greater scattering occurs when incident radiation is of a shorter wavelength. In Mie scattering (Fig. 6.3.1b), more energy is scattered in a forward than in a backward direction. Furthermore, as the particle size increases, so does forward scattering, and shape of the scattering "balloon" is altered.

Radiation scattered by one particle strikes other particles in the medium, and this process, called multiple scattering, continues in the atmosphere. In a clean dry atmosphere, about half of the energy thus scattered goes back into space and the other half reaches the ground as scattered radiation. In an atmosphere containing dust particles, more scattered energy reaches the ground because of greater forward scattering. Mathematical formulations dealing with scattering by single particles and scattering by a number of particles in a medium have been well developed in the literature. In this section we are interested in the depletion of a direct solar beam by primary scattering. Diffuse radiation arriving on the ground due to scattering will be treated in a later section.

In the next three sections, we shall study depletion of direct radiation due to the following:

(1) Scattering by air molecules (Rayleigh scattering) represented by the attenuation coefficient $k_{r\lambda}$.

(2) Scattering by water vapor and dust particles (Mie scattering) represented by the attenuation coefficients $k_{ws\lambda}$ and $k_{d\lambda}$, respectively. This procedure is somewhat outdated. However, since we shall use these coefficients in describing the ASHRAE algorithm in Chapter 7, it seems appropriate that we present the necessary basic material here.

(3) Attenuation by aerosols (Mie scattering) represented by Ångström's turbidity coefficient $k_{a\lambda}$. This will replace the coefficients $k_{ws\lambda}$ and $k_{d\lambda}$ in (2) above.

6.4 Rayleigh Scattering of Air Molecules

Scattering by air molecules described by Rayleigh's theory is based on the assumption that the scattering particles are spherical, that they are less than 0.2λ in diameter, and that the particles scatter independently of one another.

Rayleigh's mathematical formulation includes considerations such as the number of particles per unit volume and the refractive index of the particles. This theory is well described in the texts of Siegel and Howell [2] and Kondratyev [3], for example. The essence of the theory is that the monochromatic optical extinction coefficient varies approximately as λ^{-4}, and this has been verified experimentally. As air density and composition vary with altitude, so does the value of the attenuation coefficient. For the U.S.S.A. 1962, Elterman [4] has computed in detail the coefficients at kilometer intervals for altitudes up to 50 km. These coefficients are available in tabular form for wavelengths 0.27–4.00 μm and are listed in Table 6.4.1. The values in this table are valid for sea-level standard temperature and pressure. Although these coefficients are quite accurate, it is also practical to express these

Table 6.4.1

Rayleigh Spectral Attenuation Coefficients

Wavelength (μm)	Rayleigh spectral attenuation coefficients $k_{r\lambda}$ Eq. (6.4.1)	Elterman [4]
0.270	1.831	1.928
0.280	1.573	1.645
0.300	1.187	1.222
0.320	0.912	0.927
0.340	0.713	0.717
0.360	0.564	0.564
0.380	0.453	0.450
0.400	0.367	0.364
0.450	0.227	0.223
0.500	0.148	0.145
0.550	0.100	0.098
0.600	0.070	0.069
0.650	0.051	0.050
0.700	0.037	0.037
0.800	0.022	0.021
0.900	0.013	0.013
1.026	0.008	0.007
1.060	0.007	0.003
1.670	0.001	0.001
2.170	0	0
3.500	0	0
4.000	0	0

coefficients in a simple formula. On the basis of Penndorf's [5] theoretical formulation of molecular scattering, Leckner [6] has presented the following approximate formula for the scattering coefficient of dry air[5] at standard conditions:

$$k_{r\lambda} = 0.008735\lambda^{-4.08}. \tag{6.4.1}$$

The coefficients from (6.4.1) are also listed in Table 6.4.1. In the ultraviolet, there are some differences between Elterman's values and those from (6.4.1). These differences are not too important, since most of the energy in these wavelengths is absorbed by ozone before it arrives on the earth—a topic treated in Section 6.12.

Because of the variation of the Rayleigh scattering coefficient with λ^{-4}, the spectral transmittance of air molecules rapidly increases with wavelength

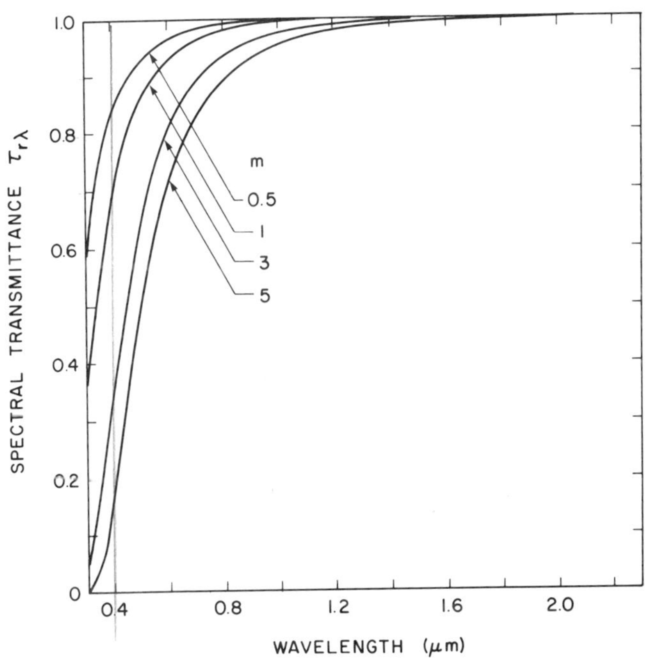

Figure 6.4.1 Rayleigh spectral transmittance as a function of air mass.

[5] The Rayleigh theory of molecular scattering applies also to atmospheric air containing water vapor in gaseous form. However, most formulations such as Eq. (6.4.1) are based on dry air at standard conditions. This results in only a slight overall error, because even a large amount of humidity does not substantially increase the number density of wet air molecules. Moreover, the amount of moisture present in the atmosphere is highly variable.

and decreases with increasing optical air mass. We rewrite the complete expression for transmittance by Rayleigh scattering

$$\tau_{r\lambda} = \exp(-0.008735\lambda^{-4.08} m_a), \tag{6.4.2}$$

where m_a is the relative optical air mass at actual pressure. Multiplication by m_a is necessitated because $k_{r\lambda}$ in Eq. (6.4.1) is evaluated at standard pressure and at $m_a = 1$. At high geographic altitudes, $m_a < 1$ and $m_a \to 0$ at the upper limits of the earth's atmosphere.

Equation (6.4.2) is plotted in Fig. 6.4.1 and shows the influence of wavelength and air mass on transmittance. At wavelengths greater than 1, scattering is negligible. At wavelength 0.5 μm and air mass 1, the atmosphere is more than 85% clear to solar radiation. It may therefore be concluded that molecular scattering is limited to short-wave radiation only.

6.5 Mie Scattering of Water Vapor and Dust Particles

We now turn our attention to scattering by the two other elements of the atmosphere: water vapor and dust particles. The size of water-vapor molecules compares to that of many dry air molecules. However, there is often coagulation of several water-vapor molecules which creates liquid particles several times bigger than their molecules. Also, it is known that under suitable conditions of humidity, small dust (or other) particles will grow in size because of water condensing on them. Therefore, the number and size of particles in the atmosphere are influenced by the presence of water vapor.

Scattering of solar radiation by dust ("dry" or "wet") particles is very difficult to handle mathematically. Van de Hulst [7] presents a detailed treatment of the Mie theory which deals with a problem of this kind. However, it is difficult to have all the required information available to employ this theory. The difficulty lies in the variability of the form, size, distribution, and nature of dust particles found in the atmosphere which are not measured and reported regularly. In this text, experimentally derived values of the attenuation coefficients are presented and are believed to be accurate enough for the present purpose.

In the literature, there are two simple but different procedures which treat scattering of radiation by particles other than dry air molecules. In this section, following Moon [8], we shall present separate coefficients for water and for dust scattering. In the next section treating the second procedure, developed by Ångström [9, 10], we shall present a combined coefficient for the two. Both approaches have been employed in the literature.

According to the first procedure, the coefficient for water "vapor" scattering is as follows:

$$k_{ws\lambda} = 0.008635\lambda^{-2}. \tag{6.5.1}$$

It is valid for $m_r = 1$ and $w = 1.0$ cm of precipitable water. It is quite apparent from the wavelength exponent of this equation that it does not apply to Rayleigh scattering of water (vapor) molecules. It is, therefore, obvious that this equation essentially treats water droplets, although in the literature it is usually referred to as water-vapor scattering.

The coefficient for dust scattering is as follows:

$$k_{d\lambda} = 0.08128\lambda^{-0.75}. \tag{6.5.2}$$

It is valid for $m_a = 1$ and 800 dust particles per cubic centimeter. It is generally considered that an atmosphere containing 200 dust particles per cubic centimeter is very clean and an atmosphere containing 800 dust particles per cubic centimeter is very polluted.

In comparing the Rayleigh scattering coefficient with those for water and dust particles, it becomes apparent that the last two are less sensitive to wavelength than the first. Let us define the spectral transmittances due to scattering by water vapor and dust particles as

$$\tau_{ws\lambda} = \exp[-0.008635\lambda^{-2}wm_r] \tag{6.5.3}$$

and

$$\tau_{d\lambda} = \exp[-0.08128\lambda^{-0.75}(d/800)m_a], \tag{6.5.4}$$

respectively.

We now examine Ångström's method, which treats attenuation by "wet" and "dry" particulate matter in a single stroke.

6.6 Ångström's Turbidity Formula for All Aerosols

In 1940, when Moon [8] presented the attenuation coefficients for particles, the attenuation was assumed to be caused by scattering effects only; however, the particulate matter absorbs as well as scatters electromagnetic radiation. Spectral investigations give good reasons for supposing that, in general, extinction by aerosol due to scattering or absorption, or by both combined, is a continuous function of wavelength without selective bands or lines. Absorption by aerosols may display some spectral selectivity, but it is not expected to be highly structured. In general, scattering is much greater than absorption. It can, therefore, be said that the coefficients in Eqs. (6.5.1)–(6.5.4) represent the effects of both scattering and absorption.

In Moon's coefficient for dust attenuation, the number of particles per unit volume can be varied; however, the coefficient is independent of the size of dust particles. We have discussed earlier that aerosols contain particles of various sizes. The next step, then, is to incorporate the particle size in the attenuation formula. Furthermore, since attenuation effects of scattering and absorption by dust are difficult to separate, Ångström [9, 10] suggests a single formula generally known as *Ångström's turbidity formula*, given by the following:

$$k_{a\lambda} = \beta\lambda^{-\alpha}. \tag{6.6.1}$$

In this formula, β is called Ångström's turbidity coefficient, α is the wavelength exponent, and the wavelength λ is in micrometers. It is called "turbidity" because scattering of solar radiation by matter other than dry air molecules is called turbidity of the atmosphere (in the optical sense). Consequently, $k_{a\lambda}$ includes attenuation due to "dry" as well as "wet" dust particles—that is, all aerosols.

In Eq. (6.6.1), β, which varies from 0.0 to 0.5 or even higher, is an index representing the amount of aerosols present in the atmosphere in the vertical direction. The wavelength exponent α is related to the size distribution of the aerosol particles. Large values of α indicate a relatively high ratio of small particles to large particles. It appears obvious that α should vary from 4 to 0; when the aerosol particles are very small, of the order of air molecules, α should approach 4, and it should approach 0 for very large particles. Generally, α has a value of between 0.5 and 2.5: a value of 1.3 is commonly employed, since it was originally suggested by Ångström. A good average value for most natural atmospheres is $\alpha = 1.3 \pm 0.5$.

The parameters β and α can be determined simultaneously with a dual-wavelength sun photometer by measuring aerosol attenuation at two wavelengths where molecular absorption is either absent or is minimal. The wavelengths usually chosen are 0.38 and 0.5 μm. At 0.38 μm there is no molecular absorption, and at 0.5 μm ozone has weak absorption (Table 6.12.1). However, β alone can be measured with a single-wavelength Volz instrument by assuming $\alpha = 1.3$. Or β can be measured at $\lambda = 1$ μm, where the exponent α does not enter into the calculations. At this wavelength only water vapor has a weak absorption band (Table 6.13.1). Examples 6.6.1 and 6.14.3 illustrate these points.

Table 6.6.1 lists combinations of the parameters β and α, which may be considered for various degrees of atmospheric cleanliness. It is necessary to keep in mind that a high value of β and a low value of α (signifying particles of large size) also represent turbid or very turbid skies. For example, $\beta = 0.3$ and $\alpha = 0.5$ signify very turbid skies.

Table 6.6.1

Parameters for Various Degrees of Atmospheric Cleanliness

Atmosphere	β	α	Visibility (km)
Clean	0.00	1.30	340
Clear	0.10	1.30	28
Turbid	0.20	1.30	11
Very turbid	0.40	1.30	<5

Like many other climatic variables, β and α can vary throughout an individual day simply because of changes in temperature that cause evaporation or condensation of moisture in the atmosphere. These changes can decrease or increase the value of these parameters. For a number of U.S. locations, long-term monthly average values of β [11] are listed in Table 6.6.2. These values of β are at $\lambda = 1\ \mu$m. Within the continental United States, β varies from 0.02 to 0.3. It is maximum during summer periods.

Routine measurements of β and α with a dual-wavelength sun photometer are carried out at a number of locations. When such measurements are not available, the turbidity parameter β may be determined from a measurement of visibility in the horizontal direction, if such a measurement exists. This visibility is also called the *meteorological range*. For visibilities greater than 5 km, the turbidity parameter can be determined from the following equation developed by McClatchey and Selby [12] and others:

$$\beta = (0.55)^{\alpha}(3.912/\mathrm{Vis} - 0.01162)[0.02472(\mathrm{Vis} - 5) + 1.132], \quad (6.6.2)$$

where Vis is in kilometers. In order to predict β, it is necessary first to guess the value of α. Unfortunately, visibility measured even with the help of a proper instrument is somewhat subjective in nature. Therefore, the above relation yields only a gross estimate of β.

The relations of α, β, and visibility are demonstrated through Fig. 6.6.1. At a fixed value of β a lower value of α signifies higher visibility—that is, higher atmospheric transparency. By inference it can be concluded that lower values of α (larger average particle size) would result in higher amounts of solar radiation reaching the ground.

Using Ångström's turbidity formula, we can write the aerosol transmittance as

$$\tau_{a\lambda} = \exp(-\beta\lambda^{-\alpha}m_a). \quad (6.6.3)$$

Figure 6.6.2 shows the variation of spectral transmittance $\tau_{a\lambda}$ with a number of βm_a combinations and two values of α, $\alpha = 1.3$ and 0.7, respectively. These

Table 6.6.2

Mean Atmospheric Turbidity β at $\lambda = 1\ \mu m$[a]

	Jan	Feb	Mar	Apr	May	Jun	Jul	Aug	Sep	Oct	Nov	Dec
Albuquerque, NM	0.027	0.035	0.043	0.061	0.053	0.043	0.040	0.045	0.032	0.026	0.022	0.016
Apalachicola, FL	0.132	0.136	0.164	0.209	0.238	0.235	0.228	0.237	0.246	0.192	0.195	0.143
Bismark, ND	0.053	0.073	0.120	0.125	0.121	0.106	0.123	0.108	0.067	0.056	0.025	0.033
Brownsville, TX	0.050	0.058	0.065	0.075	0.075	0.080	0.080	0.074	0.067	0.060	0.065	0.070
Boston, MA	0.085	0.098	0.133	0.157	0.175	0.171	0.184	0.173	0.11	0.105	0.084	0.073
Cape Hatteras, NC	0.056	0.077	0.092	0.121	0.149	0.174	0.242	0.254	0.211	0.092	0.068	0.049
Caribou, ME	0.048	0.059	0.049	0.084	0.073	0.080	0.087	0.075	0.046	0.050	0.034	0.048
Charleston, SC	0.056	0.077	0.092	0.121	0.149	0.174	0.242	0.254	0.211	0.092	0.068	0.049
Columbia, MO	0.036	0.050	0.054	0.071	0.090	0.088	0.092	0.089	0.077	0.043	0.032	0.039
Dodge City, KA	0.032	0.037	0.043	0.061	0.082	0.073	0.089	0.092	0.076	0.035	0.025	0.038
El Paso, TX	0.050	0.058	0.065	0.075	0.075	0.080	0.080	0.074	0.067	0.060	0.065	0.070
Ely, NE	0.050	0.058	0.065	0.075	0.075	0.080	0.080	0.074	0.067	0.060	0.065	0.070
Ft. Worth, TX	0.036	0.079	0.088	0.121	0.124	0.133	0.127	0.190	0.144	0.082	0.039	0.036
Fresno, CA	0.063	0.076	0.078	0.050	0.105	0.059	0.062	0.053	0.077	0.074	0.053	0.031
Great Falls, MN	0.054	0.063	0.073	0.080	0.078	0.065	0.067	0.066	0.066	0.064	0.061	0.050
Lake Charles, LA	0.132	0.136	0.164	0.209	0.238	0.235	0.228	0.237	0.246	0.192	0.195	0.143
Madison, WI	0.053	0.062	0.074	0.099	0.127	0.109	0.108	0.116	0.075	0.072	0.070	0.052
Medford, OR	0.093	0.126	0.159	0.192	0.225	0.196	0.166	0.175	0.218	0.199	0.146	0.093
Miami, FL	0.132	0.136	0.164	0.209	0.238	0.235	0.228	0.237	0.246	0.192	0.195	0.143
Nashville, TN	0.090	0.110	0.130	0.150	0.170	0.200	0.185	0.170	0.150	0.130	0.110	0.090
New York, NY	0.096	0.097	0.139	0.154	0.167	0.177	0.219	0.201	0.163	0.141	0.097	0.077
N. Omaha, NB	0.036	0.050	0.054	0.071	0.090	0.088	0.092	0.089	0.077	0.043	0.032	0.039
Phoenix, AZ	0.050	0.058	0.065	0.075	0.075	0.080	0.080	0.074	0.067	0.060	0.065	0.070
Raleigh, NC	0.065	0.083	0.100	0.125	0.150	0.250	0.230	0.230	0.165	0.100	0.085	0.070
Santa Maria, CA	0.093	0.126	0.159	0.192	0.225	0.196	0.166	0.175	0.218	0.199	0.146	0.093
Seattle, WA	0.093	0.126	0.159	0.192	0.225	0.196	0.166	0.175	0.218	0.199	0.146	0.093
Washington, DC	0.120	0.140	0.160	0.180	0.200	0.220	0.240	0.210	0.180	0.160	0.130	0.140

[a] From SOLMET Vol. 2 [11].

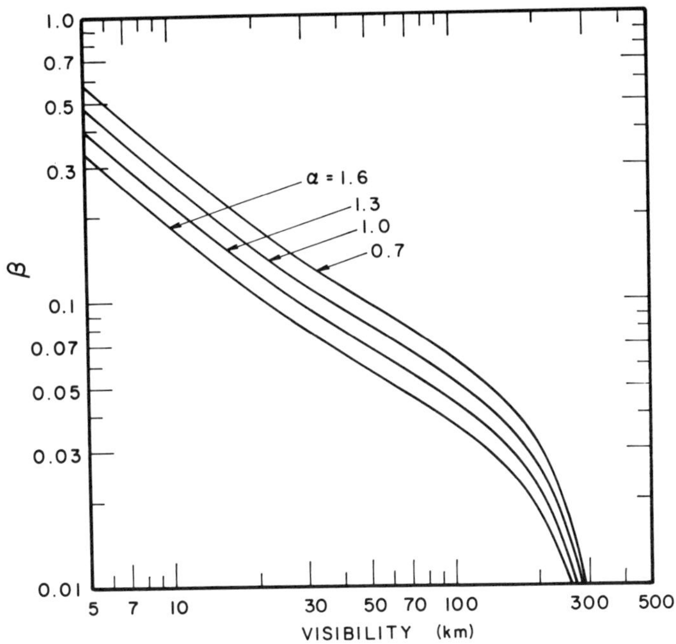

Figure 6.6.1 Variation of the horizontal visibility with α and β.

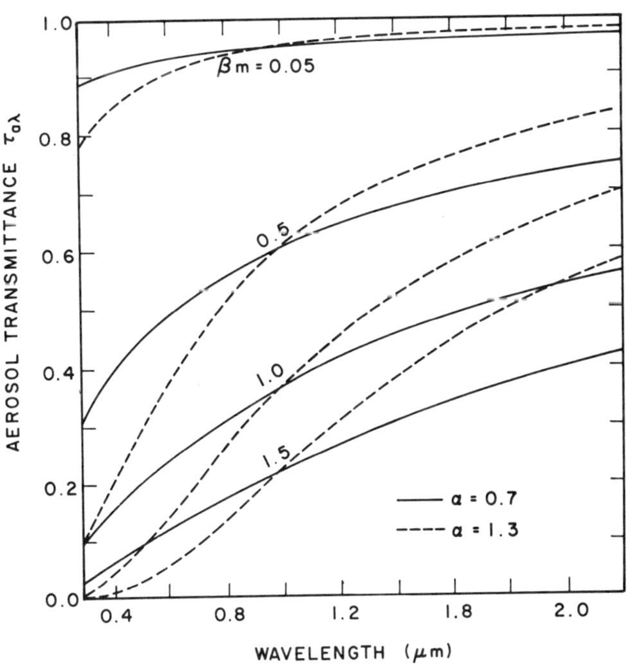

Figure 6.6.2 Aerosol spectral transmittance as a function of α and βm.

plots indicate that the wavelength has a very pronounced effect when the βm_a product is large. Furthermore, $\lambda = 1$ μm acts as a switchover point for the two wavelength exponents considered. At wavelengths shorter than 1 μm, finer particles transmit less energy than the coarser particles. At wavelengths longer than 1 μm, the reverse is true.

□ EXAMPLE 6.6.1. β, α, and the visibility determined for the U.S. standard atmosphere, in which the vertical aerosol optical depths at 0.38 and 0.5 μm are 0.3538 and 0.2661, respectively.

Solution. The optical depths can be written

$$k_{a\lambda}\big|_{\lambda=0.38\,\mu\text{m}} = \beta(0.38)^{-\alpha} = 0.3538$$

and

$$k_{a\lambda}\big|_{\lambda=0.5\,\mu\text{m}} = \beta(0.5)^{-\alpha} = 0.2661.$$

From the above two equations we can write

$$\alpha = \frac{\ln(0.3538/0.2661)}{\ln(0.50/0.38)} = 1.038.$$

We can now compute β at either of the two wavelengths,

$$\beta\big|_{\lambda=0.38} = 0.1296, \qquad \beta\big|_{\lambda=0.5} = 0.1296.$$

That is, β is independent of wavelength.

Visibility can be computed from Eq. (6.6.2):

$$0.1296 = (0.55)^{1.038}(3.91/\text{Vis} - 0.01162)[0.02472(\text{Vis} - 5) + 1.132],$$

which gives Vis = 24.18 km. (The U.S.S.A. lists a visibility of 23 km). □

6.7 Direct Transmittance due to Continuum Attenuation

We can now calculate $\tau_{c\lambda}$, the monochromatic transmittance of the atmosphere due only to continuum attenuation, as follows:

$$\tau_{c\lambda} = \tau_{r\lambda}\tau_{ws\lambda}\tau_{d\lambda} = \exp\{-[k_{r\lambda}m_a + k_{ws\lambda}wm_r + k_{d\lambda}(d/800)m_a]\}. \quad (6.7.1)$$

Using m as a generic symbol for all relative optical masses, we have the following:

$$\tau_{c\lambda} = \exp\{-[k_{r\lambda} + k_{ws\lambda}w + k_{d\lambda}(d/800)]m\}. \quad (6.7.2)$$

The attenuation coefficients $k_{ws\lambda}$ (scattering by water) and $k_{d\lambda}$ (scattering and absorption by dust) can be replaced by the single turbidity formula (6.6.1) of Ångström. Thus $\tau_{c\lambda}$ can also be written as follows:

$$\tau_{c\lambda} = \tau_{r\lambda}\tau_{a\lambda} = \exp[-(k_{r\lambda} + k_{a\lambda})m]. \qquad (6.7.3)$$

Equations (6.7.2) and (6.7.3) should give identical results, provided an exact equivalence can be established between the two groups of parameters, α and β on one side, and precipitable water and number of dust particles on the other. We shall not deal with Eqs. (6.7.1) or (6.7.2) until the end of Chapter 7, where the ASHRAE algorithm is presented.

In the preceding discussion, we have treated continuum attenuation of solar radiation due to scattering by dry air molecules and scattering and absorption by aerosols. In the next section, we shall discuss selective absorption by dry air molecules and water vapor.

□ EXAMPLE 6.7.1. Calculation of the spectral transmittances due to the continuum attenuation of direct solar irradiance at $\lambda = 0.7$ and $2.0\ \mu$m under cloudless-sky conditions and assuming zenith angle 60°, $\beta = 0.1$, and $\alpha = 1.3$.

Solution. At $\theta_z = 60°$, assume $m = 2.0$.

(a) At $\lambda = 0.7\ \mu$m.

Rayleigh scattering:

$$\tau_{r\lambda} = \exp[-0.008735(0.7)^{-4.08} \times 2.0] = 0.9279.$$

Aerosol scattering:

$$\tau_{a\lambda} = \exp[-0.1(0.7)^{-1.3} \times 2.0] = 0.7276.$$

Spectral continuum attenuation:

$$\tau_{c\lambda} = \tau_{r\lambda}\tau_{a\lambda} = 0.9279 \times 0.7276 = 0.675.$$

(b) at $\lambda = 2.0\ \mu$m.

$$\tau_{r\lambda} = 0.999, \qquad \tau_{a\lambda} = 0.9220,$$

and consequently,

$$\tau_{c\lambda} = 0.999 \times 0.922 = 0.921.$$

The transmittance at 2μm is about 50% higher than that at $0.7\ \mu$m. □

6.8 Absorption of Direct Solar Radiation by Gases

Whereas scattering of solar radiation by the atmosphere and absorption of radiation by particulate matter is a continuous function of the wavelength, absorption of solar radiation by gases (dry air molecules and water vapor) is a selective process. Among the dry air molecules, the main absorbers are CO_2, O_3, N_2O, CO, O_2, CH_4, and N_2. The minor absorbers are oxides of nitrogen NO_2, N_2O_4, N_2O_5; hydrocarbon combinations C_2H_4, C_2H_6, C_3H_8; and sulfurous gas H_2S. The contribution by isotopic formations of a number of the above-named molecules is small. Water vapor and the gases described above are generally called *molecular absorbers*, most of which are active mainly in the near- and far-infrared wavelength regions. The term *molecular absorbers* is employed to distinguish them from atomic gases such as O and N, which absorb mainly the maximum uv and shorter wavelengths.

In the previous section, we saw that when electromagnetic radiation impinges upon a particle, the particle acts like a new source of energy by scattering a portion of the energy in the incident wave. A portion of the incident energy may also be absorbed by the particle; the particle, therefore, becomes an energy sink within the field of the incident wave. As far as gases are concerned, the absorption process depends on the energy state of a molecule. Radiation is absorbed during a transition from one energy state to another. This absorption process occurs only at discrete wavelengths and is called *selective absorption.* When the absorption takes place over a number of wavelengths very close to each other, it is called *band absorption.* The limits of a band are generally loosely defined. Sometimes a number of closely spaced consecutive bands are lumped together into a single band denomination. Within a band, absorption is not constant. The various absorbers in different wavelength bands are presented in Sections 6.9 and 6.10.

6.9 Absorbers in the Ultraviolet and Visible Spectrum

Diatomic oxygen (O_2), nitrogen (N_2), atomic oxygen (O), nitrogen (N), and ozone (O_3) are the five principal absorbers in the ultraviolet and visible spectrum.

Atomic oxygen and nitrogen absorb x rays and other short-wave radiation continuously up to about 0.0850 μm. As these two gases are found high in the atmosphere (Fig. 5.3.1), no radiation of wavelength less than 0.085 μm passes through to the lower atmosphere.

Diatomic oxygen and nitrogen absorb solar radiation in a number of overlapping bands in wavelengths below 0.20 μm. Since the upper (beyond 90 km) and lower atmospheres are composed primarily of these four gases (atomic and molecular oxygen and nitrogen), no radiation below 0.2 μm reaches the surface of the earth. This is fortunate, because this far-ultraviolet radiation degrades many materials and colors. It is also harmful to human beings.[6]

Diatomic oxygen also has three weak absorption bands in the visible spectrum centered at 0.63, 0.69, and 0.76 μm. However, in this region there is some overlap by ozone bands. Ozone exhibits a number of absorption bands beyond 0.20 μm in the ultraviolet, visible, and near infrared. Ozone has a strong absorption band from 0.2 to 0.3 μm, weaker bands from 0.3 to 0.35 μm, and stronger bands again in the visible region from 0.45 to 0.77 μm. Ozone also has absorption bands below 0.2-μm wavelengths. However, since all radiation below 0.20 μm is absorbed by O, N, O_2, and N_2 before it reaches the stratosphere where ozone exists, we need not be concerned with this range as far as the calculation of irradiance on the earth is concerned.

6.10 Absorbers in the Infrared Spectrum

A number of gases absorb solar electromagnetic radiation in the infrared wavelengths. Of these gases, the important absorbers are H_2O, CO_2, O_3, N_2O, CO, O_2, CH_4, and N_2. The water-vapor bands of importance are at 0.72, 0.82, 0.94, 1.1, 1.38, 1.87, 2.7, 3.2, and 6.3 μm; those of carbon dioxide are at 1.45, 1.6, 2.0, 2.7, 4.3, 4.8, and 5.2 μm; and those of oxygen are at 0.69 and 0.76 μm. Since we are concerned with the solar spectrum at $\lambda < 4.0$ μm, we can ignore the H_2O and CO_2 bands at higher wavelengths, and the higher-wavelength bands of other gases. The concentration of N_2O, CO, and CH_4 in the atmosphere is very small, and their total effect on the solar radiation arriving on the earth is minimal.

6.11 Transmittances of Molecular Absorbers

We shall now present a numerical procedure for computing the transmittance of the entire solar spectrum due to gaseous absorption.

In order to compute direct spectral irradiance on the earth, we need the values of the monochromatic transmittances due to the various molecular absorbers. Molecular scattering depends primarily on the number density of molecules in the radiation path, whereas the molecular absorption is a function of local pressure and temperature as well. The wavelength dependence of molecular (Rayleigh) scattering is very nearly proportional to λ^{-4}.

[6] However, uv radiation is useful in the treatment of some diseases such as rickets. It is also useful in killing bacteria and some viruses.

On the other hand, the spectral variation of the molecular absorption coefficient is a highly oscillatory function of wavelength. For molecular absorption, it is necessary to know the frequencies, intensities, shapes, and widths of spectral lines in the region of interest in order to be able to evaluate the spectral transmittance. The simple Bouguer's law does not apply in this case. Nevertheless, since elaborate procedures require a great deal of computer time, it is quite common to apply this law or a variation of it to compute molecular absorption.

Before actually discussing the following procedure, it seems necessary to say a few words about a logistical difficulty in handling transmittances due to molecular absorbers. As mentioned already, these transmittances are highly variable with wavelength; within a short wavelength interval there can be several peaks. The original data are generally obtained in analog form, and the tabulated values reported in the literature are often smoothed-out average values. The wavelengths at which such values may be available do not always correspond to the wavelengths at which extraterrestrial irradiance data are available. Consequently, in matching the two, there is some inevitable loss in accuracy.

We shall now deal with the elements of calculation needed to determine transmittance due to molecular absorption. For dry air molecules, it is useful to deal with ozone separately—because its distribution is well specified—and to combine the remaining dry air molecules into one group. The concentration profiles of some of the other dry air molecules such as CO_2, N_2O, and CO are different from one another; however, a model will be presented that assumes these gases as uniformly mixed. Finally, the transmittances for water vapor will be discussed.

6.12 Transmittance of Ozone

Absorption of electromagnetic radiation by ozone has been investigated by a number of researchers. Ozone absorbs in the ultraviolet, visible, and infrared. The infrared bands are in wavelengths greater than 4 μm, a region of little interest to us. In the ultraviolet and visible bands, the results by Vigroux [13] are commonly employed. Vigroux presents attenuation coefficients in a form suitable for Bouguer's law, Eq. (6.2.1). Rewriting this law for ozone we have

$$\tau_{o\lambda} = \exp(-k_{o\lambda} l m_r), \tag{6.12.1}$$

where $k_{o\lambda}$ is the attenuation coefficient for ozone absorption, l is the amount of ozone in cm(NTP), and m_r is the optical mass for ozone. The attenuation coefficients $k_{o\lambda}$, obtained from Vigroux, are given in Table 6.12.1. Vigroux listed the coefficients with wavelengths in Ångström intervals. Unfortunately, very few of those wavelengths correspond to the ones at which we have the spectral values of extraterrestrial irradiance (Table 3.3.2 or Table C-1). Consequently, the coefficients in Table 6.12.1 are the interpolated values (by Leckner [6]). Furthermore, they are valid for 18°C. Although ozone exists mainly in the stratosphere, where the temperature may be about −40°C, the values of the attenuation coefficient at these two temperatures are within 3% of each other.

The ozone coefficients are listed for wavelengths longer than 0.29 μm because practically all radiation at shorter wavelengths is absorbed by the ozone layer. It is evident from Table 6.12.1 that attenuation is very strong in the far-ultraviolet region. It diminishes sharply with wavelength until, at $\lambda \approx 0.35$ μm, ozone becomes almost transparent to solar radiation. The extinction coefficient increases again in the visible region and reaches a

Table 6.12.1

Spectral Absorption Coefficient $k_{o\lambda}$ for Ozone[a]

λ (μm)	$k_{o\lambda}$ (cm^{-1})	λ (μm)	$k_{o\lambda}$ (cm^{-1})	λ (μm)	$k_{o\lambda}$ (cm^{-1})
0.290	38.000	0.485	0.017	0.595	0.120
0.295	20.000	0.490	0.021	0.600	0.125
0.300	10.000	0.495	0.025	0.605	0.130
0.305	4.800	0.500	0.030	0.610	0.120
0.310	2.700	0.505	0.035	0.620	0.105
0.315	1.350	0.510	0.040	0.630	0.090
0.320	0.800	0.515	0.045	0.640	0.079
0.325	0.380	0.520	0.048	0.650	0.067
0.330	0.160	0.525	0.057	0.660	0.057
0.335	0.075	0.530	0.063	0.670	0.048
0.340	0.040	0.535	0.070	0.680	0.036
0.345	0.019	0.540	0.075	0.690	0.028
0.350	0.007	0.545	0.080	0.700	0.023
0.355	0.000	0.550	0.085	0.710	0.018
0.445	0.003	0.555	0.095	0.720	0.014
0.450	0.003	0.560	0.103	0.730	0.011
0.455	0.004	0.565	0.110	0.740	0.010
0.460	0.006	0.570	0.120	0.750	0.009
0.465	0.008	0.575	0.122	0.760	0.007
0.470	0.009	0.580	0.120	0.770	0.004
0.475	0.012	0.585	0.118	0.780	0.000
0.480	0.014	0.590	0.115	0.790	0.000

[a] Reprinted with permission from Leckner [6], Copyright 1978, Pergamon Press, Ltd., and from Vigroux [13].

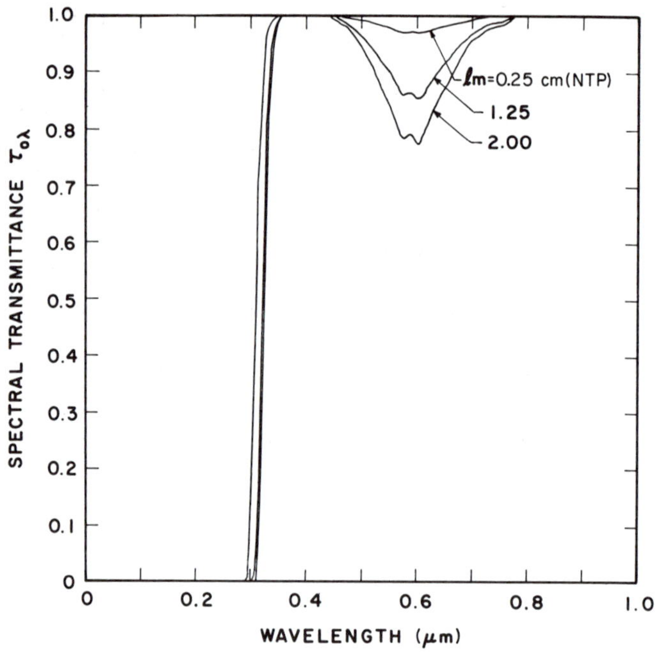

Figure 6.12.1 Ozone spectral transmittance as a function of lm.

maximum value at $\lambda = 0.605$ μm (in fact precisely at 6038 Å). Figure 6.12.1 contains the plots of $\tau_{0\lambda}$ for various values of lm. These plots show that the parameter lm has a very strong influence on the spectral transmittance at wavelengths shorter than 0.33 μm. This effect reappears in the visible range.

6.13 Transmittances of Uniformly Mixed Gases and Water Vapor

We deal here with the remaining absorbers of dry air molecules and with water vapor. Fowle [14] in 1915 determined the spectral transmittances for the various water-vapor bands, and his work remained popular for several decades. The accuracy of Fowle's work was limited, however, by the lack of sophistication of instruments available at that time, and his results did not provide a means to account for changes in pressure and temperature. An improvement on Fowle's work was provided by Howard *et al.*, who in a series of articles [15–18] presented a quantitative laboratory study of the molecular spectra of water vapor and carbon dioxide. This consisted of measurements of the absorptances of individual bands of water vapor and carbon dioxide as a function of pressure and optical path.

Another step was taken by Gates [19], and Gates and Harrop [20], who measured spectral absorptances of the actual atmosphere on a clear day in Colorado. Gates and Harrop's investigation covered absorptances due to water vapor, carbon dioxide, and other minor absorbers in the atmosphere. Thekaekara [1, 21] computed direct spectral irradiance employing Gates and Harrop's absorption coefficients. However, there are two problems with the Thekaekara/Gates–Harrop formulation: the first is one of logistics. The wavelengths for which tabulated values of extraterrestrial spectral irradiance are generally available do not correspond to the wavelengths for which Thekaekara [21] presented Gates and Harrop's spectral absorptances. The second problem is the apparently strong anomalous absorption of water vapor in the 0.8–1.0-μm region as pointed out by Hulstrom [22]. This apparent anomaly is of some concern in the area of silicon photovoltaic applications, which have peak response in this region of wavelength.

Recently the Air Force Cambridge Research Laboratories (United States) have developed rigorous methods for calculating the detailed transmittance of the atmosphere. McClatchey *et al.* [23–25] have prepared a computer code called LOWTRAN that utilizes spectral absorptances of the various molecular absorbers determined at the AFCR laboratories. Leckner [6], employing the work of McClatchey *et al.*, presented an algorithm for the calculation of direct spectral irradiance. This algorithm is simple, quite accurate for engineering calculations, and therefore appropriate for presentation in this text. The spectral transmittances are presented in order to correspond to the wavelengths in Table 3.3.2, a total of 144 intervals. The spectral transmittance for molecular absorption due to CO_2, O_2, etc., combined under a general denomination of uniformly mixed gases is given by the following:

$$\tau_{g\lambda} = \exp[-1.41k_{g\lambda}m_a/(1 + 118.93k_{g\lambda}m_a)^{0.45}]. \quad (6.13.1)$$

The corresponding spectral transmittance due to water-vapor absorption is as follows:

$$\tau_{wa\lambda} = \exp[-0.2385k_{wa\lambda}wm_r/(1 + 20.07k_{wa\lambda}wm_r)^{0.45}]. \quad (6.13.2)$$

In this formula, the variability in the vertical direction of standard pressure and temperature has been taken into account. Equations (6.13.1) and (6.13.2) are not chosen randomly but are based on what is called the "random model" for molecular absorption. This model yields more accurate results than does Bouguer's law.

For the equations given above, the attenuation coefficients $k_{g\lambda}$ and $k_{wa\lambda}$ are presented in Tables 6.13.1 and 6.13.2, respectively. Virtually all radiation is absorbed at $\lambda > 4.0$ μm. Furthermore, only 1% of extraterrestrial solar

Table 6.13.1

Spectral Absorption Coefficients of Uniformly Mixed Gases[a]

λ (μm)	$k_{g\lambda}$	λ (μm)	$k_{g\lambda}$	λ (μm)	$k_{g\lambda}$
0.76	0.300E + 01	1.75	0.100E − 04	2.80	0.150E + 03
0.77	0.210E + 00	1.80	0.100E − 04	2.90	0.130E + 00
		1.85	0.145E − 03	3.00	0.950E − 02
1.25	0.730E − 02	1.90	0.710E − 02	3.10	0.100E − 02
1.30	0.400E − 03	1.95	0.200E + 01	3.20	0.800E + 00
1.35	0.110E − 03	2.00	0.300E + 01	3.30	0.190E + 01
1.40	0.100E − 04	2.10	0.240E + 00	3.40	0.130E + 01
1.45	0.640E − 01	2.20	0.380E − 03	3.50	0.750E − 01
1.50	0.630E − 03	2.30	0.110E − 02	3.60	0.100E − 01
1.55	0.100E − 01	2.40	0.170E − 03	3.70	0.195E − 02
1.60	0.640E − 01	2.50	0.140E − 03	3.80	0.400E − 02
1.65	0.145E − 02	2.60	0.660E − 03	3.90	0.290E + 00
1.70	0.100E − 04	2.70	0.100E + 03	4.00	0.250E − 01

[a] Reprinted with permission from Leckner [6], Copyright 1978, Pergamon Press, Ltd.

Table 6.13.2

Spectral Absorption Coefficients of Water Vapor[a]

λ (μm)	$k_{wa\lambda}$ (cm^{-1})	λ (μm)	$k_{wa\lambda}$ (cm^{-1})	λ (μm)	$k_{wa\lambda}$ (cm^{-1})
0.69	0.160E − 01	0.93	0.270E + 02	1.85	0.220E + 04
0.70	0.240E − 01	0.94	0.380E + 02	1.90	0.140E + 04
0.71	0.125E − 01	0.95	0.410E + 02	1.95	0.160E + 03
0.72	0.100E + 01	0.96	0.260E + 02	2.00	0.290E + 01
0.73	0.870E + 00	0.97	0.310E + 01	2.10	0.220E + 00
0.74	0.610E − 01	0.98	0.148E + 01	2.20	0.330E + 00
0.75	0.100E − 02	0.99	0.125E + 00	2.30	0.590E + 00
0.76	0.100E − 04	1.00	0.250E − 02	2.40	0.203E + 02
0.77	0.100E − 04	1.05	0.100E − 04	2.50	0.310E + 03
0.78	0.600E − 03	1.10	0.320E + 01	2.60	0.150E + 05
0.79	0.175E − 01	1.15	0.230E + 02	2.70	0.220E + 05
0.80	0.360E − 01	1.20	0.160E − 01	2.80	0.800E + 04
0.81	0.330E + 00	1.25	0.180E − 03	2.90	0.650E + 03
0.82	0.153E + 01	1.30	0.290E + 01	3.00	0.240E + 03
0.83	0.660E + 00	1.35	0.200E + 03	3.10	0.230E + 03
0.84	0.155E + 00	1.40	0.110E + 04	3.20	0.100E + 03
0.85	0.300E − 02	1.45	0.150E + 03	3.30	0.120E + 03
0.86	0.100E − 04	1.50	0.150E + 02	3.40	0.195E + 02
0.87	0.100E − 04	1.55	0.170E − 02	3.50	0.360E + 01
0.88	0.260E − 02	1.60	0.100E − 04	3.60	0.310E + 01
0.89	0.630E − 01	1.65	0.100E − 01	3.70	0.250E + 01
0.90	0.210E + 01	1.70	0.510E + 00	3.80	0.140E + 01
0.91	0.160E + 01	1.75	0.400E + 01	3.90	0.170E + 00
0.92	0.125E + 01	1.80	0.130E + 03	4.00	0.450E − 02

[a] Reprinted with permission from Leckner [6], Copyright 1978, Pergamon Press, Ltd.

irradiance is contained in wavelengths greater than 4.0 μm; consequently, these tables are limited to this wavelength.

In Eqs. (6.13.1) and (6.13.2), m_a and m_r are the optical masses for uniformly mixed gases and water vapor, respectively, and w is precipitable water vapor in centimeters. It appears reasonable to assume that the optical mass for the uniformly mixed gases is given by Eq. (5.7.3).

The spectral transmittances due to uniformly mixed gases and water vapor are shown in Figs. 6.13.1 and 6.13.2. These figures demonstrate very low transmittance values in some of the infrared bands. In these figures an optical mass of unity is used. At higher optical masses, the transmittance will further reduce considerably.

Let us reexamine the preceding two diagrams. The transmittances are equal to unity approximately at wavelengths 1.2, 1.6, 2.2, and 3.8 μm. Practically all solar radiation at these wavelengths reaches ground unattenuated. The small bandwidths around these wavelengths are called *atmospheric windows*. This means the atmosphere is almost transparent at these wavelengths. Naturally, at these wavelengths, the atmosphere is also transparent to the energy emitted by the earth. The earth maintains itself in a certain

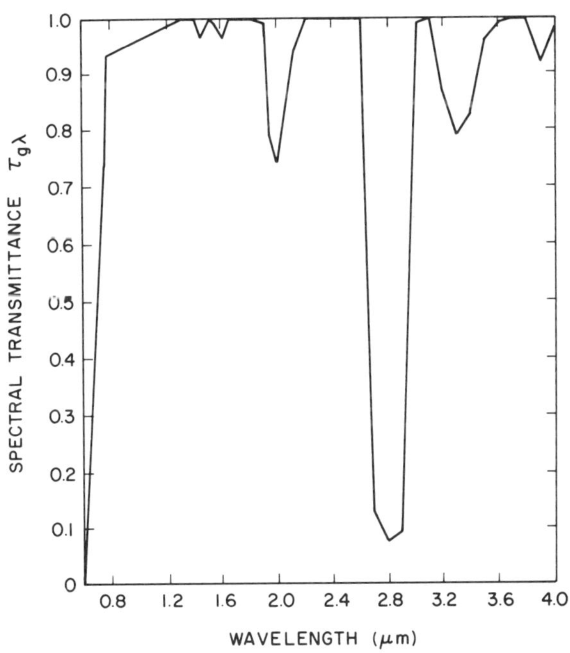

Figure 6.13.1 Spectral transmittance of the uniformly mixed gases ($m = 1$).

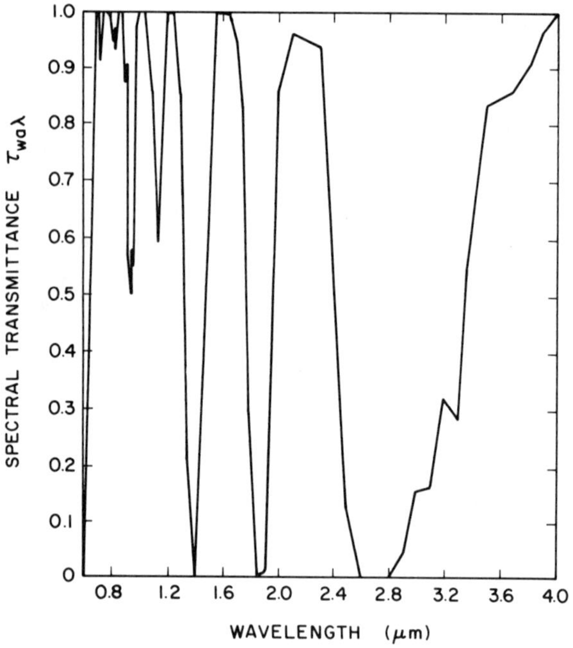

Figure 6.13.2 Spectral transmittance of water vapor ($m = 1$, $w = 2$ cm).

thermal equilibrium because of the atmosphere's ability to provide such "windows" to outer space in the infrared region.

The total transmittance due to the molecular absorbers can now be written as follows:

$$\tau_{ma\lambda} = \tau_{0\lambda}\tau_{g\lambda}\tau_{wa\lambda}. \tag{6.13.3}$$

□ EXAMPLE 6.13.1. Spectral transmittances evaluated at $\lambda = 0.7$ and 2.0 μm due to molecular absorbers under cloudless-sky conditions and assuming zenith angle 60°, precipitable water vapor 2 cm, and ozone layer thickness 0.3 cm(NTP).

Solution. At $\theta_z = 60°$, assume $m = 2.0$ throughout.
(a) At $\lambda = 0.70$ μm:

$k_{0\lambda} = 0.023$ cm^{-1} (from Table 6.12.1);

$k_{g\lambda} = 0$, and consequently $\tau_{g\lambda} = 1$;
$k_{wa\lambda} = 0.24 \times 10^{-1}$ cm^{-1} (from Table 6.13.2).

Transmittance due to absorption by ozone:

$$\tau_{0\lambda} = \exp(-0.023 \times 0.3 \times 2.0) = 0.99.$$

Transmittance due to absorption by water vapor:

$$\tau_{wa\lambda} = \exp[-0.2385 \times 0.024 \times 2 \times 2/(1 + 20.07 \times 0.024 \times 2 \times 2)^{0.45}]$$
$$= 0.99.$$

Total transmittance due to molecular absorbers:

$$\tau_{ma\lambda} = \tau_{0\lambda}\tau_{g\lambda}\tau_{wa\lambda} = 0.99 \times 1 \times 0.99 = 0.98.$$

(b) At $\lambda = 2.0\ \mu m$:

$k_{0\lambda} = 0$, and consequently $\tau_{0\lambda} = 1$;

$k_{g\lambda} = 3.0$ (From Table 6.13.1);

$k_{wa\lambda} = 2.9\ cm^{-1}$ (From Table 6.13.2).

Repeating the foregoing calculations, we have

$$\tau_{g\lambda} = 0.64, \qquad \tau_{wa\lambda} = 0.79, \qquad \text{and} \qquad \tau_{ma\lambda} = 1 \times 0.64 \times 0.79 = 0.51.$$

At 2 μm, $\tau_{ma\lambda}$ is about half of its value at 0.7 μm. □

6.14 Direct Spectral Irradiance on the Ground

Monochromatic distribution of a direct solar beam can now be computed as a function of a number of variables, including optical mass and a wide variety of atmospheric parameters—for example, water-vapor content, ozone layer thickness, and turbidity parameters. With Eqs. (6.7.3) and (6.13.3) combined, the monochromatic transmittance due to the combined effects of continuum attenuation and molecular absorption is as follows:

$$\tau_\lambda = \tau_{c\lambda}\tau_{ma\lambda} \tag{6.14.1}$$

$$= \tau_{r\lambda}\tau_{a\lambda}\tau_{o\lambda}\tau_{g\lambda}\tau_{wa\lambda}. \tag{6.14.2}$$

It is useful to restate the definitions:

$\tau_{r\lambda}$ is the spectral transmittance of the direct beam due to molecular scattering [Eq. (6.4.2)];

$\tau_{a\lambda}$ is the spectral transmittance of the direct beam due to scattering and absorption by aerosols [Eq. (6.6.3)] [this is equivalent to the combined effects of Eqs. (6.5.3) and (6.5.4); that is, $\tau_{a\lambda} = \tau_{ws\lambda}\tau_{d\lambda}$];

$\tau_{o\lambda}$ is the spectral transmittance of the direct beam due to absorption by the ozone layer [Eq. (6.12.1)];

$\tau_{g\lambda}$ is the spectral transmittance of the direct beam due to absorption by uniformly mixed gases such as CO_2 and O_2 [Eq. (6.13.1)]; and

$\tau_{wa\lambda}$ is the spectral transmittance of the direct beam due to absorption by water vapor [Eq. (6.13.2)].

In order to obtain an overall view, all these individual transmittances have been plotted simultaneously in Fig. 6.14.1. It is apparent from this figure that infrared absorption is due mainly to water vapor and uniformly mixed gases. In the ultraviolet and visible region, it is essentially ozone absorption, Rayleigh scattering, and aerosols that control attenuation of the direct beam. The transmittance by aerosols is minimum at the short wavelengths and increases slowly as the wavelength increases. However, the rate of attenuation by aerosols will change if turbidity parameters are altered. Furthermore, in this diagram the transmittance curves due to Rayleigh and aerosols are drawn only beyond that of the ozone absorption curve.

Direct spectral irradiance at any wavelength on a surface normal to the sun's rays and at mean sun–earth distance is as follows:

$$\dot{I}_{n\lambda} = \dot{I}_{0n\lambda}\tau_{\lambda}, \tag{6.14.3}$$

and on a horizontal surface,

$$\dot{I}_{b\lambda} = \dot{I}_{0n\lambda}\cos(\theta_z)\tau_{\lambda}. \tag{6.14.4}$$

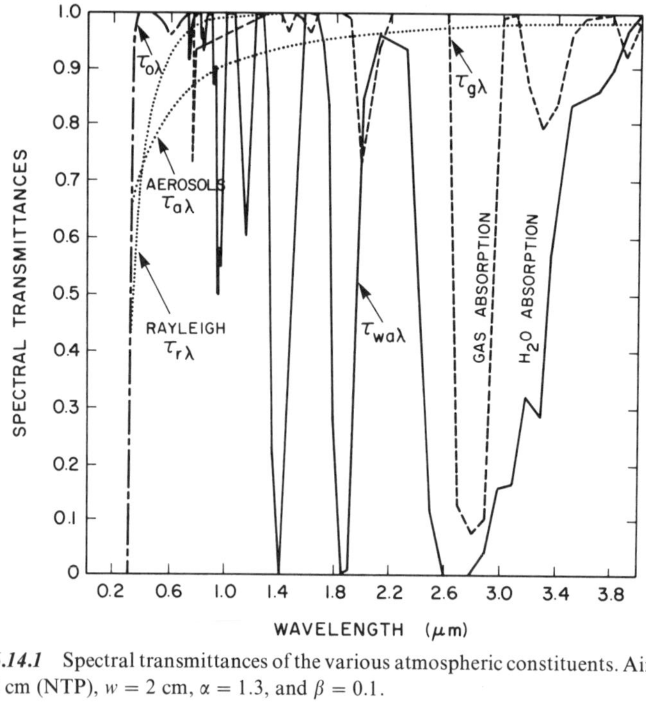

Figure 6.14.1 Spectral transmittances of the various atmospheric constituents. Air mass = 1, O_3 = 0.35 cm (NTP), w = 2 cm, α = 1.3, and β = 0.1.

Through a series of diagrams we shall now demonstrate how the extraterrestrial spectrum is modified by atmospheric parameters before it reaches the ground. We have mentioned earlier that in this chapter we shall use the extraterrestrial spectrum from the World Radiation Center only. As a start, it seems appropriate to show an example of differences between the results obtained by using the WRC and the NASA spectra. Such an example for a typical atmosphere [$\beta = 0.1$, $\alpha = 1.3$, $O_3 = 0.35$ cm(NTP), w = 2 cm, and $\theta_z = 60°$] is given in Fig. 6.14.2. The WRC spectrum results in higher amounts of irradiance, particularly in the visible portion. For this reason the WRC spectrum will be used throughout this chapter. In this particular case, the spectrally integrated (called broadband) values of the direct irradiances are 687 W m^{-2} for the WRC and 672 W m^{-2} for the NASA spectrum. This yields 2% difference, whereas the difference between the two solar constants is only 1%. A number of other calculations have shown that under most atmospheric conditions, this difference remains at about 2%. It is, therefore, interesting to note that the percentage difference between broadband direct irradiance on the ground from either spectrum is not identical to the percentage difference between the two solar constants.

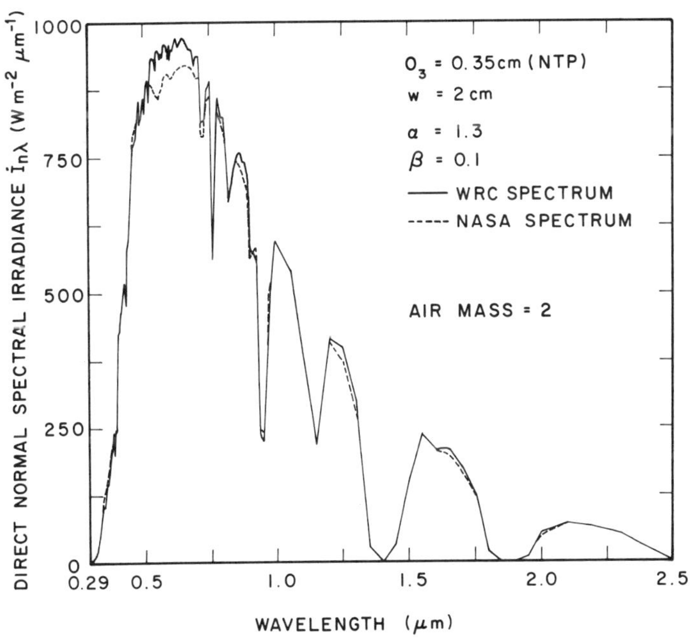

Figure 6.14.2 Comparison of the direct normal spectral irradiance using the WRC spectrum and the NASA spectrum. Air mass = 2, $O_3 = 0.35$ cm(NTP), $w = 2$ cm, $\alpha = 1.3$, and $\beta = 0.1$.

In Fig. 6.14.3 we identify the various molecular absorption bands. In this diagram we consider a clean atmosphere ($\beta = 0$) with $O_3 = 0.35$ cm(NTP) and $w = 2$ cm. The first curve under the extraterrestrial line represents depletion of the direct beam by Rayleigh scattering only. The next curve below represents the actual amount of direct solar flux reaching the ground. The amount of energy absorbed by the various gases is shown by the darkened areas. The size of the shaded areas (except those for O_2 and CO_2) depends on the amount of O_3 and H_2O present in the atmosphere. Effect of the zenith angle (air mass) applies to all areas.

The effect of a variation in ozone only is shown in Fig. 6.14.4. This diagram represents clean dry atmosphere ($\beta = w = 0$), and $O_3 = 0$, 0.2, and 0.4 cm(NTP). Attenuation by ozone is confined to the ultraviolet and the visible spectrum and is almost linear with amount of ozone.

Depletion of the extraterrestrial beam by a varying amount of water vapor is demonstrated in Fig. 6.14.5. Again, this diagram is for a clean atmosphere. Ozone is assumed zero, and absorption by oxygen and carbon dioxide is also neglected. As such, this diagram represents Rayleigh scattering and absorption by water vapor only. Under the extraterrestrial curve, the three irradiance curves are for $w = 0$, 0.4, and 4 cm of precipitable water. It is evident from this diagram that the rate of attenuation diminishes as the

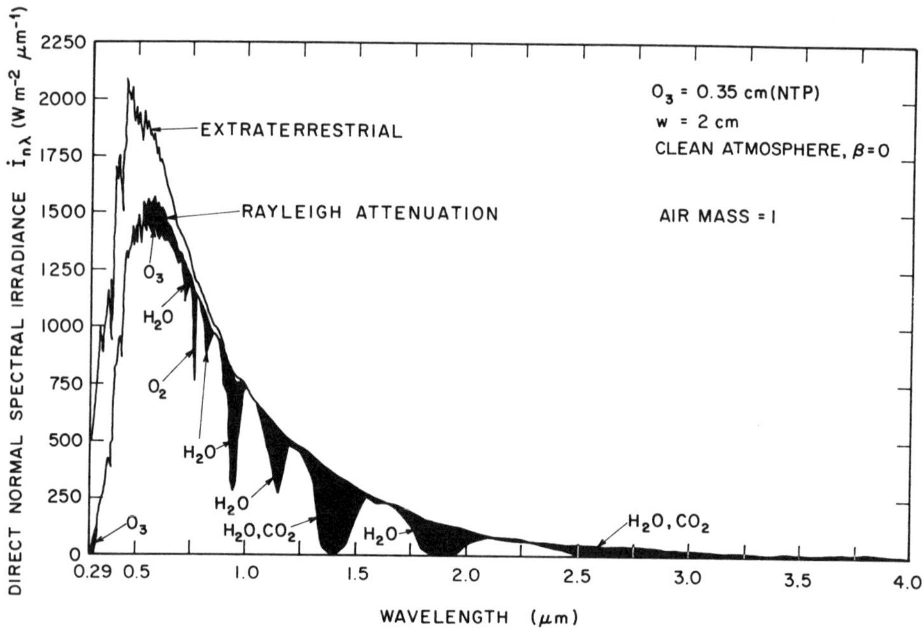

Figure 6.14.3 Identification of the various molecular absorbers.

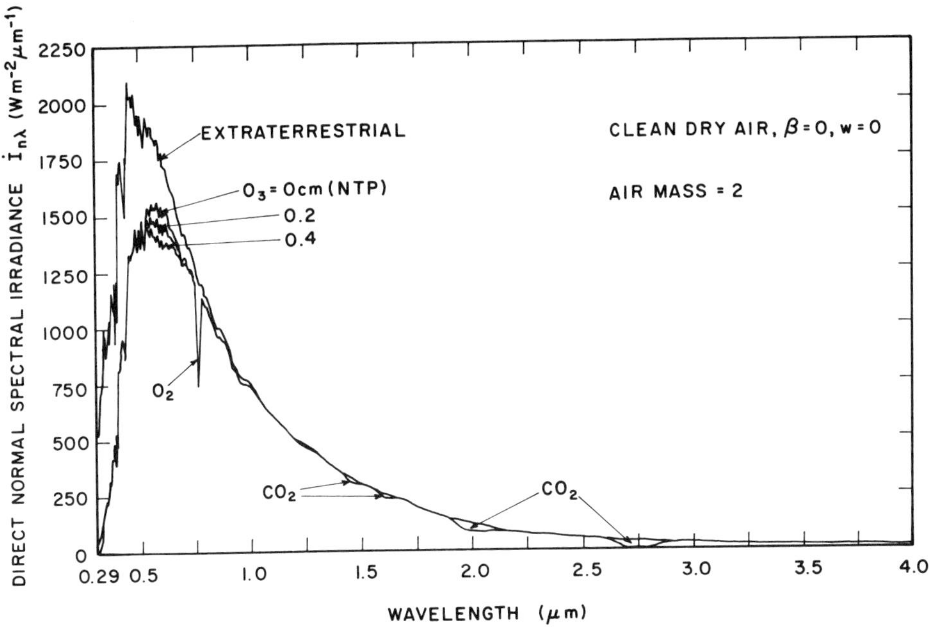

Figure 6.14.4 Direct normal spectral irradiance under varying amounts of ozone.

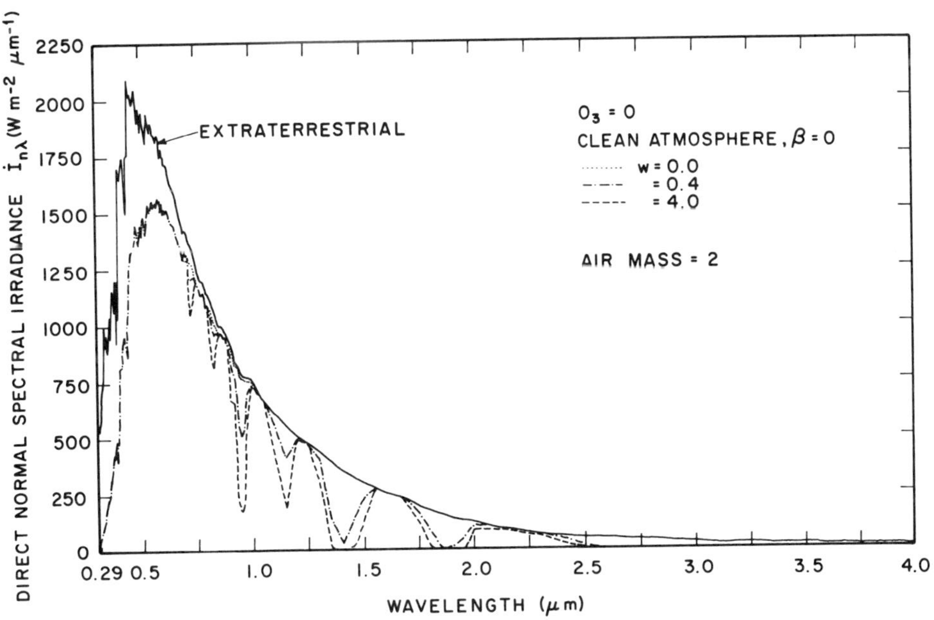

Figure 6.14.5 Direct normal spectral irradiance under varying amounts of water vapor.

amount of precipitable water increases. The first few millimeters deplete as much energy as the last few centimeters of precipitable water. Furthermore, even a small amount of moisture in the atmosphere shuts off all radiation beyond approximately 2.5 μm. In this particular example, spectrally integrated irradiances are 1142, 1025, and 934 W m^{-2} at $w = 0$, 0.4, and 4.0 cm. The first 4 mm of water vapor depletes by 10% the energy reaching ground, and it takes a further 3.6 cm of water vapor to deplete another 10% of the energy.

For a typical atmosphere [$\beta = 0.1$, $\alpha = 1.3$, $O_3 = 0.35$ cm(NTP), and $w = 2$ cm], the effect of the air mass is shown in Fig. 6.14.6. As the air mass increases, the ultraviolet and visible spectrum undergoes much stronger depletion (by absorption and scattering) than the infrared portion. For this reason the sun appears red when it is near the horizon.

The effect of variation in the turbidity parameter β (representing the amount of aerosols present in the atmosphere), and of the wavelength exponent α (representing average particle size), is shown in Figs. 6.14.7 and 6.14.8, respectively. The turbidity parameter β has a strong influence on the

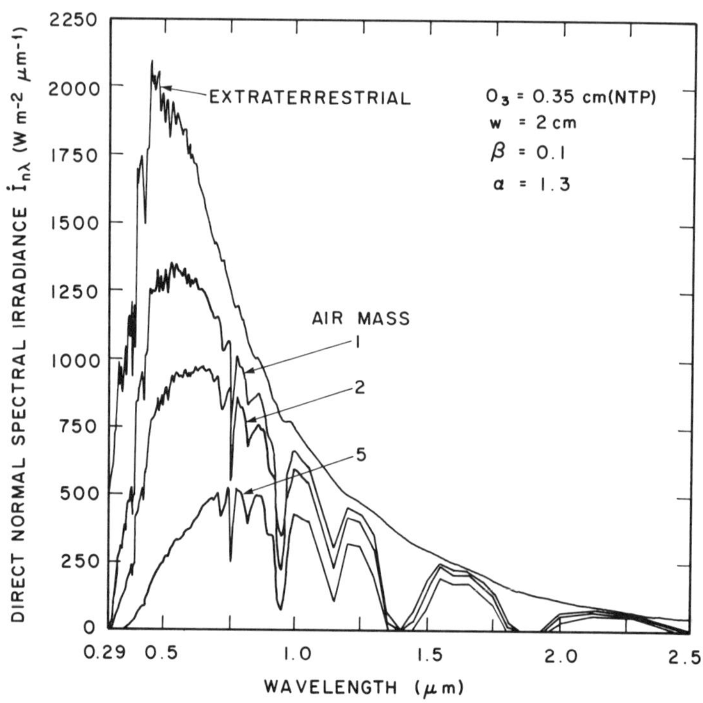

Figure 6.14.6 Direct normal spectral irradiance under different air masses.

spectral flux reaching the ground. An increase in β diminishes irradiance. The wavelength exponent α, however, has only a minor influence on spectral irradiance. Furthermore, irradiance at $\lambda > 1\ \mu m$ remains unchanged by a change in α. By keeping total amount of aerosols constant ($\beta = \text{const}$), smaller values of α (representing larger particle size and higher visibility) yield more irradiance.

In the formulation of Eq. (6.14.2), it is understood that at a single point in the vertical direction, attenuation processes occur simultaneously. However, processes dominant in the short wave such as ozone absorption first occur high in the atmosphere. The remaining energy in the direct beam then begins to be attenuated by Rayleigh scattering, which occurs predominantly in the short wave and in the upper layers of the atmosphere. Further down toward the ground, the direct beam is attenuated by water vapor, uniformly mixed gases, and aerosols, all of which occur essentially near the earth's surface. A general picture of the depletion of energy within various bandwidths and at various heights above sea level is shown in Table 6.14.1 [26].

We now give some examples of actual computations to determine direct spectral irradiance.

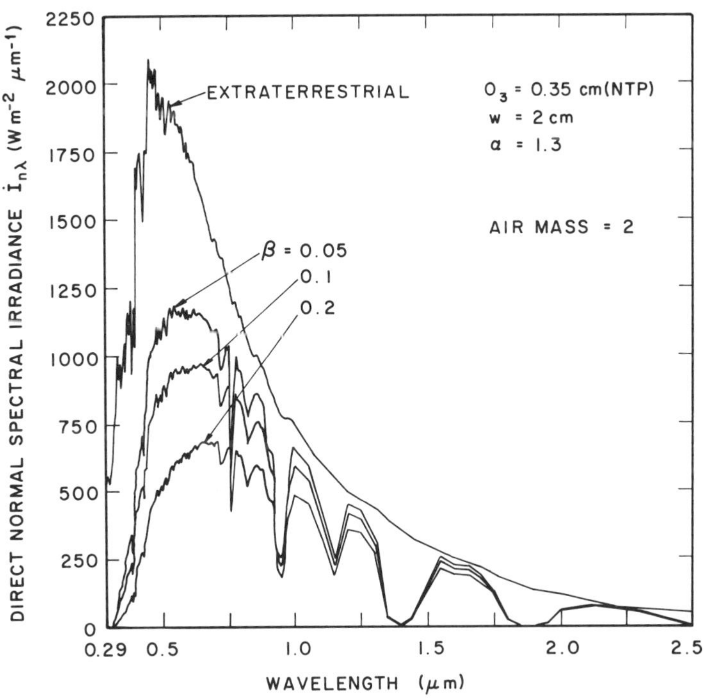

Figure 6.14.7 Direct normal spectral irradiance under different aerosol amounts.

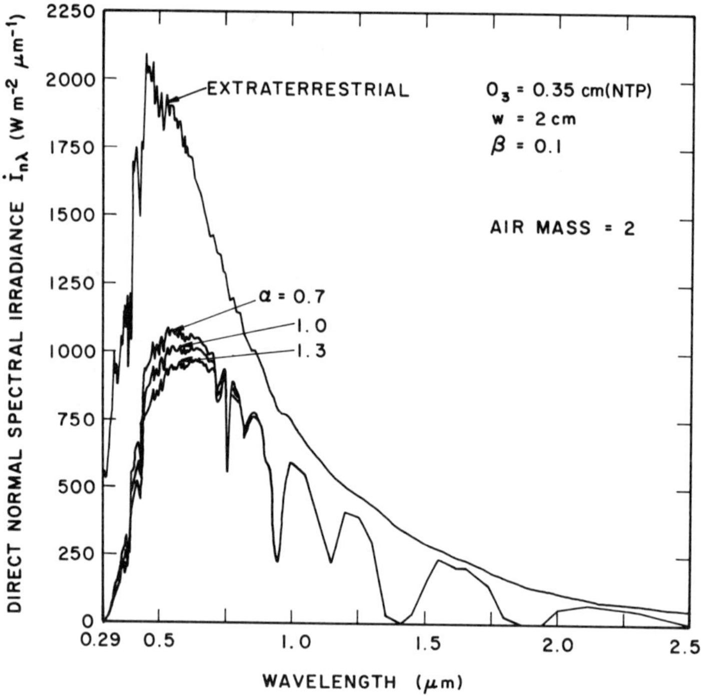

Figure 6.14.8 Direct normal spectral irradiance under varying values of α.

□ EXAMPLE 6.14.1 The direct spectral irradiance on a horizontal surface on 15 June under the conditions of Examples 6.7.1 and 6.13.1.

Solution. From Eq. (6.2.3), which applies to the mean sun–earth distance, the direct normal spectral irradiance on an individual day is given by

$$\dot{I}_{n\lambda} = E_0 \dot{I}_{0n\lambda} \prod_{i=1}^{i=j} \tau_{i\lambda}, \tag{6.2.3}$$

where E_0 for 15 June is 0.9684. We calculate the above at the two wavelengths

$$\lambda = 0.70, \quad 2.0\ \mu\text{m}.$$

(a) At $\lambda = 0.70\ \mu$m,

$$\dot{I}_{0n\lambda} = 1427.5\ \text{W m}^{-2}\ \mu\text{m}^{-1}\ \text{(from Table 3.3.2)},$$

$$\dot{I}_{n\lambda} = (0.9684) \times 1427.5\tau_{c\lambda}\tau_{ma\lambda}$$

$$= (0.9684) \times 1427.5 \times 0.675 \times 0.98$$

$$= 914.45\ \text{W m}^{-2}\ \mu\text{m}^{-1}.$$

Table 6.14.1

Attenuation of Solar Radiation by the Earth's Atmosphere[a]

Pressure (mbars)	Altitude (km)	Wavelength regions (μm) 0.12–0.20	0.20–0.29	0.29–0.32	0.32–0.35	0.35–0.55	0.55–0.9	0.9–2.5	2.5–7	7–20	Altitude (km)	
0.2	60	O_2 absorbs almost completely	Attenuation by scattering increases markedly toward shorter wavelengths								Above 60	110 km Chemosphere
7.5	33		(0.20–0.21 μm absorption by O_2) Absorption by O_3 appreciable	O_3 absorption not important					Energy small	Energy very small	60–33	
227	11		No radiation penetrates below about 11 km	O_3 absorption attenuates more than loss by scattering	O_3 absorption significantly attenuates radiation	Irradiation diminished mostly by scattering by permanent gases in atmosphere	H_2O responsible for major absorption; CO_2 absorbs slightly at 2 μm; water vapor (or ice crystals) is found up to about 21 km		No significant penetration below 2 km except in "windows" at approximately 3.8 and 4.9 μm	Strong O_3 absorption at 9.6 μm; strong CO_2 absorption 12–17 μm	33–11	20 km Stratosphere 11 km
							Energy transmitted with small loss down to 2 km	Energy penetrates to sea level only through "windows" at approximately 1.2, 1.6, and 2.2 μm		Energy transmited with moderate loss; many absorption bands due to atmospheric gases	11–2	Troposphere
795	2				Highly-variable dust, haze (H_2O and smoke) responsible for attenuation in regions 0.32–0.7 μm						2–sea level	
1013			Sea level	Appreciable penetration through "clear" atmosphere to sea level About 7% About 30%	Penetration through "clear" atmosphere to sea level about 40%	Dust may rise to more than 4 km						0 km

[a] From "Handbook of Geophysics and Space Environments" by J. N. Howard, J. I. F. King, and P. R. Gast. Copyright © 1961 Macmillan Publishing Co., Inc.

Beam irradiance on a horizontal surface is

$$\dot{I}_{b\lambda} = \dot{I}_{n\lambda} \cos \theta_z$$
$$= 914.45 \times 0.50 = 457.23 \text{ W m}^{-2} \mu\text{m}^{-1}.$$

(b) At $\lambda = 2.0\ \mu$m,

$$\dot{I}_{0n\lambda} = 118.5 \text{ W m}^{-2} \mu\text{m}^{-1} \text{ (from Table 3.3.2)}.$$

Repeating the foregoing calculations,

$$\dot{I}_{n\lambda} = 0.9684 \times 118.5 \times \tau_{c\lambda}\tau_{ma\lambda}$$
$$= 0.9684 \times 118.5 \times 0.921 \times 0.51$$
$$= 53.90 \text{ W m}^{-2} \mu\text{m}^{-1},$$
$$\dot{I}_{b\lambda} = 53.90 \times 0.5$$
$$= 26.95 \text{ W m}^{-2} \mu\text{m}^{-1}.$$

Through the following example we show the effect of a variation in α.

□ EXAMPLE 6.14.2. Example 6.14.1 with α changing from 1.3 to 0.7; all other parameters remain unchanged.

Solution. (a) At $\lambda = 0.7\ \mu$m,

$$\tau_{r\lambda} = 0.9279 \qquad \text{(see Example 6.7.1)},$$
$$\tau_{a\lambda} = \exp[-0.1(0.7)^{-0.7} \times 2.0] = 0.7736,$$
$$\tau_{c\lambda} = \tau_{r\lambda}\tau_{a\lambda} = 0.9279 \times 0.7736 = 0.7178.$$

From Example 6.13.1, $\tau_{ma\lambda} = 0.98$,

$$\dot{I}_{0n\lambda} = 1427.5 \text{ W m}^{-2} \mu\text{m}^{-1} \text{ (from Table 3.3.2)},$$
$$\dot{I}_{n\lambda} = 0.9684 \times 1427.5 \times 0.7178 \times 0.98$$
$$= 972.43 \text{ W m}^{-2} \mu\text{m}^{-1}$$
$$\dot{I}_{b\lambda} = 972.43 \times 0.50 = 486.22 \text{ W m}^{-2} \mu\text{m}^{-1}.$$

With α changing from 1.3 to 0.7, there is a slight increase in $\dot{I}_{n\lambda}$ or $\dot{I}_{b\lambda}$ at $\lambda = 0.7\ \mu$m.

(b) At $\lambda = 2.0\ \mu$m,

$$\tau_{r\lambda} = 0.999$$
$$\tau_{a\lambda} = \exp[-0.1(2.0)^{-0.7} \times 2.0] = 0.8842,$$
$$\tau_{c\lambda} = 0.999 \times 0.8842 = 0.8833,$$

$$\dot{I}_{n\lambda} = 0.9684 \times 118.5 \times 0.8833 \times 0.5 = 51.70 \text{ W m}^{-2}\ \mu\text{m}^{-1},$$

$$\dot{I}_{b\lambda} = 51.70 \times 0.50 = 25.85 \text{ W m}^{-2}\ \mu\text{m}^{-1}.$$

With α changing from 1.3 to 0.7, there is a slight decrease in $\dot{I}_{n\lambda}$ or $\dot{I}_{b\lambda}$ at $\lambda = 2.0\ \mu$m. Note, however, that the increase at 0.7 μm is much larger than the reduction at $\lambda = 2.0\ \mu$m. □

In the following example we show the method of determining turbidity by a single-wavelength sun photometer.

□ EXAMPLE 6.14.3. The turbidity parameter β determined on 15 June, when the sun is 30° above the horizon, at a location where a single-wavelength sun photometer reads 600 W m^{-2} μm^{-1} at $\lambda = 1\ \mu$m; the estimated amount of precipitable water in the atmosphere is 1.0 cm and ozone 0.30 cm (NTP); ambient temperature is 25°C and pressure is 1025 mbars.

Solution. At $\theta_z = 60°$, $m_r = 1.99$ and $m_a = 1.99(1025/1013.25) = 2.01$. For water vapor we need to carry out pressure and temperature correction,

$$w = 1\left(\frac{1025}{1013.25}\right)^{3/4}\left(\frac{273}{298}\right)^{1/2} = 0.9654 \approx 1 \text{ cm}.$$

At $\lambda = 1$, $k_{g\lambda} = k_{o\lambda} = 0$. Consequently, $\tau_{g\lambda} = \tau_{o\lambda} = 1$. Therefore, $\tau_\lambda = \tau_{r\lambda}\tau_{a\lambda}\tau_{wa\lambda}$, $\tau_{r\lambda} = \exp[-(0.008735 \times 2.01)] = 0.9826$. At $\lambda = 1$, $k_{wa\lambda} = 0.25 \times 10^{-2}$ cm^{-1}, which is very low.

$$\tau_{wa\lambda} = \exp\left(-\frac{0.2385 \times 0.25 \times 10^{-2} \times 1 \times 1.99}{(1 + 20.07 \times 0.25 \times 10^{-2} \times 1 \times 1.99)^{0.45}}\right) = 0.9967.$$

At $\lambda = 1\ \mu$m, $\dot{I}_{On\lambda} = 743.99$ W m^{-2} μm^{-1}. On 15 June, $E_0 = 0.9684$. Equation (6.14.3) can now be written

$$600 = 0.9684 \times 743.99 \times 0.9826 \times 0.9967 \times \tau_{a\lambda}$$

or

$$\tau_{a\lambda} = 0.8503.$$

At $\lambda = 1$, $k_{a\lambda} = \beta$, and

$$\tau_{a\lambda} = \exp(-\beta \times 2.01) = 0.8503$$

or

$$\beta = 0.0807.$$

The accuracy in determining β through this method depends on the accuracy in determining atmospheric water vapor. If a wavelength other than unity is used, β may be determined by assuming $\alpha = 1.3$. □

6.15 Origins of Diffuse Spectral Radiation

Diffuse radiation is generated by the scattering effects of air molecules and aerosols. Let us reconsider Fig. 6.3.1. In this diagram, diffuse radiation generated by the first impingement of the direct radiation, called primary scattering, is shown. This diffuse radiation in turn strikes other molecules and particles and thus the scattering process continues. This continuous scattering process is called multiple scattering. Figure 6.15.1 shows multiple scattering by air molecules, and Fig. 6.15.2 by aerosols. Both these diagrams show that diffuse radiation generated by primary scattering is dominant. However, it is partly dependent on the density of molecules and aerosol particles. A portion of total primary and multiply scattered radiation goes back to space, a portion is absorbed, and a portion reaches the ground. Analysis of multiple scattering is mathematically very complex and requires a great deal of computational time. Furthermore, the contribution by

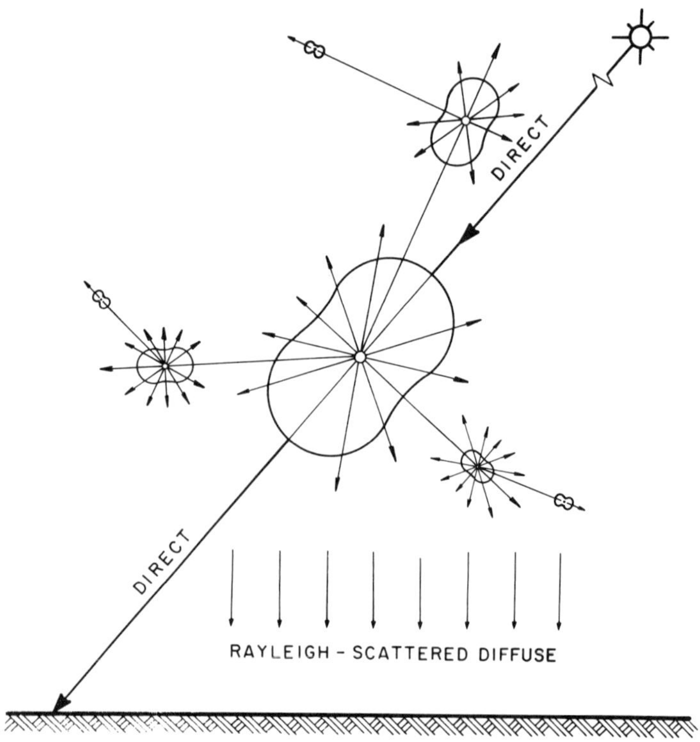

Figure 6.15.1 Multiple scattering by air molecules.

multiple scattering is often small, and any gain from an exact solution is offset by uncertainties in the optical properties of a real atmosphere, especially those of aerosols and the ground cover. Consequently, we shall present the effects of single scattering only. The analysis that follows is simple, approximate, and empirical, yet quite accurate.

The main objective of the following material is to develop simple and separate expressions for diffuse radiation produced by air molecules, and by aerosols. We shall approach this development in steps.

Consider a pure Rayleigh atmosphere irradiated by one unit of parallel monochromatic beam. Diffuse radiation produced by primary scattering is given by

$$1 - \tau_{r\lambda}. \tag{6.15.1}$$

We can now consider a pure aerosol atmosphere irradiated by one unit of parallel monochromatic beam. The transmittance $\tau_{a\lambda}$ of aerosols is due to

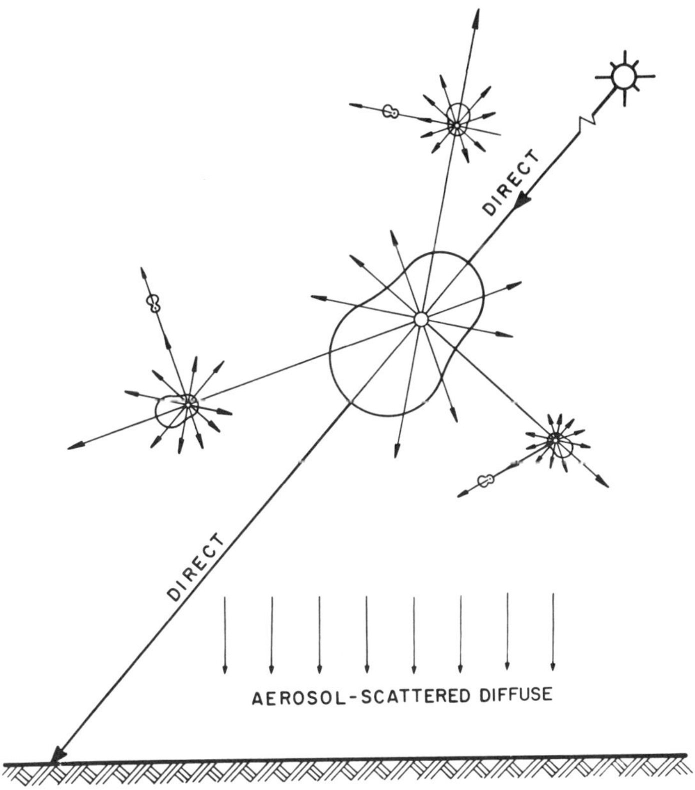

Figure 6.15.2 Multiple scattering by aerosols.

absorption and scattering effects of particulate matter, both of which take place simultaneously. Let us define

$\tau_{\mathrm{aa}\lambda}$ is the spectral transmittance of aerosols due to absorption, and
$\tau_{\mathrm{as}\lambda}$ is the spectral transmittance of aerosols due to scattering effects.

The total attenuation by the aerosols is as follows:

$$1 - \tau_{\mathrm{a}\lambda} = 1 - \tau_{\mathrm{aa}\lambda}\tau_{\mathrm{as}\lambda}. \tag{6.15.2}$$

It is difficult to separate the amount scattered and the amount absorbed. To circumvent this difficulty we define another term commonly employed by meteorologists:

ω_0 is the single scattering albedo, which is the ratio of energy scattered by aerosols to total attenuation under the first impingement by direct radiation.

From the above definition, the energy scattered by aerosols is

$$\omega_0(1 - \tau_{\mathrm{a}\lambda}). \tag{6.15.3}$$

In practice, determination of ω_0 is almost impossible. This ratio depends on the material, shape, size, and optical properties of aerosol particles. Small aerosol particles in rural environments usually scatter more than those found in urban–industrial areas. It is common to assign a fixed and arbitrary value to ω_0 invariant with wavelength, usually between 0.7 and 1.0. For urban–industrial regions $\omega_0 \approx 0.6$, and for rural–agricultural areas $\omega_0 \approx 0.9$. The former contain more carbon and as such have high absorptance.

As a continuation of the step-by-step development of this subject, we consider a homogeneous mixture of the pure Rayleigh and aerosol atmosphere. We ignore gaseous absorbers. Assuming one unit of incident energy, we determine that the amount attenuated by this atmosphere is

$$1 - \tau_{\mathrm{r}\lambda}\tau_{\mathrm{a}\lambda}. \tag{6.15.4}$$

From this expression we are interested in determining the amount of energy scattered. It is quite common in the literature (Sivkov [27], Leckner [6], and others) to ignore the absorbing effects of aerosols, and in the presence of molecular absorbers to assume that the energy scattered is

$$\tau_{\mathrm{ma}\lambda}(1 - \tau_{\mathrm{r}\lambda}\tau_{\mathrm{a}\lambda}). \tag{6.15.5}$$

Furthermore, it is also quite common to assume that one-half the total scattered energy reaches the ground and multiple reflections are ignored.

In the preceding expression, it is desirable to separate the Rayleigh-scattered and aerosol-scattered parts, which have dissimilar scattering characteristics (Fig. 6.3.1). Brine and Iqbal [28] suggest the following.

Rayleigh-scattered diffuse radiation:

$$\tau_{ma\lambda}(1 - \tau_{r\lambda})\tau_{a\lambda}, \tag{6.15.6}$$

Aerosol-scattered diffuse radiation:

$$\tau_{ma\lambda}(1 - \tau_{a\lambda})\tau_{r\lambda}\omega_0. \tag{6.15.7}$$

The preceding formulation is based on the simple assumption that scattering by molecules and by aerosols can be linearly separated. This is also called the *two-stream approximation.* Expressions (6.15.5)–(6.15.7) are all empirical and approximate. Moreover, it is implicit in these expressions that molecular absorption occurs after scattering has taken place. For zenith angles less than 60°, Brine and Iqbal's formulation shows good correspondence with some rigorous solutions of this problem by Dave and colleagues [29–32].

A portion of Rayleigh- and aerosol-scattered diffuse radiation reaches the ground after the first pass through the atmosphere. Naturally, the balance goes back to space. In Sections 6.16 and 6.17 we will present formulations for these two quantities.

Diffuse radiation arriving on the ground after the first pass through the atmosphere and direct radiation are in part reflected by the ground. This upwelling radiation is partly reflected back to the ground by the atmosphere. This process continues ad infinitum. These multiple reflections between the ground and the atmosphere (not to be confused with multiple scattering) add to the diffuse radiation reaching the ground after the first pass through the atmosphere. Let

$\dot{I}_{d\lambda}$ be the diffuse spectral irradiance on a horizontal surface. This diffuse irradiance is composed of three parts, as follows:

$\dot{I}_{dr\lambda}$ is the diffuse spectral irradiance produced by Rayleigh scattering that arrives on the ground after the first pass through the atmosphere;

$\dot{I}_{da\lambda}$ is the diffuse spectral irradiance produced by aerosols that arrives on the ground after the first pass through the atmosphere, and

$\dot{I}_{dm\lambda}$ is the diffuse spectral irradiance produced by multiple reflections.

Consequently,

$$\dot{I}_{d\lambda} = \dot{I}_{dr\lambda} + \dot{I}_{da\lambda} + \dot{I}_{dm\lambda}. \tag{6.15.8}$$

This downward-diffuse radiation is also called sky-diffuse or sky radiation, and should not be confused with the long-wave radiation emitted by the earth's atmosphere.

In Sections 6.16–6.19 we shall study methods of determining the three quantities given above.

6.16 Rayleigh-Scattered Spectral Diffuse Irradiance

Consider the Rayleigh scattering diagrams (Figs. 6.3.1a and 6.15.1). It appears obvious that, irrespective of the angle of incidence, about half the diffuse radiation is directed toward the ground and the remainder goes back to space. This assumption is commonly made in literature treating spectrally integrated diffuse radiation originating from Rayleigh scattering. Spectral studies seem to indicate that this assumption is quite good for $\lambda > 0.5\ \mu m$. However, because of multiple scattering, at shorter wavelengths a little less than half the diffuse radiation is directed toward the ground.

Consider a plane-parallel layer representing the earth's atmosphere containing all the attenuators we have treated so far. From Eq. (6.15.6) it can be seen that the Rayleigh-scattered diffuse radiation generated by an irradiance of $\dot{I}_{0n\lambda} \cos \theta_z$ on a horizontal surface at the top of the atmosphere is as follows:

$$\dot{I}_{0n\lambda} \cos(\theta_z)\tau_{ma\lambda}[(1 - \tau_{r\lambda})\tau_{a\lambda}]. \qquad (6.16.1)$$

If we assume that half the above amount is directed downwards, the diffuse radiation reaching the ground after the first pass through the atmosphere is as follows:

$$\dot{I}_{dr\lambda} = \dot{I}_{0n\lambda} \cos(\theta_z)\tau_{ma\lambda}[0.5(1 - \tau_{r\lambda})\tau_{a\lambda}]. \qquad (6.16.2)$$

It is appropriate at this moment to explain two phrases often employed by meteorologists: *forward scatterance* and *backscatterance.* The former implies the fraction of the *total scattered* energy propagated in the direction of the incident wave. This direction is usually toward the ground. Backscatterance is the fraction of the total scattered energy propagated in a direction opposite to the incident wave, and this direction is usually toward outer space. In the Rayleigh atmosphere treated above, we have assumed that the forward and the backward scatterances are each equal to 0.5.

In Fig. 6.16.1 we show some results of the Rayleigh-scattered spectral diffuse irradiance. We assume $\beta = 0.1$ and $\alpha = 1.3$. In this diagram the shaded areas represent the effects of molecular absorption, with ozone increasing from 0 to 0.35 cm(NTP), and water vapor from 0 to 2 cm. The effect of absorption is small and is mainly in the ultraviolet and a portion of the visible region. The absorption effect increases, however, with an increase in the zenith angle. Furthermore, the peak in the curves shifts toward higher wavelengths as the zenith angle increases. As expected, an increase in the zenith angle reduces the amount of radiation reaching the ground.

We have observed in Fig. 6.4.1 that Rayleigh scattering is negligible at $\lambda > 1\ \mu m$, and this is substantiated by Fig. 6.16.1. The small amount of

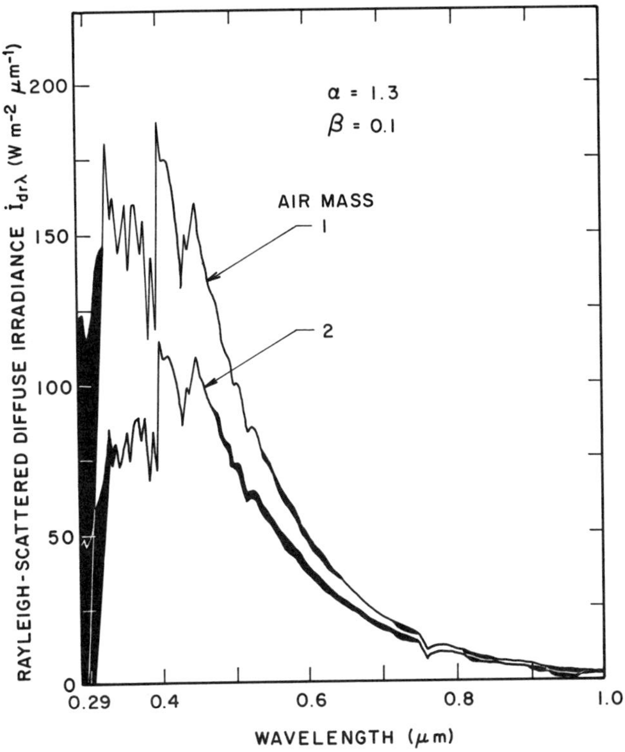

Figure 6.16.1 Rayleigh-scattered spectral diffuse irradiance. The shaded areas indicate increase in attenuation when ozone is increased from 0 to 0.35 cm(NTP) and water vapor is increased from 0 to 2 cm.

diffuse irradiance at $\lambda > 1$ μm is due to the fact that the Rayleigh and the aerosol transmittances are slightly less than one at this wavelength.

6.17 Aerosol-Scattered Spectral Diffuse Irradiance

Aerosols are the biggest single atmospheric element creating severe problems in calculating solar irradiance. The frustration is nowhere as acutely felt as in the study of diffuse irradiance. One part of the difficulty lies in specifying their number, size, distribution, and optical properties. The second part is the difficulty in solving the integrodifferential equations that govern this subject. In general, aerosol-generated diffuse radiation is substantial and hence cannot be ignored. In many industrialized and/or dusty parts of the world, it is much greater than the Rayleigh-scattered diffuse radiation.

Let us consider again the earth's atmosphere as a plane-parallel layer containing all the attenuating elements, and irradiated by a flux of $\dot{I}_{0n\lambda} \cos \theta_z$ on a horizontal surface. Again, following Eq. (6.15.7), the total amount of aerosol-scattered diffuse radiation is given by the following:

$$\dot{I}_{0n\lambda} \cos(\theta_z)\tau_{ma\lambda}[\omega_0(1 - \tau_{a\lambda})\tau_{r\lambda}]. \tag{6.17.1}$$

This formulation represents the amount of diffuse radiation streaming out in all directions. We are interested in that portion which is directed toward the ground. We have already seen in Figs. 6.3.1b and 6.15.2 that aerosol scattering is mainly in the forward direction. Therefore, we need to specify a ratio of forward to total energy scattered, that is, forward scatterance. Let this ratio be called F_c such that

$$F_c = \frac{\text{energy scattered in the forward direction}}{\text{total energy scattered}}.$$

As F_c is forward scatterance, $1 - F_c$ is backscatterance.

Determination of F_c is difficult. This ratio depends on particle size, shape, and wavelength. Since the word "forward" implies the direction of beam radiation, F_c is a function of the zenith angle as well. Next to ω_0, F_c is the most poorly specified of the aerosol properties. For the type of aerosols prevailing over the British Isles, Robinson [33] determined the value of F_c as a function of the zenith angle. Table 6.17.1 lists the values of F_c which may be used until more precise studies are available.

It follows that aerosol-scattered diffuse radiation reaching the ground after the first pass through the atmosphere is as follows:

$$\dot{I}_{da\lambda} = \dot{I}_{0n\lambda} \cos(\theta_z)\tau_{ma\lambda}[F_c\omega_0(1 - \tau_{a\lambda})\tau_{r\lambda}]. \tag{6.17.2}$$

The factors F_c and ω_0 are both assumed invariant with wavelength.

In Fig. 6.17.1 we show the spectral variation of aerosol-scattered diffuse irradiance [Eq. (6.17.2)]. Nonabsorbing aerosols are assumed—that is, $\omega_0 = 1$. Molecular absorbers are included and their effect, which is small, is

Table 6.17.1

F_c, the Ratio of Forward Scattering to Total Scattering[a]

θ_z	0	10	20	30	40	50	60	70	80	85
F_c	0.92	0.92	0.90	0.90	0.90	0.85	0.78	0.68	0.60	0.50

[a] Adapted from Robinson [33].

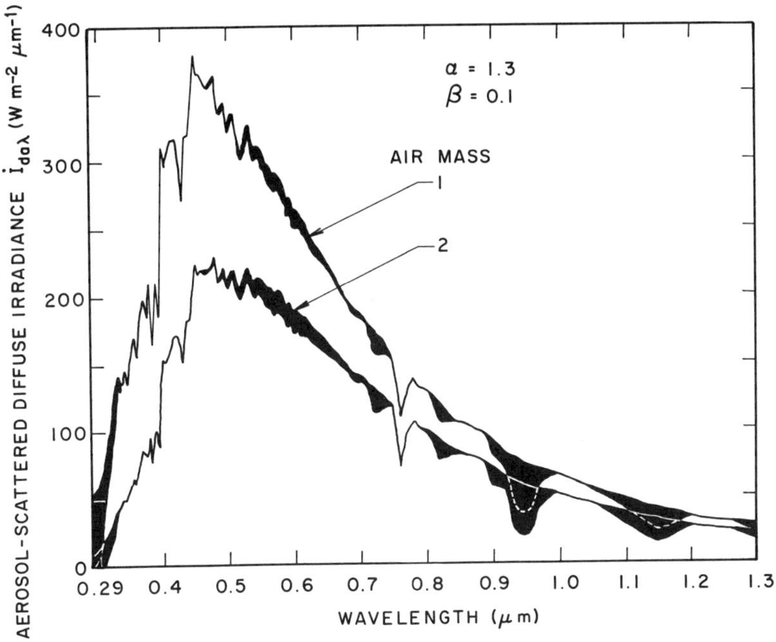

Figure 6.17.1 Aerosol-scattered diffuse irradiance. The shaded areas indicate increase in attenuation when ozone is increased from 0 to 0.35 cm(NTP) and water vapor is increased from 0 to 2 cm.

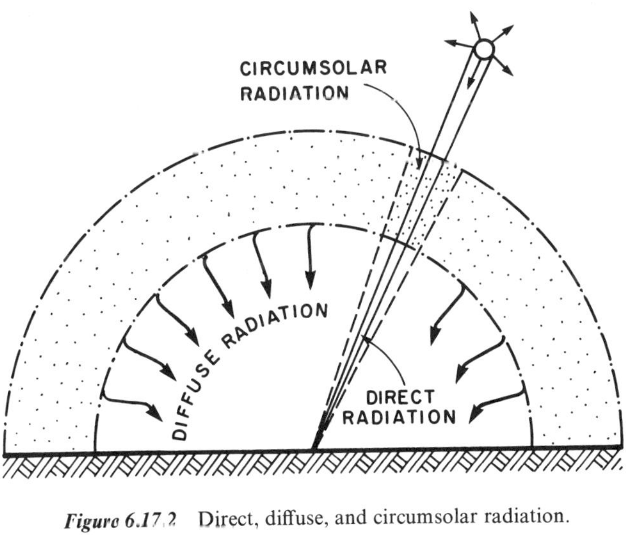

Figure 6.17.2 Direct, diffuse, and circumsolar radiation.

shown by the shaded areas, which represent increasing ozone from 0 to 0.35 cm(NTP) and water vapor from 0 to 2 cm. Radiation is further attenuated at larger air masses, and molecular absorption becomes more important. Aerosol-scattered radiation extends to longer wavelengths than does Rayleigh-scattered radiation.

Because the aerosols have a strong forward scattering component, a substantial portion of the aerosol-scattered diffuse radiation appears to come from a small annular area around the solar disk; this radiation is called circumsolar radiation or the solar aureole (Fig. 6.17.2). The size of the annular area varies with the amount of aerosol loading and aerosol particle size distribution. It is necessary to point out that $\dot{I}_{\mathrm{da}\lambda}$ includes circumsolar radiation.

6.18 Atmospheric Albedo

For the development of material in the next section, we need some information regarding reflectance properties of the earth and its atmosphere. In meteorology, reflectance is also called albedo. Chapter 9 is devoted entirely to ground albedo. In this section we develop a mathematical expression for the albedo of the cloudless atmosphere. Albedo is defined as the ratio of the energy reflected back to the incident energy. Consider the atmosphere as a plane-parallel layer, irradiated by one unit of parallel monochromatic beam (Fig. 6.18.1). In this diagram, the partition of the incident energy into various components is shown. Albedo of the atmosphere is the diffuse component reflected back to space. In an approximate form it is as follows:

$$\rho_{\mathrm{a}\lambda} = \tau_{\mathrm{ma}\lambda}[0.5(1 - \tau_{\mathrm{r}\lambda})\tau_{\mathrm{a}\lambda} + (1 - F_{\mathrm{c}})\omega_0(1 - \tau_{\mathrm{a}\lambda})\tau_{\mathrm{r}\lambda}]. \qquad (6.18.1)$$

The first term on the right-hand side of this equation represents the albedo of the Rayleigh atmosphere, and the second that of aerosols.

In the event that the incident radiation is of diffuse nature, for instance the upwelling radiation from the ground, the angle of incidence will vary from 0° to 90°. In such a situation, calculation of atmospheric albedo is quite complicated. It is common to assume a single average angle of incidence corresponding to air mass 1.6–1.9. Here we shall assume $m = 1.9$, the value that yields the best correspondence with the results of Dave and colleagues [31, 32]. We rewrite the preceding equation in order to prevent possible confusion:

$$\rho'_{\mathrm{a}\lambda} = \tau'_{\mathrm{ma}\lambda}[0.5(1 - \tau'_{\mathrm{r}\lambda})\tau'_{\mathrm{a}\lambda} + (1 - F'_{\mathrm{c}})\omega_0(1 - \tau'_{\mathrm{a}\lambda})\tau'_{\mathrm{r}\lambda}] \qquad (6.18.2)$$

The primes over various quantities indicate that they are to be evaluated at $m = 1.9$, $\theta_z \approx 60°$.

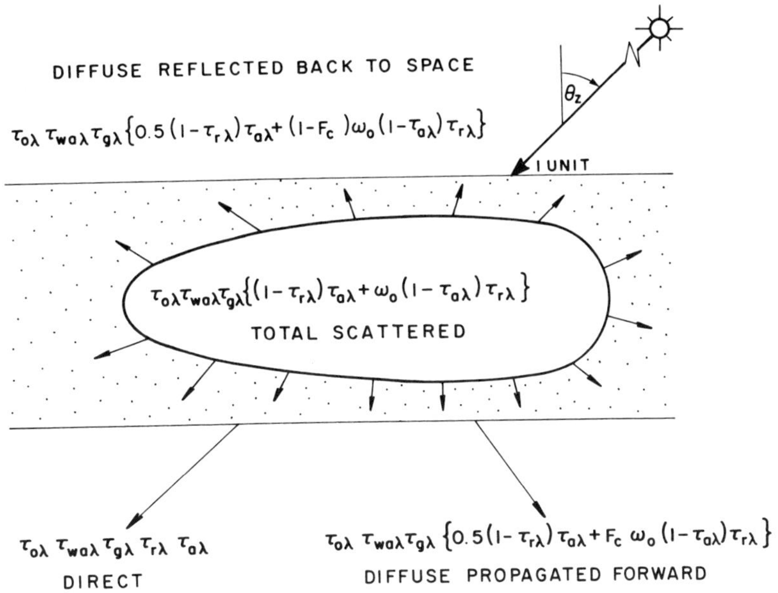

Figure 6.18.1 Partition of radiation into direct, diffuse propagated forward, and diffuse reflected back to space.

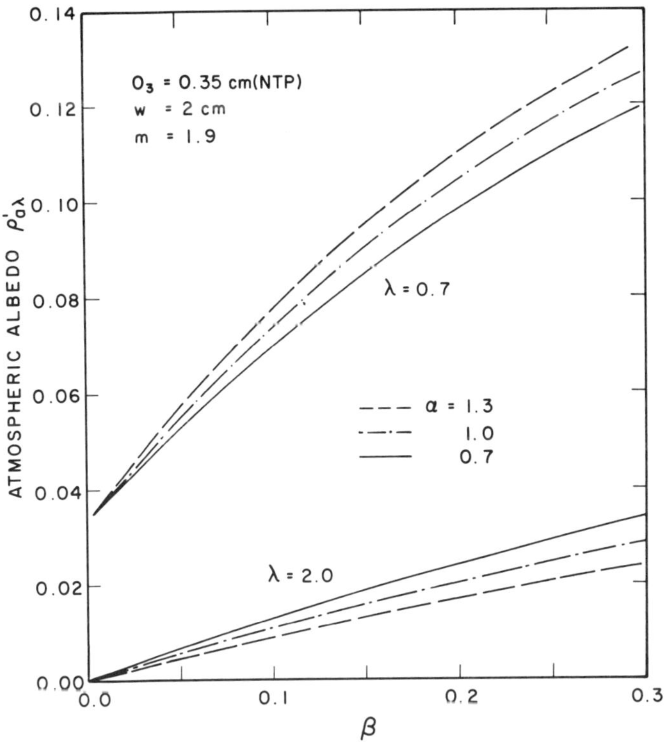

Figure 6.18.2 Variation of the atmospheric albedo with wavelength and β.

It is necessary to distinguish between the atmospheric albedo described above and the *planetary albedo*. The latter indicates the albedo of the earth and its atmosphere combined.

The variation of atmospheric albedo $\rho'_{a\lambda}$ with α and β is shown in Fig. 6.18.2 for two specific wavelengths. The albedo is greater at shorter wavelengths because more energy is scattered in the shorter than in the longer wavelengths. The albedo increases with β and α. Larger values of α represent small average particle size, and we know that smaller particles scatter more energy than do bigger particles. Consequently, the atmospheric albedo increases with α.

6.19 Multiply Reflected Spectral Diffuse Irradiance

Let us consider diffuse irradiance reaching the ground after the first pass, and beam irradiance on a horizontal surface. Let

$$Q_\lambda = (\dot{I}_{dr\lambda} + \dot{I}_{da\lambda}) + \dot{I}_{n\lambda} \cos\theta_z. \qquad (6.19.1)$$

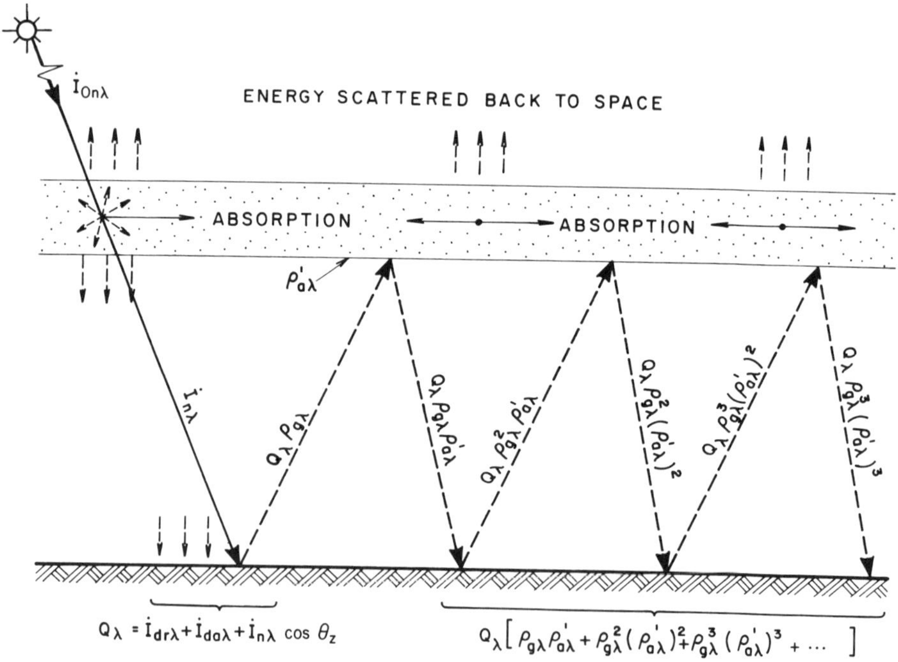

Figure 6.19.1 The process of multiple reflection of radiation between the atmosphere and the ground.

We are interested in determining the diffuse irradiance resulting from multiple reflections of Q_λ between the ground and the atmosphere. Since Q_λ contains a diffuse and a beam component, in theory we should have two separate values for ground albedo. Here, we assume a single value for this purpose, and call it $\rho_{g\lambda}$. Furthermore, we assume that this albedo produces a perfectly isotropic diffuse reflection. That is, the upwelling radiation $Q_\lambda \rho_{g\lambda}$ has a uniform intensity in all directions.

Figure 6.19.1 shows the process of multiple reflections between the ground and the atmosphere. The upwelling diffuse radiation $Q_\lambda \rho_{g\lambda}$ is reflected back to the ground, and the process of multiple reflections between the ground and the sky continues. The radiation is continuously attenuated at each impact.

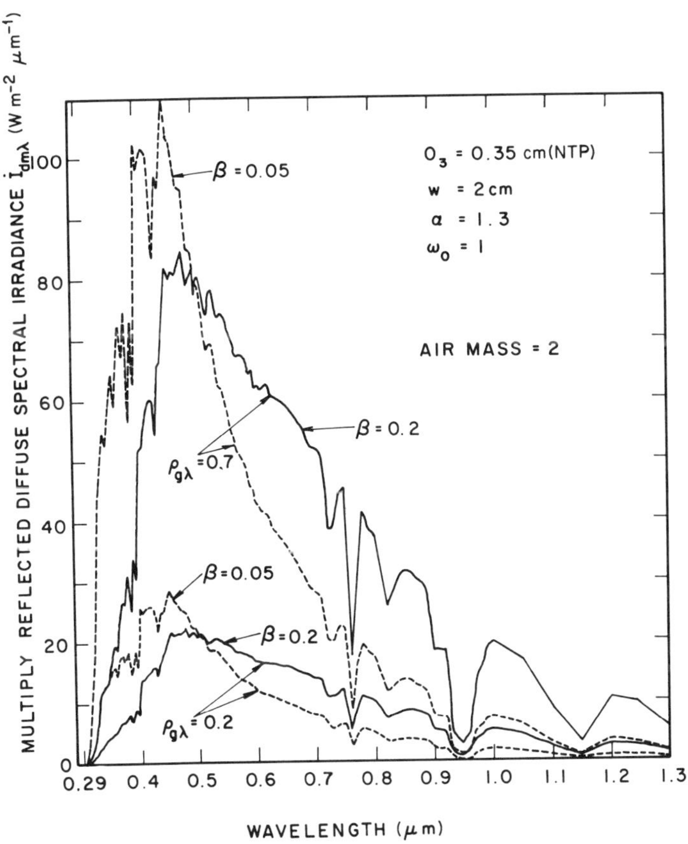

Figure 6.19.2 Multiply reflected diffuse spectral irradiance on a horizontal surface as a function of β and $\rho_{g\lambda}$.

From Fig. 6.19.1, it can be seen that the total amount of multiply reflected radiation reaching ground is as follows:

$$\dot{I}_{\mathrm{dm}\lambda} = Q_\lambda(\rho_{\mathrm{g}\lambda}\rho'_{\mathrm{a}\lambda} + \rho^2_{\mathrm{g}\lambda}\rho'^2_{\mathrm{a}\lambda} + \rho^3_{\mathrm{g}\lambda}\rho'^3_{\mathrm{a}\lambda} + \cdots)$$

or

$$\dot{I}_{\mathrm{dm}\lambda} = Q_\lambda(\rho_{\mathrm{g}\lambda}\rho'_{\mathrm{a}\lambda}/(1 - \rho_{\mathrm{g}\lambda}\rho'_{\mathrm{a}\lambda}). \tag{6.19.2}$$

The value of $\rho'_{\mathrm{a}\lambda}$ is very small, and $\rho_{\mathrm{g}\lambda}$ is usually about 0.2. Consequently, the effect of multiple reflections is minor. However, in the event of a snow cover, $\rho_{\mathrm{g}\lambda}$ is approximately 0.7 and the effect of multiple reflections assumes important dimensions. In this chapter we assume $\rho_{\mathrm{g}\lambda}$ invariant with wavelength.

Variations of multiply reflected irradiance with ground albedos of 0.2 and 0.7 are shown in Fig. 6.19.2. Naturally, higher albedos increase the level of multiply reflected radiation. In the same diagram, the effect of turbidity is also shown. The effect of turbidity appears to vary with wavelength. At lower turbidity ($\beta = 0.05$), $\dot{I}_{\mathrm{dm}\lambda}$ is higher at shorter wavelengths ($\lambda < 0.5$ μm). At higher turbidity ($\beta = 0.2$), $\dot{I}_{\mathrm{dm}\lambda}$ is higher at longer wavelengths ($\lambda > 0.5$ μm).

6.20 Diffuse Spectral Irradiance on the Ground

If we combine Eqs. (6.19.1) and (6.19.2), the diffuse spectral irradiance on a horizontal surface [Eq. (6.15.8)]

$$\dot{I}_{\mathrm{d}\lambda} = \dot{I}_{\mathrm{dr}\lambda} + \dot{I}_{\mathrm{da}\lambda} + \dot{I}_{\mathrm{dm}\lambda}$$

can be written as follows:

$$\dot{I}_{\mathrm{d}\lambda} = (\dot{I}_{\mathrm{dr}\lambda} + \dot{I}_{\mathrm{da}\lambda})\frac{1}{1 - \rho_{\mathrm{g}\lambda}\rho'_{\mathrm{a}\lambda}} + \dot{I}_{\mathrm{n}\lambda}\cos\theta_{\mathrm{z}}\left(\frac{\rho_{\mathrm{g}\lambda}\rho'_{\mathrm{a}\lambda}}{1 - \rho_{\mathrm{g}\lambda}\rho'_{\mathrm{a}\lambda}}\right). \tag{6.20.1}$$

We can now examine the variation of diffuse irradiance with some climatic and physical parameters. Air mass, turbidity parameters α and β, and ground albedo are the dominant factors that modify diffuse irradiance. We have already observed that an increase in air mass reduces diffuse irradiance. Figure 6.20.1 shows the effect of a change in turbidity parameter β. Naturally, greater turbidity results in higher amounts of diffuse radiation. The effect of a variation in α is demonstrated in Fig. 6.20.2. A small average particle size represented by high values of α produces greater amounts of diffuse irradiance. The influence of ground albedo on spectral diffuse irradiance can be examined in Fig. 6.20.3. An increase in ground albedo from 0.2 to 0.7 substantially increases diffuse irradiance. However, this increase is not confined to the ultraviolet and the visible portion of the spectrum. Due to the

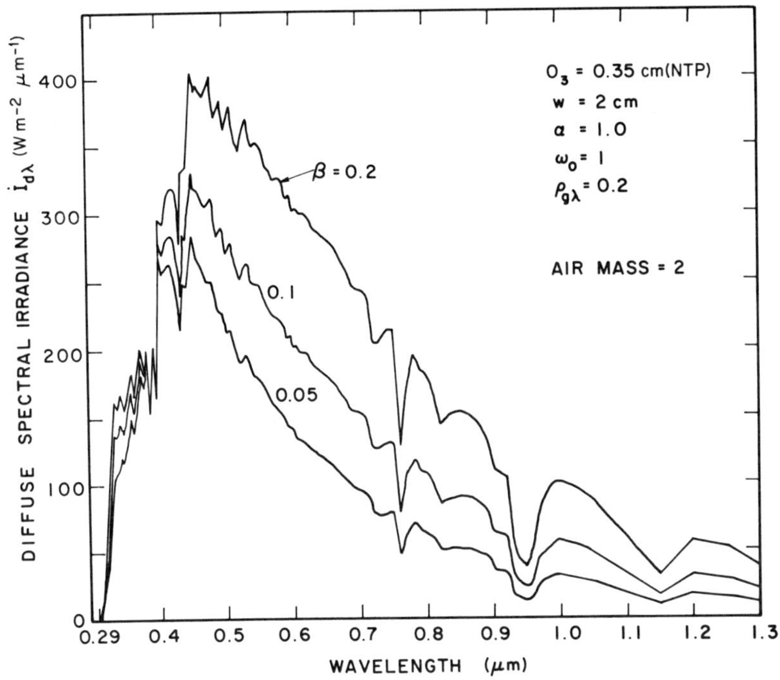

Figure 6.20.1 Diffuse spectral irradiance on a horizontal surface as a function of β.

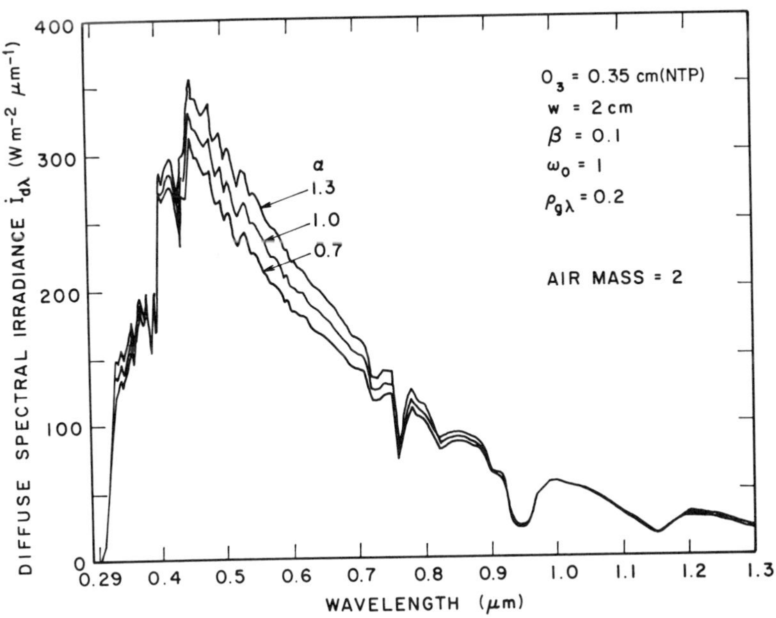

Figure 6.20.2 Diffuse spectral irradiance on a horizontal surface as a function of α.

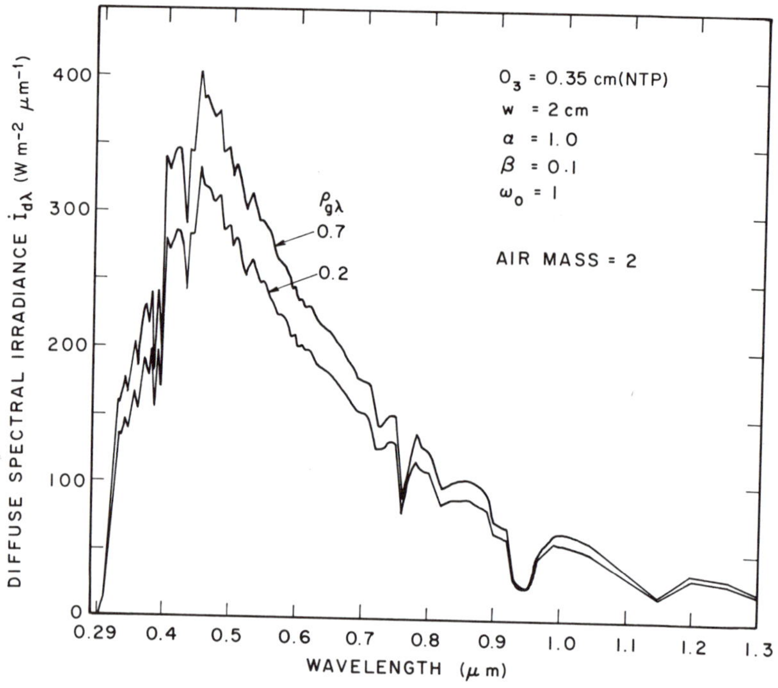

Figure 6.20.3 Diffuse spectral irradiance on a horizontal surface as a function of $\rho_{g\lambda}$.

increase in the visible spectrum, on a clear day immediately following a fresh snow fall, it becomes necessary to wear sunglasses.

In Fig. 6.20.4, a comparison between $\dot{I}_{dr\lambda}$, $\dot{I}_{da\lambda}$, and $\dot{I}_{dm\lambda}$ is given. At $\beta = 0.1$ and $\rho_{g\lambda} = 0.2$, the multiply reflected component is the smallest and the aerosol-scattered component the largest in magnitude. The flux $\dot{I}_{da\lambda}$ depends primarily on the magnitude of β, while $\dot{I}_{dm\lambda}$ depends on both β and $\rho_{g\lambda}$. Naturally, at $\beta = \rho_{g\lambda} = 0$, $\dot{I}_{da\lambda}$ and $\dot{I}_{dm\lambda}$ both reduce to zero.

6.21 Global Spectral Irradiance on the Ground

Global irradiance is the sum of the beam plus diffuse irradiance on a horizontal surface. Let $\dot{I}_{\lambda}$ is the global spectral irradiance on a horizontal surface and

$$\dot{I}_{\lambda} = \dot{I}_{n\lambda} \cos\theta_z + \dot{I}_{d\lambda} \tag{6.21.1}$$

or

$$\dot{I}_{\lambda} = (\dot{I}_{n\lambda} \cos\theta_z + \dot{I}_{dr\lambda} + \dot{I}_{da\lambda})/(1 - \rho_{g\lambda}\rho'_{a\lambda}). \tag{6.21.2}$$

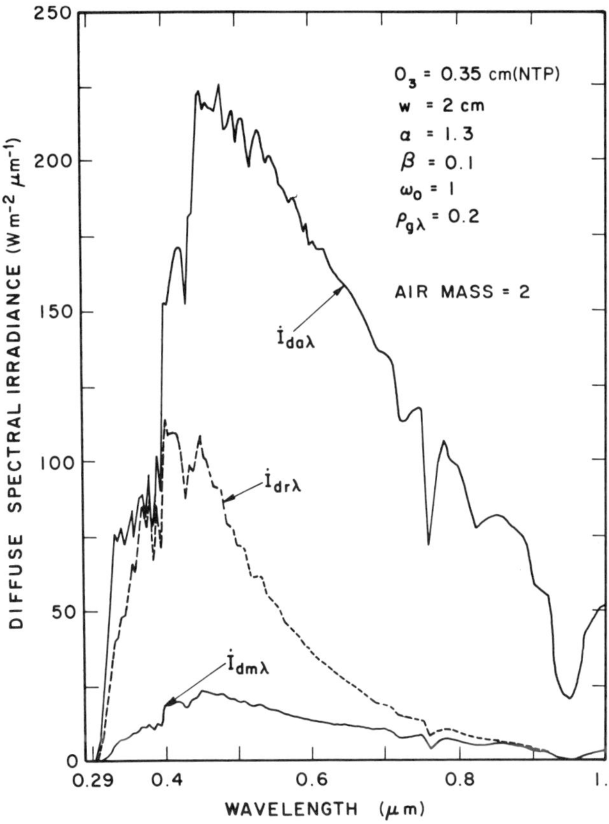

Figure 6.20.4 The various components of the spectral diffuse irradiance on a horizontal surface.

The variation of global solar irradiance with β, α, and ground albedo is shown in Figs. 6.21.1–6.21.3. In the study of the direct normal radiation we have observed (see Fig. 6.14.7) that an increase in the magnitude of β decreases the amount of direct spectral irradiance. On the other hand, the amount of diffuse irradiance increases as β increases (see Fig. 6.20.1). However, the decrease in direct irradiance is far greater than the increase in diffuse irradiance. Consequently, global radiation decreases with an increase in β (Fig. 6.21.1). However, this change is not as drastic as in the case of direct or diffuse irradiance. Furthermore, the effect of a change in β on global radiation is minimal at $\theta_z = 0$, and this effect increases as the zenith angle increases.

The effect of a variation in α is demonstrated through Fig. 6.21.2. The global radiation remains almost insensitive to changes in α. Small variations that do occur in global irradiance are confined to the ultraviolet and the visible portion of the spectrum. An increase in α results in slightly lower values of global irradiance.

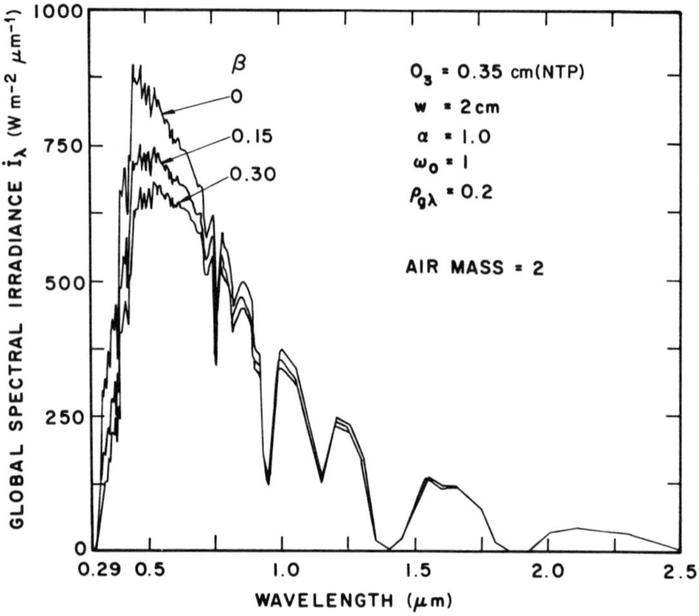

Figure 6.21.1 Global spectral irradiance on a horizontal surface as a function of β.

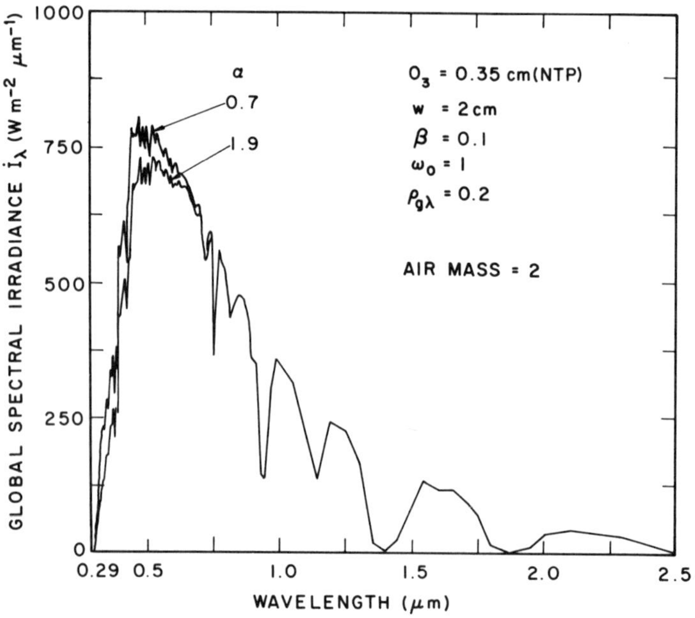

Figure 6.21.2 Global spectral irradiance on a horizontal surface as a function of α.

An increase in ground albedo increases the amount of global spectral irradiance because of the multiple reflection effects (Fig. 6.21.3). Again, small changes in global irradiance are confined mainly to $\lambda < 1$ μm.

It is instructive to look at the relative magnitudes of diffuse and direct components of global radiation. In Fig. 6.21.4 we present these plots for a typical group of the atmospheric parameters. At wavelengths longer than 1 μm, global radiation is almost entirely composed of the direct component. At wavelengths shorter than 1 μm, the fractional contribution of diffuse radiation to global radiation depends on air mass, turbidity, and ground albedo. Other atmospheric parameters such as ozone and water vapor have negligible effect.

The algorithm to calculate beam radiation is quite accurate. On the other hand, the formulation to compute diffuse radiation is approximate. Because (in most instances) beam radiation is far greater than diffuse, the global radiation values should be quite accurate.

□ Example 6.21.1. The spectral diffuse and the global irradiance, under standard sea-level conditions, on a horizontal surface calculated at (a) $\lambda = 0.7$ and (b) $\lambda = 2.0$ μm on 15 June, at a location where the skies are cloudless and the sun is at 60° from its zenith position, for $\beta = 0.1$, $\alpha = 1.3$, $O_3 = 0.3$ cm(NTP), $w = 2$ cm, and ground albedo $\rho_{g\lambda} = 0.2$, assuming $\omega_0 = 0.95$.

Solution. In order to solve this problem, we shall employ the results already obtained in Examples 6.7.1, 6.13.1, and 6.14.1. The spectral diffuse irradiance will be obtained through Eq. (6.20.1) and the global through Eq. (6.21.2). We assume $m = 2.0$ throughout.

(a) At $\lambda = 0.7$ μm, from Eq. (6.16.2),

$$\dot{I}_{dr\lambda} = (0.9684)(1427.5)(0.5)(0.98)(0.7276)[0.5(1 - 0.9279)]$$
$$= 17.77 \text{ W m}^{-2}\ \mu\text{m}^{-1}.$$

From Eq. (6.17.2) and Table 6.17.1,

$$\dot{I}_{da\lambda} = (0.9684)(1427.5)(0.5)(0.98)(0.9279)[0.78 \times 0.95(1 - 0.7276)]$$
$$= 126.54 \text{ W m}^{-2}\ \mu\text{m}^{-1}.$$

In order to obtain the multiply reflected component of the diffuse irradiance, we should first obtain the albedo of the atmosphere $\rho'_{a\lambda}$ from Eq. (6.18.2). In this equation all the transmittances are to be computed at $m = 1.9$. We list the transmittances:

$$\tau'_{r\lambda} = 0.9313, \qquad \tau'_{a\lambda} = 0.7393, \qquad \tau'_{o\lambda} = 0.9870, \qquad \tau'_{wa\lambda} = 0.9865,$$

and

$$\tau'_{g\lambda} = 1 \qquad \text{since} \quad k_{g\lambda} = 0.$$

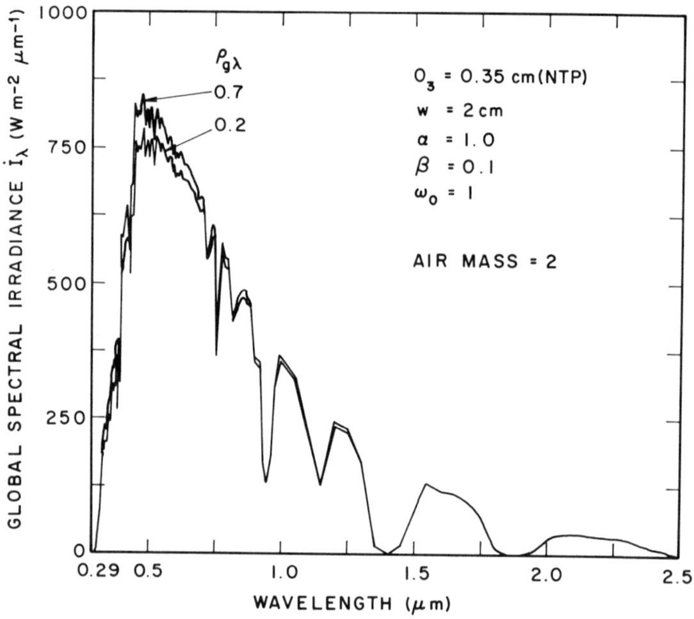

Figure 6.21.3 Global spectral irradiance on a horizontal surface as a function of $\rho_{g\lambda}$.

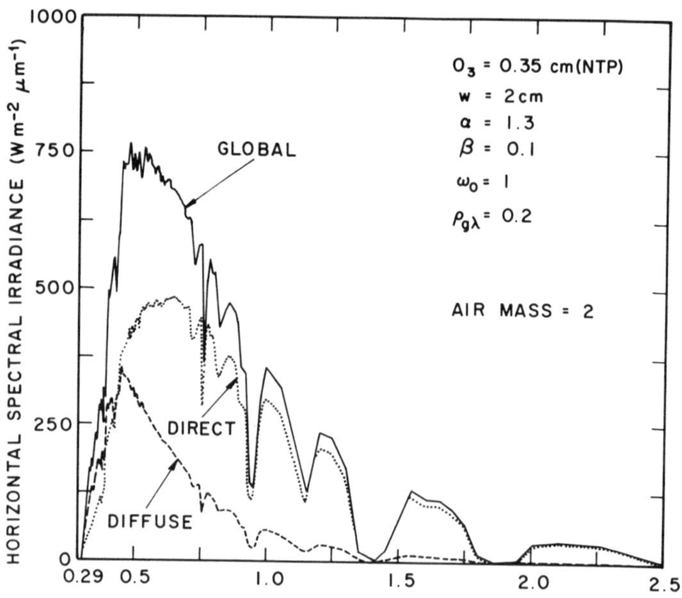

Figure 6.21.4 Global, direct, and diffuse spectral irradiance on a horizontal surface.

Therefore

$$\tau'_{\mathrm{ma}\lambda} = \tau'_{\mathrm{o}\lambda}\tau'_{\mathrm{wa}\lambda}\tau'_{\mathrm{g}\lambda} = 0.9736.$$

From the above, the albedo of the atmosphere is

$$\begin{aligned}\rho'_{\mathrm{a}\lambda} &= 0.9736[0.5(1 - 0.9313)(0.7393) + (1 - 0.78)(0.95)(1 - 0.7393)(0.9313)] \\ &= 0.07413.\end{aligned}$$

Therefore,

$$\frac{\rho_{\mathrm{g}\lambda}\rho'_{\mathrm{a}\lambda}}{1 - \rho_{\mathrm{g}\lambda}\rho'_{\mathrm{a}\lambda}} = \frac{0.2(0.07413)}{1 - (0.2)(0.07413)} = 0.01505,$$

$$\begin{aligned}\dot{I}_{\mathrm{dm}\lambda} &= (17.77 + 126.54 + 457.23)(0.01505) \\ &= 9.05 \text{ W m}^{-2}\ \mu\text{m}^{-1}.\end{aligned}$$

Finally,

$$\dot{I}_{\mathrm{d}\lambda} = 17.77 + 126.54 + 9.05 = 153.36 \text{ W m}^{-2}\ \mu\text{m}^{-1}.$$

The global spectral irradiance is

$$\dot{I}_{\lambda} = 153.36 + 457.23 = 610.59 \text{ W m}^{-2}\ \mu\text{m}^{-1}.$$

(b) At $\lambda = 2.0\ \mu$m, we present the main results:

$$\dot{I}_{\mathrm{dr}\lambda} = 0.0139 \text{ W m}^{-2}\ \mu\text{m}^{-1}, \qquad \dot{I}_{\mathrm{da}\lambda} = 1.74 \text{ W m}^{-2}\ \mu\text{m}^{-1};$$

$$\tau'_{\mathrm{ma}\lambda} = 0.5177, \qquad \tau'_{\mathrm{r}\lambda} = 0.999, \qquad \tau'_{\mathrm{a}\lambda} = 0.9257;$$

$$\dot{I}_{\mathrm{dm}\lambda} = 0.0476 \text{ W m}^{-2}\ \mu\text{m}^{-1}.$$

Finally,

$$\dot{I}_{\mathrm{d}\lambda} = 1.80 \text{ W m}^{-2}\ \mu\text{m}^{-1}, \qquad \dot{I}_{\lambda} = 28.75 \text{ W m}^{-2}\ \mu\text{m}^{-1}.$$ □

6.22 Further Reading

In this chapter we have referred to a large amount of literature. Sufficient ground has been covered on direct spectral irradiance. The subject of diffuse radiation is difficult as well as fascinating. The mathematical groundwork on multiple scattering was laid down by Chandrasekhar [34]. Deirmendjian [35] and Deirmendjian and Sekera [36] present some solutions to this problem. Dave and colleagues [29–31] have treated extensively the subject of multiple scattering for a variety of atmospheric models and parametric variations. These solutions yield also the hemispherical variation of diffuse radiation intensity for every wavelength.

Measurement of the diffuse spectral irradiance has been reported by Kok and colleagues [37–40], McCartney [41], Murai *et al.* [42], and Bird and Hulstrom [43]. Dave [44] presented tables of numerical values of spectral diffuse irradiance derived from solution of the basic transfer equations. Tables of direct and diffuse spectral irradiance have also been given by Boer [45] but from a less rigorous procedure. The theory of gaseous absorption is covered in a classic text by Goody [46]. Tiwari [47] presents an updated review of this subject.

Methods of measuring the atmospheric turbidity parameters have been described by Ångström [48, 49], Robinson [50], and Hansen [51]. In the modern literature, Ångström's turbidity factors or visibility are used to account for the attenuation by aerosols. In the older (and also very good) literature, attenuation by aerosols has been expressed through Schüepp's turbidity coefficient B and the turbidity factor T of Linke (pronounced Linkay). B is easily related to β; however, T is a spectrally integrated average turbidity coefficient the theoretical evaluation of which is somewhat complicated. See Robinson [50] for further reading on this subject.

It seems appropriate to add that anyone interested in detailed data on solar spectral irradiance should obtain the computer program LOWTRAN from the National Climatic Center, Asheville, North Carolina.

6.23 Historical Notes

Over the years, confusion has arisen as to the true author of the law of attenuation of light through a medium [Eq. (6.2.1)]. It seems appropriate to clear the air on this point. Pierre Bouguer[7] (1698–1758), born in France, a physicist and hydrographer, experimentally developed this law in 1729 [52]. About 30 years later, Johann Heinrich Lambert[8] (1728–1777), born in Alsace, a mathematician and philosopher, theoretically developed it in 1760 [53]. Middleton [54], in an annotated translation of Bouguer's publications, throws an interesting light on the almost simultaneous studies of the two researchers.

The law of attenuation of light [Eq. (6.2.1)] is sometimes also ascribed to August Beer (1825–1863), born in Germany. In fact, Beer's law [55] states that the absorption of radiation depends only on the concentration of the absorbing species along the path. It is, therefore, a restrictive form of Bouguer's law.

[7] Bouguer is pronounced Bôgεr, with g hard as in go.

[8] Lambert was a Swiss-German of French descent. Therefore, whether or not the t in his name should be mute is debatable.

It seems of interest to add here that Bouguer and Lambert also developed, independently of each other, expressions for air mass [56].

Nomenclature

Å	Angstrom, 10^{-10} m
D	Particle diameter (μm)
d	Number of dust particles per cubic centimeter
F_c	Ratio of energy scattered in the forward direction to total energy scattered (dimensionless)
$\dot{I}_b$	Total broadband direct irradiance on a horizontal surface (W m^{-2})
$\dot{I}_{b\lambda}$	Direct spectral irradiance on a horizontal surface (W m^{-2} μm^{-1})
$\dot{I}_{da\lambda}$	Aerosol-scattered spectral diffuse irradiance arriving on a horizontal surface after the first pass through the atmosphere (W m^{-2} μm^{-1})
$\dot{I}_{dm\lambda}$	Multiply reflected spectral diffuse irradiance arriving on a horizontal surface (W m^{-2} μm^{-1})
$\dot{I}_{dr\lambda}$	Rayleigh-scattered spectral diffuse irradiance arriving on a horizontal surface after the first pass through the atmosphere (W m^{-2} μm^{-1})
$I_{d\lambda}$	Diffuse spectral irradiance on a horizontal surface (W m^{-2} μm^{-1})
$\dot{I}_n$	Total broadband direct normal irrdiance (W m^{-2})
$\dot{I}_{n\lambda}$	Direct normal spectral irradiance (W m^{-2} μm^{-1})
$I_{0n\lambda}$	Extraterrestrial spectral irradiance at mean sun–earth distance on a surface normal to the solar rays (W m^{-2} μm^{-1})
I_λ	Global spectral irradiance on a horizontal surface (W m^{-2} μm^{-1})
$k_{a\lambda}$	Coefficient of attenuation due to scattering and absorption by aerosols (dimensionless)
$k_{d\lambda}$	Coefficient of attenuation due to scattering by dust particles (dimensionless)
$k_{g\lambda}$	Coefficient of attenuation due to absorption by mixed gases (dimensionless)
$k_{o\lambda}$	Coefficient of attenuation due to absorption by ozone (cm^{-1})
$k_{r\lambda}$	Coefficient of attenuation due to Rayleigh scattering by air molecules (dimensionless)
$k_{wa\lambda}$	Coefficient of attenuation due to absorption by water vapor (cm^{-1})
$k_{ws\lambda}$	Coefficient of attenuation due to scattering by water vapor (cm^{-1})
k_λ	Coefficient of attenuation due to all processes combined (dimensionless)
l	Ozone layer thickness [cm(NTP)]
m	Air mass (dimensionless)
m_a	Air mass at actual pressure (dimensionless)
m_r	Air mass at standard pressure (dimensionless)
n	Refractive index (dimensionless)
Q_λ	See Eq. (6.19.1)
r	Actual sun–earth distance (AU)
r_0	Mean sun–earth distance, 1 AU
Vis	Horizontal visibility (km)
w	Thickness of precipitable water (cm)
α	Wavelength exponent in Ångström's turbidity formula (dimensionless)
β	Ångström's turbidity parameter (dimensionless)
θ_z	Zenith angle (degrees)
λ	Wavelength (μm)
$\rho_{a\lambda}$	Spectral albedo of the cloudless atmosphere (dimensionless)
$\rho_{g\lambda}$	Spectral albedo of the ground (dimensionless)
$\tau_{a\lambda}$	Transmittance due to aerosol attenuation (dimensionless)

$\tau_{aa\lambda}$ Transmittance due to absorption by the aerosols (dimensionless)
$\tau_{as\lambda}$ Transmittance due to scattering by the aerosols (dimensionless)
$\tau_{c\lambda}$ Transmittance due to Rayleigh and aerosol effects (continuum attenuation) (dimensionless)
$\tau_{d\lambda}$ Transmittance due to dust scattering (dimensionless)
$\tau_{g\lambda}$ Transmittance due to absorption by mixed gases (dimensionless)
$\tau_{ma\lambda}$ Transmittance due to all molecular absorbers (dimensionless)
$\tau_{wa\lambda}$ Transmittance due to absorption by water vapor (dimensionless)
$\tau_{ws\lambda}$ Transmittance due to scattering by water vapor (dimensionless)
ω_0 Single-scattering albedo, ratio of energy scattered to total attenuation by the aerosols (dimensionless)

References

1. M. P. Thekaekara (ed.), The Energy Crisis and the Energy from the Sun, pp. 21–49, Data on incident solar energy, Institute of Environmental Sciences, Mount Prospect, Illinois, (1974).
2. R. Siegel and J. R. Howell, "Thermal Radiation Heat Transfer." McGraw-Hill, New York, 1981.
3. K. Ya. Kondratyev, "Radiation in the Atmosphere." Academic Press, New York, 1969.
4. L. Elterman, U.V., Visible and IR attenuation for altitudes to 50 km. ARCRL-68-0153, Environmental Research Paper No. 285 (1968).
5. R. Penndorf, Tables of the refractive index for standard air and the Rayleigh scattering coefficient for the spectral region between 0.2 and 20.0 μm and their application to atmospheric optics. *J. Opt. Soc. Am.* **47**(2), 176–182 (1957).
6. B. Leckner, The spectral distribution of solar radiation at the earth's surface—elements of a model. *Sol. Energy* **20**(2), 143–150 (1978).
7. H. C. van de Hulst, "Light Scattering by Small Particles." Chapman and Hall, New York, 1957. See also "Multiple Light Scattering, Tables, Formulas and Applications," Vols. 1 and 2. Academic Press, New York (1980).
8. P. Moon, Proposed standard solar radiation curves for engineering use. *J. Franklin Inst.* **230,** 583–617 (1940).
9. A. Ångström, On the atmospheric transmission of sun radiation and on dust in the air. *Geografis. Annal.* **2,** 156–166 (1929).
10. A. Ångström, On the atmospheric transmission of sun radiation. *Geografis. Annal.* **2** and **3,** 130–159 (1930).
11. SOLMET Vol. 2, Hourly solar radiation—surface meteorological observations—Final Rep. TD-9724. National Climatic Center, Asheville, North Carolina (1979).
12. R. A. McClatchey and J. E. Selby, Atmospheric transmittance from 0.25 to 38.5 μm: computer code LOWTRAN-2. Air Force Cambridge Research Laboratories, AFCRL-72-0745, Environ. Res. Paper 427 (1927).
13. E. Vigroux, Contribution a l'étude expérimentale de l'absorption de l'ozone. *Ann. Phys.* **8,** 709–762 (1953).
14. F. E. Fowle, The transparency of aqueous vapor. *J. Astrophys.* **42,** 394–411 (1915).
15. J. N. Howard, D. E. Burch, and D. Williams, Infrared transmission of synthetic atmospheres. II. Absorption by carbon dioxide. *J. Opt. Soc. Am.* **46**(4), 237–241 (1956).
16. J. N. Howard, D. E. Burch, and D. Williams, Infrared transmission of synthetic atmospheres. III. Absorption by water vapor. *J. Opt. Soc. Am.* **46**(4), 242–245 (1956).
17. J. N. Howard, D. E. Burch, and D. Williams, Infrared transmission of synthetic atmosperes. IV. Application of theoretical band models. *J. Opt. Soc. Am.* **46**(5), 334–338 (1956).

18. J. N. Howard, D. E. Burch, and D. Williams, Infrared transmission of synthetic atmospheres. V. Absorption laws for overlapping bands. *J. Opt. Soc. Am.* **46**(6), 452–455 (1956).
19. D. M. Gates, Near infrared atmospheric transmission to solar radiation. *J. Opt. Soc. Am.* **50**(12), 1299–1304 (1960).
20. D. M. Gates and W. H. Harrop, Infrared transmission of the atmosphere to solar radiation. *Appl. Opt.* **2**(9), 887–898 (1963).
21. M. P. Thekaekara, Solar irradiance, total and spectral, "Solar Energy Engineering" (A. A. M. Sayigh, ed.), Chapter 3. Academic Press, New York, 1977.
22. R. L. Hulstrom, Insolation models, data and algorithms, SERI/TR-36-110, Solar Energy Research Institute, Golden, Colorado (1978).
23. R. A. McClatchey, R. W. Fenn, J. E. A. Selby, F. E. Volz, and J. S. Goring, Optical Properties of the Atmosphere (3rd ed.), AFCRL-72-0497 (1972).
24. J. E. A. Selby and R. A. McClatchey, Atmospheric transmittance from 0.25 to 28.5 microns. Computer Code Lowtran 3. Air Force Cambridge Research Lab., AFCRL-TR-75-0255, AD-A017734 (1975).
25. J. E. A. Selby, F. Y. Kneizys, J. H. Chetwind Jr., and R. A. McClatchey, Atmospheric Transmittance/Radiance: Computer Code Lowtran 4, AFGL-TR-78-0053 (1978).
26. J. N. Howard, J. I. F. King, and P. R. Gast, Thermal radiation, "Handbook of Geophysics," Chapter 16. MacMillan, New York, 1961.
27. S. I. Sivkov, Computation of Solar Radiation Characteristics. Israeli Program of Scientific Translations, Jerusalem (1971).
28. D. T. Brine and M. Iqbal, Diffuse and global solar spectral irradiance under cloudless skies. *Sol. Energy* **30**(5), 447–453 (1983).
29. J. V. Dave, A direct solution of the spherical harmonics approximation to the radiative transfer equation for an arbitrary solar elevation. Part I: Theory. *J. Atmos. Sci.* **32,** 790–798 (1975).
30. J. V. Dave, a direct solution of the spherical harmonics approximation to the radiative transfer equation for an arbitrary solar elevation. Part II: Results. *J. Atmos. Sci.* **32,** 1463–1474 (1975).
31. J. V. Dave, P. Halpern, and N. Braslau, Spectral distribution of the direct and diffuse solar energy received at sea-level of a model atmosphere. Rep. No. G320-3332, IBM Palo Alto Scientific Center (1975).
32. J. V. Dave, Private communication, a magnetic tape of the direct and diffuse fluxes at the ground-level of the Dave–Braslaw models, obtained from Dr. Dave. IBM Palo Alto Scientific Center (1981).
33. G. D. Robinson, Absorption of solar radiation by atmospheric aerosol as revealed by measurements from the ground. *Arch. Meteorol. Geophys. Bioklimatol. Ser. B* **12,** 19–40 (1962).
34. S. Chandrasekhar, "Radiative Transfer." Oxford Univ. Press (Clarendon), London and New York, 1950.
35. D. Deirmendjian, "Electromagnetic Scattering on Spherical Polydispersions." American Elsevier, New York, 1969.
36. D. Deirmendjian and Z. Sekera, Global radiation resulting from multiple scattering in a Rayleigh atmosphere. *Tellus* **6**(4), 382–398 (1954).
37. C. J. Kok, Spectral irradiance of daylight at Pretoria. *J. Phys. D; Appl. Phys.* **5,** 1513–1520 (1972).
38. C. J. Kok, Spectral irradiance of daylight for air mass 2. *J. Phys. D; Appl. Phys.* **5,** L85–L88 (1972).
39. C. J. Kok and A. N. Chalmers, Spectral irradiance of daylight at Durban. C.S.I.R., Res. Rep. 339, Pretoria, South Africa (1978).

40. C. J. Kok, A. N. Chalmers, and R. Turner, Spectroradiometry of daylight at sea level in the Southern Hemisphere: Durban. *S-Afr. Tydskr. Fis.* **2**(2), 47–53 (1979).
41. H. A. McCartney, Spectral distribution of solar radiation. II: global and diffuse. *Q. J. R. Meteorol. Soc.* **104,** 911–926 (1978).
42. K. Murai, M. Kobayashi, R. Goto and T. Yamauchi, Ground-based spectral measurements of solar radiation (II)—global and diffuse sky radiation. *Papers Meteorol. Geophys.* **30**(2), 83–91 (1979).
43. R. E. Bird and R. L. Hulstrom, Solar spectral measurements and modelling. SERI/TR-642-1013 Solar Energy Research Institute, Golden, Colorado (1981).
44. J. V. Dave, Extensive datasets of the diffuse radiation in realistic atmosphere models with aerosols and common absorbing gases. *Sol. Energy* **21**(5), 361–369 (1979).
45. K. W. Boer, The solar spectrum at typical clear weather days. *Sol. Energy* **19**(5), 525–538 (1977).
46. R. M. Goody, "Atmospheric Radiation. I: Theoretical Basis." Oxford Univ. Press, London and New York, 1964.
47. S. N. Tiwari, Models for infrared atmospheric radiation. *Adv. Geophys.* **20,** 1–80 (1978).
48. A. Ångström, Techniques of determining the turbidity of the atmosphere. *Tellus* **13**(2), 214–223 (1961).
49. A. Ångström, The parameters of atmospheric turbidity. *Tellus* **16**(1), 64–75 (1964).
50. N. Robinson, "Solar Radiation." American Elsevier, New York, 1966.
51. V. Hansen, Determination of atmospheric turbidity parameters from spectral solar radiation measurements. *Arch. Meteorol. Geophys. Bioklimatol. Ser B* **22,** 301–308 (1974).
52. Pierre Bouguer, "Essai d'Optique sur la gradation de la lumière," Paris (1729).
53. Johann Heinrich Lambert, "Photometria sive de mensura et gradibus luminis, colorum, et umbrae." Augsburg (1760).
54. W. E. K. Middleton, "Pierre Bouguer's Optical Treatise on the Gradation of Light." Univ. of Toronto Press, Toronto, 1961.
55. "Grand Larousse Encyclopédique," Vol. 2. Librairie Larousse, Paris, 1960.
56. F. Linke, "Handbuch der Geophysik." Verlag von Gebrüder Borntraeger, Berlin-Zehlendorf, 1942.

Chapter 7

TOTAL[1] (BROADBAND) RADIATION UNDER CLOUDLESS SKIES

7.1 Introduction

For most engineering problems, determining the amount of radiation over a certain bandwidth or over the complete solar spectrum is required. Radiation over a certain bandwidth can be evaluated only by adding up the spectral values. On the other hand the radiation over the entire solar spectrum can be computed either by integrating over the complete solar spectrum or by what may be called a broadband approach. Integration of the monochromatic values can be carried out with the help of the material presented in Chapter 6. However, such a procedure is very time consuming. Consequently, simple broadband equations have been developed to account for the attenuation by each of the atmospheric constituents.

Under the broadband approach, we discuss the parameterization method and the ASHRAE algorithm. For each method, formulations to compute

[1] In this chapter the terms *total* and *broadband* will be used synonymously and interchangeably. Both will mean integrated values over the complete solar spectrum, which for this study covers 0.29–4.0 μm. The term broadband applies to this broad band.

direct, diffuse, and global radiation are presented. We present first the equations treating spectral integration of irradiance.

7.2 Spectral Integration Model

In order to provide a means of checking or comparing the results of the parameterization models discussed in the following sections, it will be useful to give here the basic equations obtained from the spectral integration method.

The spectrally integrated direct irradiance on a surface normal to the sun's rays (also called direct normal radiation) and at mean sun–earth distance is given thus:

$$\dot{I}_{\mathrm{n}}\Big|_{\lambda=\lambda_1}^{\lambda=\lambda_2} = \sum_{\lambda=\lambda_1}^{\lambda=\lambda_2} \dot{I}_{0\mathrm{n}\lambda}\tau_\lambda\,\Delta\lambda, \tag{7.2.1}$$

or

$$\dot{I}_{\mathrm{n}} = \sum_{\lambda=0}^{\infty} \dot{I}_{0\mathrm{n}\lambda}\tau_\lambda\,\Delta\lambda. \tag{7.2.2}$$

Equation (7.2.1) applies only to a specified bandwidth, whereas Eq. (7.2.2) applies to the total solar spectrum. The former is used in applications where determining energy in a certain bandwidth is required, as in photovoltaics. In both equations the monochromatic transmittance τ_λ is obtained from Eq. (6.14.2). The wavelength intervals $\Delta\lambda$ employed above vary, as Table 3.3.2 indicates.

From Eq. (7.2.1) some interesting information regarding the energy contained in various color bands can be computed. The fraction of direct energy in different colors is shown in Table 7.2.1. As the zenith angle increases, the fraction in the infrared band increases. Further, as the turbidity increases, the fraction in the infrared band increases correspondingly, showing that the direct radiation in the visible portion has been dissipated by scattering. In any event, even for a clear atmosphere, energy in the infrared band accounts for a minimum of approximately 50% of the total direct energy.

Parallel to the foregoing procedure, the broadband diffuse irradiance on the ground, at mean sun–earth distance, is determined by summation of $\dot{I}_{\mathrm{d}\lambda}$ [Eq. (6.20.1)] over the required wavelength. Thus

$$\dot{I}_{\mathrm{d}}\Big|_{\lambda=\lambda_1}^{\lambda=\lambda_2} = \sum_{\lambda=\lambda_1}^{\lambda=\lambda_2} \dot{I}_{\mathrm{d}\lambda}\,\Delta\lambda, \tag{7.2.3}$$

Table 7.2.1

Direct Normal Irradiance within Various Bands[a]

θ_z (degrees)	β	$\dot{I}_n$ ($W\ m^{-2}$)	uv <0.39	Violet 0.39–0.455	Blue 0.455–0.492	Green 0.492–0.577	Yellow 0.577–0.597	Orange 0.597–0.622	Red 0.622–0.77	ir >0.77
			Fraction of direct energy in different colors, λ (μm)							
0.0	0.0	1053.30	0.04	0.08	0.06	0.13	0.03	0.04	0.18	0.45
60.0	0.0	934.01	0.02	0.06	0.06	0.13	0.03	0.04	0.19	0.47
70.0	0.0	853.01	0.01	0.05	0.05	0.13	0.03	0.04	0.19	0.49
80.0	0.0	689.18	0.00	0.03	0.04	0.11	0.03	0.04	0.20	0.55
85.0	0.0	525.18	0.00	0.01	0.02	0.07	0.02	0.03	0.20	0.64
0.0	0.1	895.12	0.03	0.07	0.06	0.12	0.03	0.04	0.18	0.48
60.0	0.1	688.19	0.01	0.05	0.05	0.11	0.03	0.04	0.18	0.53
70.0	0.1	558.62	0.01	0.03	0.04	0.10	0.03	0.04	0.18	0.58
80.0	0.1	334.38	0.00	0.01	0.02	0.06	0.02	0.03	0.17	0.69
85.0	0.1	166.73	0.00	0.00	0.00	0.02	0.01	0.01	0.12	0.83
0.0	0.2	766.07	0.02	0.06	0.05	0.12	0.03	0.04	0.17	0.51
60.0	0.2	518.89	0.01	0.03	0.04	0.10	0.02	0.03	0.17	0.59
70.0	0.2	381.62	0.00	0.02	0.03	0.08	0.02	0.03	0.17	0.66
80.0	0.2	181.35	0.00	0.00	0.01	0.03	0.01	0.02	0.13	0.80
85.0	0.2	67.60	0.00	0.00	0.00	0.01	0.00	0.00	0.06	0.93
0.0	0.3	659.98	0.02	0.05	0.04	0.11	0.03	0.04	0.17	0.55
60.0	0.3	399.40	0.01	0.02	0.03	0.08	0.02	0.03	0.16	0.65
70.0	0.3	270.33	0.00	0.01	0.02	0.06	0.02	0.02	0.15	0.72
80.0	0.3	107.15	0.00	0.00	0.00	0.02	0.01	0.01	0.09	0.88
85.0	0.3	32.36	0.00	0.00	0.00	0.00	0.00	0.00	0.02	0.97

[a] $w = 2$ cm, $O_3 = 0.35$ cm(NTP), $\alpha = 1.3$, $\dot{I}_{sc} = 1367\ W\ m^{-2}$, extraterrestrial spectrum from Table 3.3.2.

Table 7.2.2

Diffuse Horizontal Irradiance within Various Bands[a]

θ_z (degrees)	β	$\dot{I}_d$ (W m^{-2})	Fraction of diffuse energy in different colors, λ (μm)							
			uv <0.39	Violet 0.39–0.455	Blue 0.455–0.492	Green 0.492–0.577	Yellow 0.577–0.597	Orange 0.597–0.622	Red 0.622–0.77	ir >0.77
0.0	0.0	70.06	0.29	0.26	0.13	0.16	0.02	0.03	0.07	0.04
60.0	0.0	52.96	0.25	0.26	0.14	0.17	0.03	0.03	0.08	0.04
0.0	0.1	205.03	0.12	0.16	0.10	0.18	0.03	0.04	0.16	0.21
60.0	0.1	129.69	0.08	0.14	0.10	0.18	0.04	0.05	0.17	0.24
0.0	0.2	316.94	0.09	0.14	0.10	0.18	0.04	0.05	0.17	0.25
60.0	0.2	185.70	0.05	0.11	0.09	0.18	0.04	0.05	0.19	0.30
0.0	0.3	409.95	0.07	0.13	0.09	0.18	0.04	0.05	0.18	0.27
60.0	0.3	226.90	0.04	0.10	0.08	0.17	0.04	0.05	0.20	0.33

[a] $w = 2$ cm, $O_3 = 0.35$ cm(NTP), $\alpha = 1.3$, $\omega_0 = 1$, $\rho = 0.2$, $\dot{I}_{sc} = 1367$ W m^{-2}, extraterrestrial spectrum from Table 3.3.2.

or

$$\dot{I}_d = \sum_{\lambda=0}^{\infty} \dot{I}_{d\lambda}\,\Delta\lambda. \tag{7.2.4}$$

Table 7.2.2 shows the fraction of diffuse energy in different color bands. When the atmosphere is clean, maximum energy is in the ultraviolet and violet bands. However, with increase in turbidity, the fraction in the infrared increases. Yet even at high values of turbidity, the fraction in the visible spectrum accounts for about 75% of the total diffuse radiation arriving on the ground.

The global irradiance on a horizontal surface at the mean sun–earth distance can be written as follows:

$$\dot{I} = \dot{I}_n \cos\theta_z + \dot{I}_d. \tag{7.2.5}$$

The global irradiance can also be obtained by integrating Eq. (6.21.2). Thus

$$\dot{I} = \sum_{\lambda=0}^{\infty} \left((\dot{I}_{n\lambda} \cos\theta_z + \dot{I}_{dr\lambda} + \dot{I}_{da\lambda}) \frac{1}{1 - \rho_{g\lambda}\rho'_{a\lambda}} \right) \Delta\lambda. \tag{7.2.6}$$

In the foregoing, we have laid down the spectral integration method to compute instantaneous values of the direct, diffuse, and global radiation. We look now at the parameterization methods.

7.3 Parameterization Method

In many applications of solar energy such as heating and cooling, only the total (spectrally integrated) quantity of the incident radiation needs to be determined. It is therefore necessary to develop methods to calculate rapidly the radiant energy under specified atmospheric parameters. One such method is called the parameterization method. In parameterization, an overall spectrally integrated transmittance for each atmospheric constituent is employed to compute the total transmittance of the atmosphere. Let us define the total transmittance as follows:

$$\begin{aligned} \tau &= \text{transmittance of the direct normal irradiance} \\ &= \dot{I}_n/\dot{I}_{sc}, \end{aligned} \tag{7.3.1}$$

or

$$\tau = \left(\sum_{\lambda=0}^{\infty} \dot{I}_{0n\lambda}\tau_\lambda\,\Delta\lambda \right) \Big/ \dot{I}_{sc}. \tag{7.3.2}$$

The right-hand side of the above equation is a function of the ozone layer thickness l, precipitable water w, the turbidity parameters β and α, and the air mass m. Approximate algebraic expressions can be developed through regression analysis to fit Eq. (7.3.2) such that we can say

$$\tau = f(l, w, \beta, \alpha, m). \tag{7.3.3}$$

Let us now define certain quantities called spectrally integrated individual transmittances of the direct irradiance due to the various atmospheric constituents. These transmittances can be listed as follows:

τ_o is the transmittance to direct irradiance due to absorption by ozone,
τ_g is the transmittance to direct irradiance due to absorption by uniformly mixed gases,
τ_w is the transmittance to direct irradiance due to absorption by water vapor,
τ_r is the transmittance to direct irradiance due to Rayleigh scattering effects of air molecules, and
τ_a is the transmittance to direct irradiance due to attenuation by aerosols.

Each of the above is a spectrally integrated transmittance; for instance, τ_r is defined as follows:

$$\tau_r = \left(\sum_{\lambda=0}^{\infty} \dot{I}_{0n\lambda} \exp(-0.008735\lambda^{-4.08} m_a)\, \Delta\lambda \right) \Big/ \dot{I}_{sc}. \tag{7.3.4}$$

It is often assumed that the transmittances are multiplicative, such that we can write

$$\tau = \tau_o \tau_g \tau_w \tau_r \tau_a. \tag{7.3.5}$$

In brief, the above may be written

$$\tau = \prod_{i=1}^{i=j} \tau_i, \tag{7.3.6}$$

where j stands for the total number of attenuating constituents considered. Equation (7.3.6), or a certain variation of it, is very useful in rapid calculation of the direct normal irradiance. Of course, Eq. (7.3.6) is not mathematically identical to Eq. (7.3.2) and hence may be considered incorrect. It is, however, widely used in this or a similar form.

Before going further, it is necessary to state that some of the transmittances on the right-hand side of Eq. (7.3.5) can be represented by either

$$1 - \alpha_i \qquad \text{or} \qquad 1 - S_i,$$

where α_i and S_i are the corresponding depletions by absorption and scattering, respectively. Atmospheric physicists, interested in the radiative heating of the atmosphere by absorption of radiation by the atmospheric constituents, have laid down individual expressions for the absorptance of the various absorbing gases, and these expressions can be used here.

Many parameterization models are available in the literature. In fact, there are so many of them that it can be confusing to decide which model to adopt. Further, the forms of the algebraic equations representing various transmittances or absorptances differ widely. We present below three such models, which we shall compare with one another. We shall name these Model A, Model B, and Model C. These models have been chosen for their simplicity and accuracy. It is hoped that in the near future accurate insolation data accompanied by accurate measurements of such atmospheric parameters as ozone, water vapor, and turbidity will become available and help us to decide which one of the three models is most accurate. Such data at present are not available. As we shall see, the models, with their different forms, ultimately yield results not too different from one another.

7.4 Parameterization Model A

A number of authors have contributed to the development of this model [1–6]. We shall present separate expressions for (a) direct, (b) diffuse, and (c) global irradiance.

A. *Direct Irradiance*

The basic expression for the total direct transmittance was suggested by Paltridge and Platt [1]. From it the direct normal irradiance at mean sun–earth distance can be written

$$\dot{I}_{n} = \dot{I}_{0n}(\tau_o \tau_r - \alpha_w)\tau_a, \tag{7.4.1}$$

where α_w is the absorptance of direct irradiance by water vapor and

$$\alpha_w = 1 - \tau_w. \tag{7.4.2}$$

In Eq. (7.4.1) it is implied that the absorptance by water vapor should be subtracted after the attenuation of direct irradiance by ozone and Rayleigh scattering. This is reasonable considering the fact that absorption by ozone and scattering by air molecules occur high up in the atmosphere and are confined to short wavelengths. In this model, the slight additional absorptance by carbon dioxide and oxygen is included in α_w. Furthermore, it is assumed that the attenuation (absorption and scattering) by aerosols occurs over the entire solar spectrum.

The transmittance τ_o due to absorption by ozone can be written

$$\tau_o = 1 - \alpha_o, \tag{7.4.3}$$

where α_o is the absorptance of direct irradiance by ozone. Absorptance of direct irradiance by ozone has been studied by a number of researchers. Paltridge and Platt recommend the correlation developed by Lacis and Hansen [7]. From Lacis and Hansen, the ozone absorptance is given by the following:

$$\alpha_0 = \frac{0.021\,18U_3}{1 + 0.042U_3 + 3.23 \times 10^{-4}U_3^2} + \frac{1.082U_3}{(1 + 138.6U_3)^{0.805}} + \frac{0.0658U_3}{1 + (103.6U_3)^3}. \tag{7.4.4}$$

In (7.4.4), the first term on the right-hand side represents the absorptance of the visible spectrum, and the last two terms account for the absorptance of the ultraviolet portion of solar radiation. U_3 is the ozone relative optical path length given by

$$U_3 = lm_r, \tag{7.4.5}$$

where l is the vertical ozone layer thickness in cm(NTP) and m_r is the relative optical mass.

For absorptance by water vapor, Paltridge and Platt recommend the use of a correlation by Lacis and Hansen that is based on Yamamoto's study [8]. Lacis and Hansen present the following correlation to fit Yamamoto's curve to within 1% accuracy for $10^{-2} < U_1 < 10$ cm:

$$\alpha_w = \frac{2.9U_1}{(1 + 141.5U_1)^{0.635} + 5.925U_1}, \tag{7.4.6}$$

where U_1 is the pressure-corrected relative optical path length of precipitable water given by

$$U_1 = wm_r. \tag{7.4.7}$$

When precipitable water is obtained under conditions other than $p_0 =$ 1013.25 mbars and $T_0 = 273$ K, it should be corrected through Eq. (5.4.5). However, no such correction is applied to m_r.

The attenuation due to Rayleigh scattering can be evaluated through Eq. (7.3.4). A regression-type correlation that fits such an equation is given by Davies *et al.* [2] as below:

$$\tau_r = 0.972 - 0.082\,62m_a + 0.009\,33m_a^2 - 0.000\,95m_a^3 + 0.000\,437m_a^4. \tag{7.4.8}$$

Davies and Hay [4] have presented tabulated values of τ_r that are said to be 2% more accurate than those obtained through Eq. (7.4.8).

The transmittance due to aerosol attenuation is difficult to express in simple terms. Following Houghton [9], in its simplest form, it can be written[2]

$$\tau_a = K^{m_a}, \tag{7.4.9}$$

where K, a function of local turbidity usually obtained through measured direct normal irradiance, ranges from 0.6 to 1.0. Once the value of K is determined at one location, it can be applied to other locations with identical turbidity conditions. Henceforth, we shall not use this procedure. A better approach, however, is to obtain an equation for τ_a through a regression analysis of

$$\tau_a = \left(\sum_{\lambda=0}^{\infty} \dot{I}_{0n\lambda} \exp(-\beta\lambda^{-\alpha} m_a) \, \Delta\lambda \right) \Big/ \dot{I}_{sc}. \tag{7.4.10}$$

From Mächler [6], the following expression fits the preceding equation:

$$\tau_a = (0.124\,45\alpha - 0.016\,2) + (1.003 - 0.125\alpha) \times \exp[-\beta m_a(1.089\alpha + 0.512\,3)], \qquad \beta < 0.5. \tag{7.4.11}$$

It will be very useful to express (7.4.11) in terms of visibility also [visibility is related to α, β through (6.6.2)]. Again, from [6]

$$\tau_a = [0.97 - 1.265(\text{Vis})^{-0.66}]^{m_a^{0.9}}, \qquad 5 < \text{Vis} < 180 \text{ km}. \tag{7.4.12}$$

The above equation has been developed in such a manner that within $0.9 < \alpha < 1.5$, τ_a is a function only of the visibility and air mass.

B. *Diffuse Irradiance*

We now deal with diffuse irradiance, utilizing the above transmittances. Following the empirical procedures laid down for the spectral diffuse irradiance in Sections 6.15–6.20, the broadband diffuse irradiance on a horizontal surface can also be written

$$\dot{I}_d = \dot{I}_{dr} + \dot{I}_{da} + \dot{I}_{dm} \tag{7.4.13}$$

where

$\dot{I}_{dr}$ is the broadband diffuse irradiance on the ground due to Rayleigh scattering,

[2] In the literature some authors use m_r and some use m_a for the aerosol optical mass. That is, some authors consider it necessary to include pressure correction in the air mass to compute aerosol attenuation, and some do not. This author recommends that the pressure-corrected air mass m_a be used consistently in computing τ_a, τ_r, and τ_g. On the other hand, as the ozone thickness l and the precipitable water w are pressure-corrected quantities, further pressure correction is not required in computing their optical masses.

$\dot{I}_{da}$ is the broadband diffuse irradiance on the ground due to scattering by aerosols, and

$\dot{I}_{dm}$ is the broadband diffuse irradiance on the ground due to the multiple reflections between the earth's surface and its atmosphere.

The first two terms on the right-hand side of Eq. (7.4.13) represent radiant flux reaching the ground after the first pass through the atmosphere.

The diffuse irradiance because of Rayleigh scattering is as follows:

$$\dot{I}_{dr} = \dot{I}_{sc} \cos \theta_z \tau_0 [0.5(1 - \tau_r)] \tau_a. \quad (7.4.14)$$

It is assumed that one-half of the Rayleigh-scattered radiation reaches the ground. In other words, molecular scattering is assumed symmetric in the forward and backward directions. In the above equation, τ_w, the transmittance due to absorption by water vapor, does not appear because water vapor attenuates radiation in the spectral band where $\tau_r \simeq 1$. In this respect it is useful to direct our attention again to the corresponding spectral equation (6.16.2), where $\tau_{wa\lambda}$ and $\tau_{g\lambda}$ were included because, in spectral calculations, these quantities automatically adjust their values.

Corresponding to Eq. (6.17.2), the diffuse irradiance on a horizontal surface due to scattering of aerosols is

$$\dot{I}_{da} = \dot{I}_{sc} \cos \theta_z (\tau_o \tau_r - \alpha_w)[F_c \omega_0 (1 - \tau_a)], \quad (7.4.15)$$

where the quantities F_c and ω_0 have the same meaning they have in Sections 6.15 and 6.17. It may be noted that absorptance by water vapor is subtracted before aerosol scattering.

Following the development of Eq. (6.19.2), the downward irradiance due to multiple reflections between the ground and the atmosphere is

$$\dot{I}_{dm} = (\dot{I}_n \cos \theta_z + \dot{I}_{dr} + \dot{I}_{da}) \rho_g \rho'_a / (1 - \rho_g \rho'_a), \quad (7.4.16)$$

where ρ_g is the ground albedo and ρ'_a is the albedo of the cloudless sky. The latter is computed from

$$\rho'_a = 0.0685 + 0.17(1 - \tau'_a)\omega_0, \quad (7.4.17)$$

where the prime over τ_a indicates that the aerosol transmittance should be computed at air mass $= 1.66p/p_0$.[3] The factor 0.17 represents $1 - F_c$ at $\theta_z = 53°$, and $1 - F_c$ is backscatterance. Consequently, the second term on the right-hand side of the above equation represents the albedo of cloudless skies due to the presence of aerosols, whereas the first term represents the albedo of clean air.

[3] The air mass at 1.66 times the minimum air mass for direct radiation accounts for an overall air mass for the upwelling diffuse irradiance. Various authors have recommended slightly different values for this air mass. Note: in Eq. (6.18) we have used $m = 1.9$.

C. Global Irradiance

The global (direct plus diffuse) irradiance on a horizontal surface can be written as follows:

$$\dot{I} = \dot{I}_n \cos\theta_z + \dot{I}_d$$
$$= (\dot{I}_n \cos\theta_z + \dot{I}_{dr} + \dot{I}_{da})\left(\frac{1}{1-\rho_g\rho'_a}\right). \tag{7.4.18}$$

We illustrate the method of computing the instantaneous values through the following example.

☐ EXAMPLE 7.4.1. Estimation of direct, diffuse, and global irradiance on a horizontal surface at solar noon in Atlantic City, New Jersey (39°27′ N), for 21 January 1977, when the aerosol optical thickness at $\lambda = 0.38\ \mu$m is listed as 0.087 and at $\lambda = 0.5\ \mu$m as 0.069 under a cloudless sky (assuming precipitable water vapor is 1 cm, O_3 is 0.30 cm (NTP), barometric pressure is 990 mbars, ambient temperature is 0°C, ground albedo is 0.2, and $\omega_0 = 0.95$).

Solution. For 21 January, $E_0 = 1.0334$ (Table 1.2.1), $\delta = -20.05°$ (Table 1.3.1); $\phi = 39°27' = 30.45°$, $\omega = 0°$ at solar noon.

From Eq. (1.5.1),

$$\cos\theta_z = \sin(-20.05)\sin(39.45) + \cos(-20.05)\cos(39.45)\cos(0)$$
$$= 0.5075$$
$$\theta_z = 59.5°.$$

From Eq. (5.7.2),

$$m_r = [0.5075 + 0.15(93.885 - 59.5)^{-1.253}]^{-1} = 1.9635.$$

From Eq. (5.7.3),

$$m_a = 1.9635(990/1013.25) = 1.919.$$

From Eq. (5.4.5),

$$w = 1(990/1013.25)^{3/4} = 0.9827 \text{ cm};$$
$$U_3 = lm_r = 0.3(1.9635) = 0.5891.$$

From Eqs. (7.4.3) and (7.4.4),

$$\tau_o = 1 - \left(\frac{0.02118(0.589\,1)}{1 + 0.042(0.589\,1) + 3.23 \times 10^{-4}(0.589\,1)^2} + \frac{1.082(0.589\,1)}{[1 + 138.6(0.589\,1)]^{0.805}} + \frac{0.0658\,(0.589\,1)}{1 + (103.6 \times 0.589\,1)^3}\right)$$
$$= 0.9696;$$

$U_1 = wm_r = 0.9827(1.9635) = 1.9295.$

From Eq. (7.4.6),

$$\alpha_w = \frac{2.9(1.929\,5)}{[1 + 141.5(1.929\,5)]^{0.635} + 5.925(1.929\,5)} = 0.1197$$

From Eq. (7.4.8),

$$\tau_r = 0.972 - 0.082\,62(1.919) + 0.009\,33(1.919)^2 - 0.000\,95(1.919)^3 + 0.000\,437(1.919)^4 = 0.8470.$$

To obtain aerosol attenuation, we require α and β. Following Example 6.6.1,

$$k_{a\lambda|\lambda=0.38\ \mu m} = \beta(0.38)^{-\alpha} = 0.087,$$
$$k_{a\lambda|\lambda=0.5\ \mu m} = \beta(0.5)^{-\alpha} = 0.069;$$

$$\alpha = \frac{\ln(0.087/0.069)}{\ln(0.50/0.38)} = 0.8446; \qquad \beta = 0.087(0.38)^{0.844\,6} = 0.038\,4.$$

From Eq. (6.6.2) these values of α and β yield Vis = 160 km.

We can employ either Eq. (7.4.11) or (7.4.12). Both should yield the same result. From Eq. (7.4.11) we have

$$\tau_a = [0.124\,45(0.844\,6) - 0.016\,2] + [1.003 - 0.125(0.844\,6)] \times \exp\{-0.038\,4(1.919) \times [1.089(0.844\,6) + 0.512\,3]\}$$
$$= 0.896\,5;$$

from Eq. (7.4.12),

$$\tau_a = [0.97 - 1.265(160)^{-0.66}]^{(1.919)^{0.9}} = 0.87.$$

From Eq. (7.4.1), with $E_o = 1.033\,4$,

$$I_n = (1.033\,4)(1\,367)[0.969\,6)(0.847\,0) - 0.119\,7](0.896\,5) = 888.48\ \mathrm{W\,m^{-2}},$$
$$I_b = I_n \cos\theta_z = 888.48(0.507\,5) = 450.90\ \mathrm{W\,m^{-2}}.$$

From Eq. (7.4.14), with $E_o = 1.033\,4$,

$$\dot{I}_{dr} = 1.033\,4(1\,367)(0.507\,5)(0.969\,6)[1 - 0.8470]0.896\,5(0.5) = 47.67 \text{ W m}^{-2}.$$

Similarly, from Eq. (7.4.15),

$$\begin{aligned} \dot{I}_{da} &= 1.033\,4(1\,367)(0.507\,5)[(0.969\,6)(0.847) - 0.119\,7] \\ &\quad \times [0.778(0.95)(1 - 0.896\,5)] \\ &= 38.47 \text{ W m}^{-2}. \end{aligned}$$

$\tau'_a = \tau_a$, computed at air mass $1.66p/p_0 = 1.66(990/1\,013.25) = 1.621\,9$,

$$\begin{aligned} \tau'_a &= [0.124\,45(0.844\,6) - 0.016\,2] + [1.003 - 0.125(0.844\,6)] \\ &\quad \times \exp\{-0.038\,4(1.621\,9)[1.08\,9(0.844\,6) + 0.512\,3]\} \\ &= 0.909\,8. \end{aligned}$$

From Eq. (7.4.17),

$$\rho'_a = 0.068\,5 + 0.17(1 - 0.909\,8)(0.95) = 0.083\,1.$$

From Eq. (7.4.16),

$$\begin{aligned} \dot{I}_{dm} &= (450.90 + 47.67 + 38.47)\{0.2(0.083\,1)/[1 - 0.2(0.083\,1)]\} \\ &= 9.08 \text{ W m}^{-2}; \end{aligned}$$

$$\dot{I}_d = \dot{I}_{dr} + \dot{I}_{da} + \dot{I}_{dm} = 47.67 + 38.47 + 9.08 = 95.22 \text{ W m}^{-2};$$

$$\dot{I} = \dot{I}_b + \dot{I}_d = 450.90 + 95.22 = 546.12 \text{ W m}^{-2}$$ □

7.5 Parameterization Model B

Sasamori *et al.* [10] and Hoyt [11, 12] have presented a formulation based on separate values of transmittances (or absorptances) for precipitable water, carbon dioxide, oxygen, ozone, Rayleigh scattering, and scattering plus absorption by aerosols. This model has been validated by data from some U.S. stations [12].

A. *Direct Irradiance*

In this model the direct normal irradiance is given by

$$\dot{I}_n = \dot{I}_{sc}\left(1 - \sum_{i=1}^{i=4} \alpha_i\right)\tau_r \tau_{as}, \tag{7.5.1}$$

where

$\alpha_1 = \alpha_w$, the absorptance of direct irradiance by water vapor,
$\alpha_2 = \alpha_g$, the absorptance of direct irradiance by uniformly mixed gases,
$\alpha_3 = \alpha_o$, the absorptance of direct irradiance by ozone,
$\alpha_4 = \alpha_a$, the absorptance of direct irradiance by aerosols, and
τ_{as} is the transmittance due to scattering by aerosols.

It should be noted that in Eqs. (7.5.1) and (7.4.1) the absorption effects have been accounted for differently. In Eq. (7.5.1), the term within the parenthesis represents $\tau_1\tau_2\tau_3\tau_4$, and it can be shown that

$$\tau_1\tau_2\tau_3\tau_4 = (1-\alpha_1)(1-\alpha_2)(1-\alpha_3)(1-\alpha_4)$$
$$= 1-\alpha_1-\alpha_2-\alpha_3-\alpha_4$$

when the higher-order terms are neglected.

We list below all the necessary equations for absorptances and transmittances. The recommended equation for absorptance by water vapor is

$$\alpha_w = 0.110(U_1 + 6.31 \times 10^{-4})^{0.3} - 0.012\,1, \tag{7.5.2}$$

where $U_1 = wm_r$ in centimeters; w is the pressure- and temperature-corrected precipitable water. When $U_1 = 0$, α_w reduces to near zero. The above is a correlation exclusively for water vapor and, like Eq. (7.4.6), is based on Yamamoto's study.

In this model, two separate expressions for absorptance by carbon dioxide and oxygen are recommended. When these are combined, we have the absorptance α_g of the uniformly mixed gases, where

$$\alpha_g = 0.002\,35(126m_a + 0.012\,9)^{0.26} - 7.5 \times 10^{-4} + 7.5 \times 10^{-3}m_a^{0.875}. \tag{7.5.3}$$

In this equation, the first two terms account for carbon dioxide absorptance and the third represents oxygen absorptance.

The ozone absorptance obtained from Manabe and Strickler [13] is

$$\alpha_o = 0.045(U_3 + 8.34 \times 10^{-4})^{0.38} - 3.1 \times 10^{-3}, \tag{7.5.4}$$

where $U_3 = lm_r$ in centimeters.

In the previous section, the transmittance due to Rayleigh scattering was given by an algebraic expression. This model presents it in tabular form such that

$$\tau_r = [f(m_a)]^{m_a}; \tag{7.5.5}$$

$f(m_a)$ is a function of the air mass and is listed in Table 7.5.1. In this table, $m_a = 0.0$ simply indicates the extraterrestrial condition. As m_a increases,

Table 7.5.1

Variation of the Function $f(m_a)$ in the Expression for Rayleigh Transmittance τ_r, Eq. (7.5.5)[a]

m_a	$f(m_a)$
0.0	1.000
0.5	0.909
1.0	0.917
1.5	0.921
2.0	0.925
2.5	0.929
3.0	0.932
3.5	0.935
4.0	0.937

[a] Reprinted with permission from Hoyt [11], copyright 1978, Pergamon Press, Ltd.

so does $f(m_a)$, reflecting the initial depletion of shorter-wavelength radiation and therefore the decreasing fractional scattering per unit air mass for the longer path lengths. The following formula [6] fits the values in Table 7.5.1 to better than 0.2%:

$$\tau_r = 0.615\,958 + 0.375\,566 \exp(-0.221\,185 m_a), \qquad 0 < m_a < 8. \quad (7.5.6)$$

The attenuation due to the presence of aerosols is treated through Ångström's turbidity parameters β and α; the wavelength exponent α is assumed equal to one, and β is obtained from $\lambda = 1\ \mu$m.

This model separates aerosol attenuation into two parts, one part represents scattering and the other absorption. The transmittance due to scattering by aerosols is given by

$$\tau_{as} = [g(\beta)]^{m_a}, \quad (7.5.7)$$

where $g(\beta)$ is a function of β. This is listed in Table 7.5.2. The following formula [6], with an accuracy of better than 0.32%, can replace this table:

$$g(\beta) = -0.914\,000 + 1.909\,267 \exp(-0.667\,023\beta), \qquad 0 < \beta < 0.5. \quad (7.5.8)$$

It is pertinent to emphasize here that the functions $f(m_a)$ and $g(\beta)$ are both spectrally integrated transmittances. For this purpose, Labs and Neckel's [14] extraterrestrial solar spectrum is employed in Model B.

Table 7.5.2

Variation of the Function $g(\beta)$ in the Expression for Attenuation due to Scattering by Aerosols, Eq. (7.5.7)[a]

β	$g(\beta)$
0.00	1.000
0.02	0.972
0.04	0.945
0.06	0.919
0.08	0.894
0.10	0.870
0.12	0.824
0.14	0.846
0.16	0.802
0.18	0.780
0.20	0.758
0.24	0.714
0.28	0.670
0.32	0.626

[a] Reprinted with permission from Hoyt [11], copyright 1978, Pergamon Press, Ltd.

The absorptance of aerosols, α_a, is given by

$$\alpha_a = (1 - \omega_0)[g(\beta)]^{m_a}, \tag{7.5.9}$$

where ω_0 has the same meaning as in Chapter 6: it is the ratio of energy scattered to total attentuation by aerosols as a result of primary scattering. A value of 0.95 is recommended by Hoyt.

B. *Diffuse Irradiance*

The procedure to compute diffuse irradiance is treated slightly differently from that in Section 7.4. We rewrite the three components of diffuse irradiance on a horizontal surface [Eq. (7.4.4)]:

$$\dot{I}_d = \dot{I}_{dr} + \dot{I}_{da} + \dot{I}_{dm}.$$

The first term on the right-hand side of the above equation (representing diffuse irradiance from the Rayleigh atmosphere after the first pass) is given by the following:

$$\dot{I}_{\mathrm{dr}} = \dot{I}_{\mathrm{sc}} \cos\theta_{\mathrm{z}} \left(1 - \sum_{i=1}^{i=4} \alpha_i\right) 0.5(1 - \tau_{\mathrm{r}}). \tag{7.5.10}$$

The diffuse irradiance originating from aerosol scattering after the first pass is as follows:

$$\dot{I}_{\mathrm{da}} = \dot{I}_{\mathrm{sc}} \cos\theta_{\mathrm{z}} \left(1 - \sum_{i=1}^{i=4} \alpha_i\right) 0.75(1 - \tau_{\mathrm{as}}). \tag{7.5.11}$$

Comparing the above two equations with the corresponding Eqs. (7.4.15) and (7.4.16), it is obvious that there are only minor differences between these two pairs of equations. In Eq. (7.5.11), the factor 0.75 representing forward scattering is equivalent to the factor $F_{\mathrm{c}}\omega_0$ in Eq. (7.4.16). Values of F_{c} are listed in Table 6.17.1. With $\omega_0 = 0.95$, and taking as an average $F_{\mathrm{c}} = 0.8$, we obtain $F_{\mathrm{c}}\omega_0 = 0.76$, which is close to 0.75.

In this model, the diffuse irradiance arising from multiple reflections between ground and cloudless-sky atmosphere is also treated somewhat differently from that in Section 7.4. Here, the multiply reflected radiation, which includes only the first term of the infinite series, is given by

$$\dot{I}_{\mathrm{dm}} = \rho_{\mathrm{g}} Q \left(1 - \sum_{i=1}^{i=4} \alpha_i'\right) [0.5(1 - \tau_{\mathrm{r}}') + 0.25(1 - \tau_{\mathrm{as}}')], \tag{7.5.12}$$

where

$$Q = \dot{I}_{\mathrm{n}} \cos\theta_{\mathrm{z}} + \dot{I}_{\mathrm{dr}} + \dot{I}_{\mathrm{da}}. \tag{7.5.13}$$

The factor 0.25 represents backscatterance; that is, $1 - F_{\mathrm{c}}$. The prime over the transmittances indicates that

$$\tau_{\mathrm{r}}' = \tau_{\mathrm{r}}\big|_{\text{air mass} = 1.66p/p_0} \tag{7.5.14a}$$

and

$$\tau_{\mathrm{as}}' = \tau_{\mathrm{as}}\big|_{\text{air mass} = 1.66p/p_0}. \tag{7.5.14b}$$

The air mass at 1.66 times the minimum air mass for beam radiation accounts for an overall air mass for the upwelling diffuse irradiance. The prime over the absorptances indicates that

$$\alpha_{\mathrm{i}}' = \alpha_{\mathrm{i}}\big|_{\text{air mass} = m_{\mathrm{a}} + 1.66p/p_0}, \tag{7.5.15}$$

where p is station pressure and p_0 is the standard pressure, 1013.25 mbars.

C. *Global Irradiance*

The global irradiance on a horizontal surface is written

$$\dot{I} = \dot{I}_\mathrm{n} \cos\theta_\mathrm{z} + \dot{I}_\mathrm{dr} + \dot{I}_\mathrm{da} + \dot{I}_\mathrm{dm}. \tag{7.5.16}$$

☐ EXAMPLE 7.5.1. **Example 7.4.1 repeated using Model B.**

Solution. From Example 7.4.1,

$$m_\mathrm{r} = 1.963\,5, \qquad m_\mathrm{a} = 1.919, \qquad \cos\theta_\mathrm{z} = 0.507\,5,$$
$$U_1 = 1.929\,5, \qquad U_3 = 0.589\,1, \qquad \beta = 0.038\,4.$$

From Eq. (7.5.2),

$$\alpha_\mathrm{w} = 0.110(1.929\,5 + 6.31 \times 10^{-4})^{0.3} - 0.012\,1$$
$$= 0.121\,9.$$

From Eq. (7.5.3),

$$\alpha_\mathrm{g} = 0.002\,35[126(1.919) + 0.012\,9]^{0.26} - 7.5 \times 10^{-4}$$
$$+ 7.5 \times 10^{-3}(1.919)^{0.875}$$
$$= 0.022\,3.$$

From Eq. (7.5.4),

$$\alpha_\mathrm{o} = 0.045(0.589\,1 + 8.34 \times 10^{-4})^{0.38} - 3.1 \times 10^{-3}$$
$$= 0.033\,7.$$

From Eq. (7.5.6),

$$\tau_\mathrm{r} = 0.615\,958 + 0.375\,566 \exp[-0.221\,185(1.919)]$$
$$= 0.861\,6.$$

This model assumes $\alpha = 1$ and β is determined at $\lambda = 1$. However, because α is close to unity, very little error is expected in this case.

From Eqs. (7.5.7) and (7.5.8),

$$\tau_\mathrm{as} = \{-0.914 + 1.909\,267 \exp[-0.667\,023(0.038\,4)]\}^{1.919}$$
$$= 0.9007.$$

From Eq. (7.5.9),

$$\alpha_\mathrm{a} = (1 - 0.95)0.900\,7 = 0.045\,0.$$

From τ_{as} and α_a, τ_a can be obtained as

$$\tau_a = \tau_{as}(1 - \alpha_a) = 0.900\,7(1 - 0.045) = 0.860\,2.$$

From Eq. (7.5.1), with $E_o = 1.033\,4$,

$$\begin{aligned} I_n &= 1.033\,4(1\,367)[1 - (0.121\,9 + 0.022\,3 + 0.033\,7 + 0.045\,0)] \\ &\quad \times (0.861\,6)(0.900\,7) \\ &= 851.92 \text{ W m}^{-2}, \\ I_b &= I_n \cos\theta_z = 851.92(0.507\,5) = 432.35 \text{ W m}^{-2}. \end{aligned}$$

From Eq. (7.5.10), with $E_o = 1.033\,4$,

$$\begin{aligned} I_{dr} &= 1.033\,4(1\,367)(0.507\,5)[1 - (0.121\,9 + 0.022\,3 + 0.033\,7 + 0.045)] \\ &\quad \times [0.5(1 - 0.861\,6)] \\ &= 38.55 \text{ W m}^{-2}. \end{aligned}$$

From Eq. (7.5.11), with $E_o = 1.033\,4$,

$$\begin{aligned} I_{da} &= 1.033\,4(1\,367)(0.507\,5)[1 - (0.121\,9 + 0.022\,3 + 0.033\,7 + 0.045)] \\ &\quad \times [0.75(1 - 0.900\,7)] \\ &= 41.49 \text{ W m}^{-2}. \end{aligned}$$

From Eq. (7.5.13),

$$Q = 432.35 + 38.55 + 41.49 = 512.39 \text{ W m}^{-2}.$$

From Eq. (7.5.14a),

$$\begin{aligned} \tau_r' &= 0.615\,958 + 0.375\,566 \exp\{(-0.221\,185)[1.66(990/1\,013.25)]\} \\ &= 0.878\,3. \end{aligned}$$

From Eq. (7.5.14b),

$$\begin{aligned} \tau_{as}' &= \{-0.914 + 1.909\,267 \exp[-0.667\,023(0.038\,4)]\}^{1.66(990/1\,013.25)} \\ &= 0.915\,4. \end{aligned}$$

From Eq. (7.5.15),

$$\begin{aligned}\alpha'_w &= 0.11\{0.982\,7[1.9190 + 1.66(990/1\,013.25)] + 6.31 \times 10^{-4}\}^{0.3} - 0.012\,1\\ &= 0.147\,8,\\ \alpha'_g &= 0.002\,35\{126[1.919 + 1.66(990/1\,013.25)] + 0.012\,9\}^{0.26} - 7.5 \times 10^{-4}\\ &\quad + 7.5 \times 10^{-3}[1.919 + 1.66(990/1\,013.25)]^{0.875}\\ &= 0.033\,4,\\ \alpha'_o &= 0.045\{0.3[1.919 + 1.66(990/1\,013.25)] + 8.34 \times 10^{-4}\}^{0.38}\\ &\quad - 3.1 \times 10^{-3}\\ &= 0.043\,0,\\ \alpha'_a &= (1 - 0.95)0.915\,4\\ &= 0.045\,8.\end{aligned}$$

From Eq. (7.5.12),

$$\begin{aligned}\dot{I}_{dm} &= 0.2(512.39)[1 - (0.147\,8 + 0.033\,4 + 0.043\,0 + 0.045\,8)]\\ &\quad \times [0.5(1 - 0.087\,3) + 0.25(1 - 0.915\,4)]\\ &= 6.13 \text{ W m}^{-2},\\ \dot{I}_d &= \dot{I}_{dr} + \dot{I}_{da} + \dot{I}_{dm} = 38.55 + 41.49 + 6.13 = 86.17 \text{ W m}^{-2},\\ \dot{I} &= \dot{I}_b + \dot{I}_d = 432.35 + 86.17 = 518.5 \text{ W m}^{-2}.\end{aligned}$$

□

We proceed now to present Model C.

7.6 Parameterization Model C

This model is based on two studies by Bird and Hulstrom [15, 16]. In the first study [15] Bird and Hulstrom carried out a detailed comparison of a large number of direct insolation models. They compared individual elements of the various models with the rigorous model, called SOLTRAN, which was constructed from LOWTRAN [17, 18]. On the basis of this comparison, they developed new equations for the transmittances of the individual atmospheric constituents. They retained the individual transmittances in forms more or less similar to those we have studied in Models A and B.

A. *Direct Irradiance*

From this model the direct normal irradiance is given by[4]

$$\dot{I}_{\mathrm{n}} = 0.975\,1\dot{I}_{\mathrm{sc}}\tau_{\mathrm{r}}\tau_{0}\tau_{\mathrm{g}}\tau_{\mathrm{w}}\tau_{\mathrm{a}} \tag{7.6.1}$$

where the factor 0.975 1 is included because the spectral interval considered by SOLTRAN is 0.3–3.0 μm.

It is instructive to compare Eqs. (7.4.1), (7.5.1), and (7.6.1). In each the absorptances are taken into account differently. We list below the various transmittances.

The transmittance by Rayleigh scattering is as follows:

$$\tau_{\mathrm{r}} = \exp[-0.0903m_{\mathrm{a}}^{0.84}(1.0 + m_{\mathrm{a}} - m_{\mathrm{a}}^{1.01})]. \tag{7.6.2}$$

The transmittance by ozone is obtained from the following:

$$\begin{aligned}\tau_{\mathrm{o}} &= 1 - \alpha_{\mathrm{o}}\\ &= 1 - [0.161\,1\,U_3(1.0 + 139.48U_3)^{-0.3035}\\ &\quad -0.002\,715U_3(1.0 + 0.044U_3 + 0.000\,3U_3^2)^{-1}],\end{aligned} \tag{7.6.3}$$

where U_3 is defined by (7.4.5).

The transmittance by uniformly mixed gases is given by

$$\tau_{\mathrm{g}} = \exp(-0.0127m_{\mathrm{a}}^{0.26}). \tag{7.6.4}$$

The transmittance by water vapor is given by

$$\begin{aligned}\tau_{\mathrm{w}} &= 1 - \alpha_{\mathrm{w}}\\ &= 1 - 2.4959U_1[(1.0 + 79.034U_1)^{0.682\,8} + 6.385U_1]^{-1},\end{aligned} \tag{7.6.5}$$

where U_1 is defined by Eq. (7.4.7).

An expression for the aerosol transmittance of this model is based on spectral attenuation at the two wavelengths commonly used by meteorological networks, 0.38 and 0.5 μm, wavelengths at which there is minimal molecular (ozone) absorption. The aerosol transmittance is given by

$$\tau_{\mathrm{a}} = \exp[-k_{\mathrm{a}}^{0.873}(1.0 + k_{\mathrm{a}} - k_{\mathrm{a}}^{0.708\,8})m_{\mathrm{a}}^{0.910\,8}], \tag{7.6.6}$$

where

$$k_{\mathrm{a}} = 0.275\,8k_{\mathrm{a}\lambda|\lambda=0.38\,\mu\mathrm{m}} + 0.35k_{\mathrm{a}\lambda|\lambda=0.5\,\mu\mathrm{m}}. \tag{7.6.7}$$

[4] In the original publication [15] the multiplying factor in this equation is 0.9662 and is based on the NASA/ASTM solar constant and its spectral distribution. This author has changed this factor to 0.9751 to conform to the solar constant 1367 W m^{-2} and its spectral distribution. This represents a net increase of 2%.

The aerosol thicknesses at the two wavelengths are measured on a regular basis by the U.S. National Weather Service. If measurement at one of the wavelengths is unknown, it can be ignored in (7.6.7).

At locations where either turbidity or visibility is recorded, it will be useful to have an equation for τ_a in terms of Ångström's turbidity parameters β and α, or visibility. For this purpose, Eqs. (7.4.11) or (7.4.12) can be employed. It can be shown that Eqs. (7.6.6), (7.4.11), and (7.4.12) yield identical results.

B. *Diffuse Irradiance*

The procedure to compute diffuse irradiance on a horizontal surface is very similar to that in Model A. However, the following equations are based on the results of two rigorous studies [19, 20]. The Rayleigh-scattered diffuse irradiance after the first pass through the atmosphere is given by

$$\dot{I}_{dr} = 0.79\dot{I}_{sc} \cos \theta_z \, \tau_o \tau_g \tau_w \tau_{aa} 0.5(1 - \tau_r)/(1 - m_a + m_a^{1.02}), \qquad (7.6.8)$$

where τ_{aa} is the transmittance of direct radiation due to aerosol absorptance and

$$\tau_{aa} = 1 - (1 - \omega_0)(1 - m_a + m_a^{1.06})(1 - \tau_a). \qquad (7.6.9)$$

Bird and Hulstrom recommend that unless a more precise figure for ω_0 is available, it should be taken as 0.9.

The aerosol-scattered diffuse irradiance after the first pass through the atmosphere is given by

$$\dot{I}_{da} = 0.79\dot{I}_{sc} \cos \theta_z \, \tau_o \tau_g \tau_w \tau_{aa} F_c(1 - \tau_{as})/(1 - m_a + m_a^{1.02}), \qquad (7.6.10)$$

where τ_{as} has been defined earlier, and

$$\tau_{as} = \tau_a/\tau_{aa}. \qquad (7.6.11)$$

This model recommends that a constant value of 0.84 be used for F_c unless better information on the aerosols is available.

To obtain multiply reflected irradiance, the following expression for the atmospheric albedo is suggested:

$$\rho_a' = 0.068\,5 + (1 - F_c)(1 - \tau_{as}). \qquad (7.6.12)$$

Note that the small difference between the two expressions for albedo in Eqs. (7.6.12) and (7.4.17), and τ_{as} in (7.6.12) is not based on the air-mass expression $1.66(p/1\,013.25)$.

The diffuse irradiance arising from multiple reflections between the earth and its atmosphere is given by Eq. (7.4.16) of Model A. This completes the diffuse-irradiance algorithm of Model C.

C. *Global Irradiance*

Again, the global irradiance on a horizontal surface is given by Eq. (7.4.18) of Model A.

☐ EXAMPLE 7.6.1. Example 7.4.1 repeated using Model C.

Solution. From Example 7.4.1,

$$m_a = 1.919, \qquad U_1 = 1.929\,5, \qquad U_3 = 0.589\,1, \qquad \cos\theta_z = 0.5075.$$

From Eq. (7.6.2),

$$\tau_r = \exp\{-0.090\,3(1.919)^{0.84}[1.0 + 1.919 - (1.919)^{1.01}]\}$$
$$= 0.8571.$$

From Eq. (7.6.3),

$$\tau_o = 1 - \alpha_o$$
$$= 1 - \{0.161\,1(0.589\,1)[1.0 + 139.48(0.589\,1)]^{-0.303\,5}$$
$$- 0.002\,715(0.589\,1)[1.0 + 0.044(0.589\,1) + 0.000\,3(0.589\,1)^2]^{-1}\}$$
$$= 0.976\,8.$$

From Eq. (7.6.4),

$$\tau_g = \exp[-0.012\,7(1.919)^{0.26}] = 0.985\,1.$$

From Eq. (7.6.5),

$$\tau_w = 1 - \alpha_w$$
$$= 1 - 2.495\,9(1.929\,5)\{[1.0 + 79.034(1.929\,5)]^{0.682\,8} + 6.385(1.929\,5)\}^{-1}$$
$$= 0.889\,1.$$

From Eqs. (7.6.6) and (7.6.7), with $k_{a\lambda}(0.38) = 0.087$, and $k_{a\lambda}(0.5) = 0.069$, we have

$$k_a = 0.275(0.087) + 0.35(0.069) = 0.048\,1$$
$$\tau_a = \exp\{-(0.048\,1)^{0.873}[1.0 + 0.048\,1 - (0.048\,1)^{0.708\,8}](1.919)^{0.910\,8}\}$$
$$= 0.887\,5.$$

From Eq. (7.6.1), with $E_o = 1.033\,4$,

$$I_n = 1.033\,4(0.975\,1)(1\,367)(0.857\,1)(0.976\,8)(0.985\,1)(0.889\,1)(0.887\,5)$$
$$= 896.44 \text{ W m}^{-2},$$
$$I_b = I_n \cos\theta_z = 896.44(0.507\,5) = 454.9 \text{ W m}^{-2}.$$

From Eq. (7.6.9),

$$\tau_{aa} = 1 - (1 - 0.9)[1 - 1.919 + (1.919)^{1.06}](1 - 0.887\,5)$$
$$= 0.987\,9.$$

From Eq. (7.6.8), with $E_o = 1.033\,4$,

$$\dot{I}_{dr} = 1.033\,4(0.79)(1\,367)(0.507\,5)(0.976\,8)(0.985\,1)(0.889\,1)(0.987\,9)$$
$$\times [0.5(1 - 0.857\,1)]/[1 - 1.919 + (1.919)^{1.02}]$$
$$= 33.66 \text{ W m}^{-2},$$

$$\tau_a = \tau_{aa}\tau_{as},$$

$$\tau_{as} = 0.887\,5/0.987\,9 = 0.898\,4.$$

From Eq. (7.6.10), with $E_o = 1.033\,4$,

$$\dot{I}_{da} = 1.033\,4(0.79)(1\,367)(0.507\,5)(0.976\,8)(0.985\,1)(0.889\,1)(0.987\,9)$$
$$\times [0.84(1 - 0.898\,4)]/[1 - 1.919 + (1.919)^{1.02}]$$
$$= 39.85 \text{ W m}^{-2}.$$

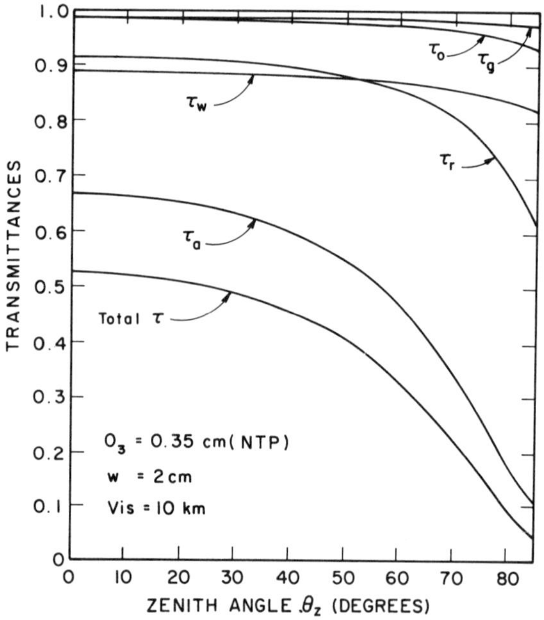

Figure 7.6.1 Transmittances of the various atmospheric constituents using Model C.

From Eq. (7.6.12),

$$\rho_a' = 0.068\,5 + (1 - 0.84)(1 - 0.898\,4) = 0.08.$$

The multiply reflected radiation is

$$\dot{I}_{dm} = (455.0 + 39.85 + 33.66)\{0.2(0.08)/[1 - 0.2(0.08)]\}$$
$$= 8.59 \text{ W m}^{-2}.$$

The diffuse irradiance can now be written

$$\dot{I}_d = 33.66 + 39.85 + 8.59 = 81.80 \text{ W m}^{-2}.$$

Finally, the global irradiance on a horizontal surface is

$$\dot{I} = 455.0 + 81.80 = 536.80 \text{ W m}^{-2}.$$

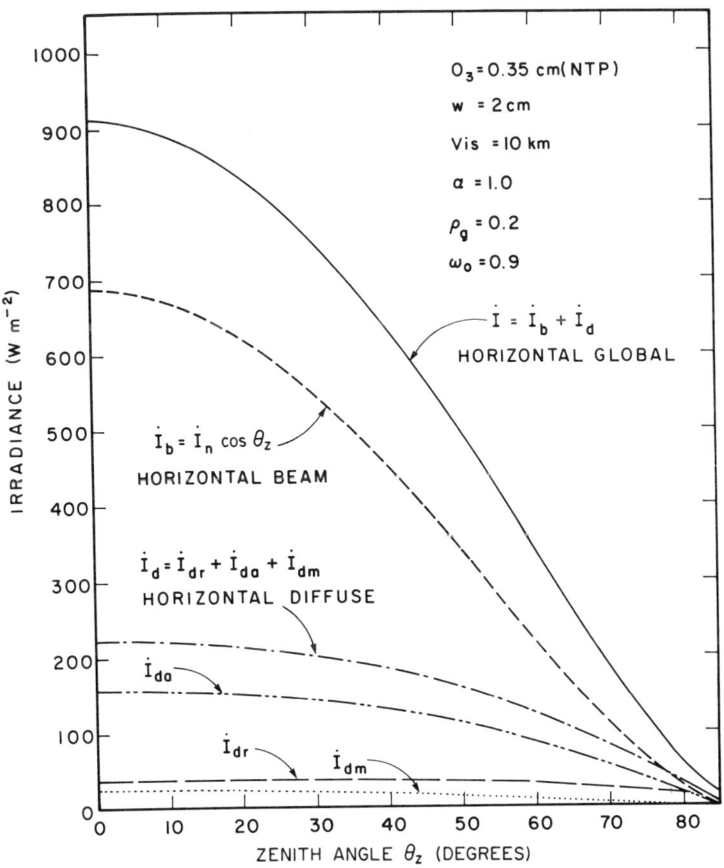

Figure 7.6.2 The various irradiances on a horizontal surface.

In review, the respective transmittances and the irradiances of the three models agree quite well. □

Before we go further and compare the various models, it seems useful to examine the individual transmittances of one of the models. In Fig. 7.6.1, the transmittances of the individual atmospheric constituents of Model C are plotted. The atmospheric parameters employed are as indicated on this diagram. Minimum attenuation is by the uniformly mixed gases. Because of this, in some models, a separate absorption term by the uniformly mixed gases is ignored. Ozone attenuates slightly more than the uniformly mixed gases. Of course, this depends on the amount of ozone in the atmosphere. Water vapor, aerosols, and (scattering by) air molecules are the three main attenuators. The relative importance of water vapor and aerosols again depends on the extent to which they are present in the atmosphere. Further, the absorptance of the aerosols depends on ω_0, the single-scattering albedo. Transmittances of the molecular absorbers remain, more or less, independent of the zenith angle. Rayleigh and aerosol transmittances decrease significantly with zenith angle.

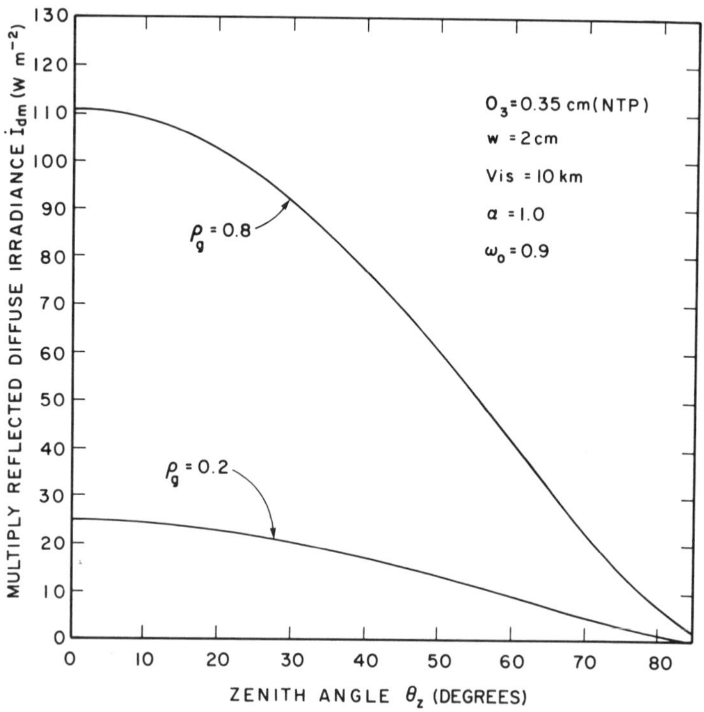

Figure 7.6.3 The effect of ground albedo on the multiply reflected diffuse irradiance.

A similar examination of the individual transmittances in the other two models indicates that the remarks made above apply equally to the other models.

In the next section we compare the individual and the overall transmittances and the irradiances resulting from the three parameterization models, and the spectral model discussed in Section 7.2.

Before we go further, it is useful to examine the magnitude of the various irradiances under a given set of the atmospheric parameters. These irradiances are shown in Fig. 7.6.2. The global and the beam irradiances on a horizontal surface decrease sharply as the zenith angle increases. However, the diffuse irradiance remains rather insensitive until the zenith angle goes beyond 30°. Under the given atmospheric parameters, the aerosol-scattered diffuse irradiance is far greater than the Rayleigh-scattered diffuse irradiance. The latter constitutes only about 4% of the global irradiance. The multiply reflected irradiance accounts for 3% of the global value.

The multiply reflected irradiance varies almost linearly with ground albedo [Eq. (7.4.16)]. The effect of ground albedo is shown in Fig. 7.6.3. An almost fourfold increase in the multiply reflected irradiance results as ρ_g is increased from 0.2 to 0.8.

7.7 Comparison of the Parameterization Models

We now compare some of the individual transmittances and the irradiances obtained from the three models.

The ozone transmittances are compared in Fig. 7.7.1. Although the algebraic expressions for τ_o in the three models are quite different, the numerical values obtained are within 1% of each other. The greatest difference is between Models B and C. In this particular diagram, $l = 0.35$ cm(NTP). These differences are a function of the amount of ozone and of the zenith angle. Because ozone transmittance is very high, its effect on the overall transmittance is minimal.

A comparison of the water-vapor transmittances is presented in Fig. 7.7.2. In this aspect Models A and B are very similar. Model C results in approximately 1%-higher values at $\theta_z = 0$. The differences vary with zenith angle. This diagram is for $w = 2$ cm. The differences will also vary with the amount of water vapor.

In a manner similar to the above, differences between the other transmittances can also be studied. Comparisons of the total irradiances for (a) direct normal radiation, (b) diffuse radiation on a horizontal surface, and (c) global radiation on a horizontal surface are presented below.

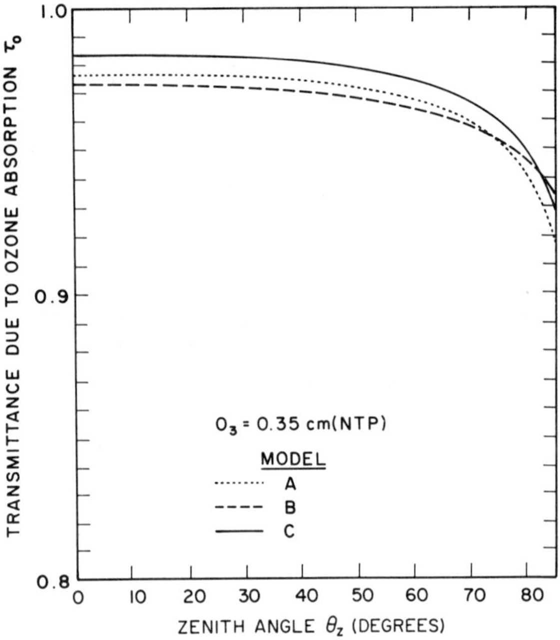

Figure 7.7.1 Transmittance of ozone as a function of the zenith angle

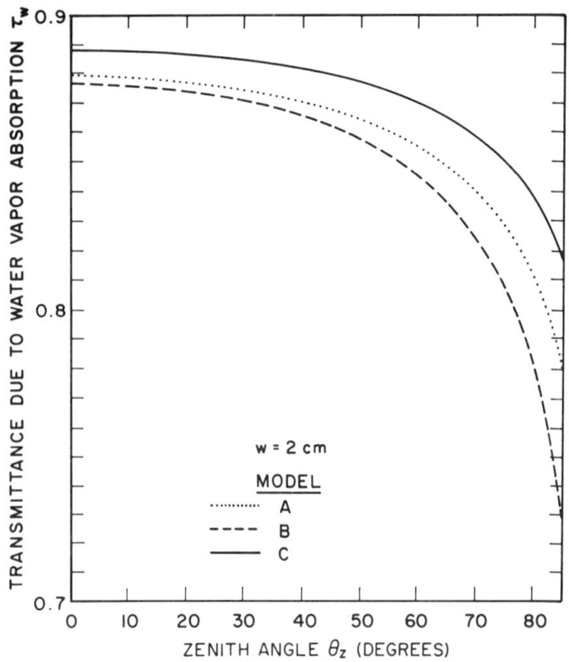

Figure 7.7.2 Transmittance of water vapor as a function of the zenith angle.

A. *Direct Normal Irradiance*

Because the three models have utilized different approaches to account for aerosol attenuation, we first present results for a clean atmosphere.

For a clean atmosphere with $w = 2$ cm and $O_3 = 0.35$ cm(NTP), the direct normal irradiances obtained from the three models and from the spectrally integrated approach are shown in Fig. 7.7.3. The correspondence between Model C and the spectrally integrated results is very good at $\theta_z <$ 60°. Model B gives lower values than all the rest. The maximum difference between any two models is not more than 10%.

The effect of turbidity has been included in Fig. 7.7.4. This diagram is for 10-km visibility, $w = 2$ cm, $O_3 = 0.35$ cm(NTP). At zenith angles less than 60° Models A, B, and C yield almost identical results. At higher zenith angles Model B shows some disagreement.

In the last two diagrams it has been noted that among the three parameterization models, Model C yields results closest to those of the spectrally integrated approach. Similar comparisons carried out with different atmospheric parameters have substantiated the foregoing observation. It can therefore be concluded that Model C is very accurate. and its use is

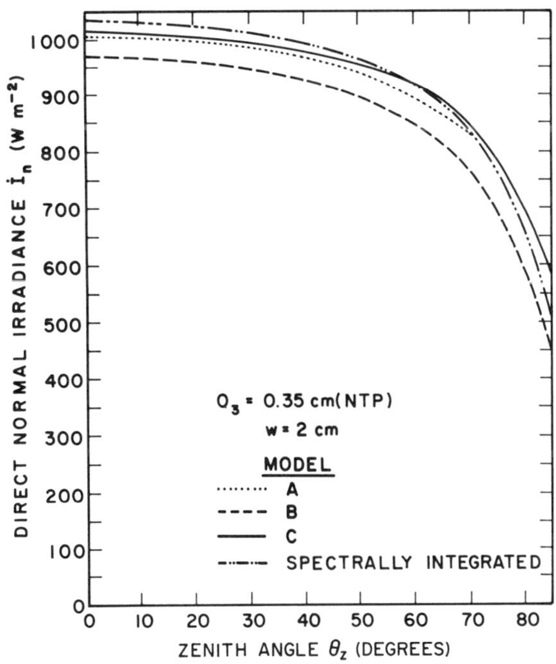

Figure 7.7.3 Direct normal irradiance of a clean atmosphere obtained from the various models.

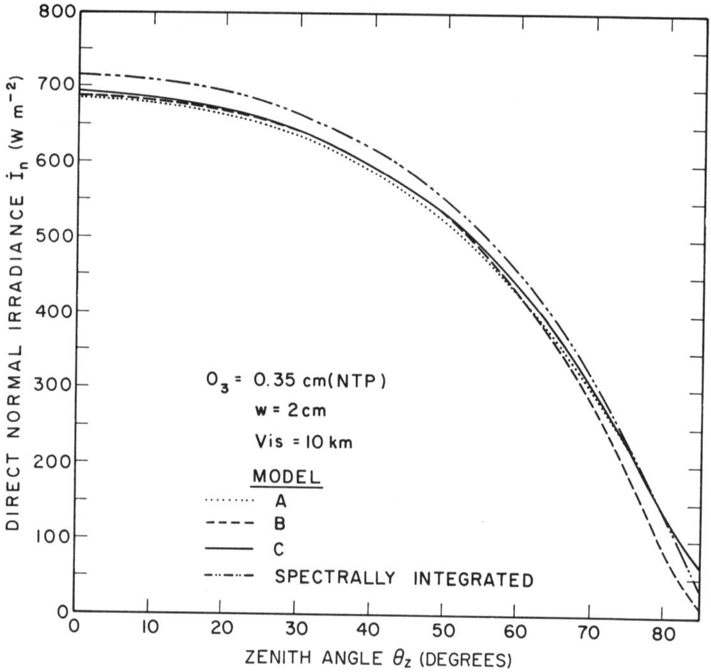

Figure 7.7.4 Direct normal irradiance at Vis = 10 km obtained from the various models.

recommended. However, as we have observed, the other two models are not too far behind Model C.

In Chapter 6 (Fig. 6.14.4) we observed that the first few millimeters of precipitable water in the atmosphere attenuate as much radiation as the next few centimeters. Figure 7.7.5 shows essentially the same results under broadband calculations. In this figure, $O_3 = 0.35$ and visibility is 10 km. It is obvious from this diagram that 0.35 cm of the precipitable water yields results that will produce the same degree of error, approximately 10%, whether the atmosphere is dry or is laden with 4 cm of water vapor. It can be shown that this statement remains essentially unaffected by changing the visibility parameter. This result is useful in developing simpler parameterization equations in which a constant amount of water vapor may be assumed.

The expressions for the transmittances τ in the three models are based almost exclusively on the NASA/ASTM extraterrestrial spectral distribution and older data. What would be the result if the current values of the extraterrestrial solar spectrum recommended by the World Radiation Center were used? Naturally, it is a difficult task to redo the algebraic equations representing the various τs in order to carry out any comparison. However, it is a simple matter to compute the spectrally integrated values from Eq.

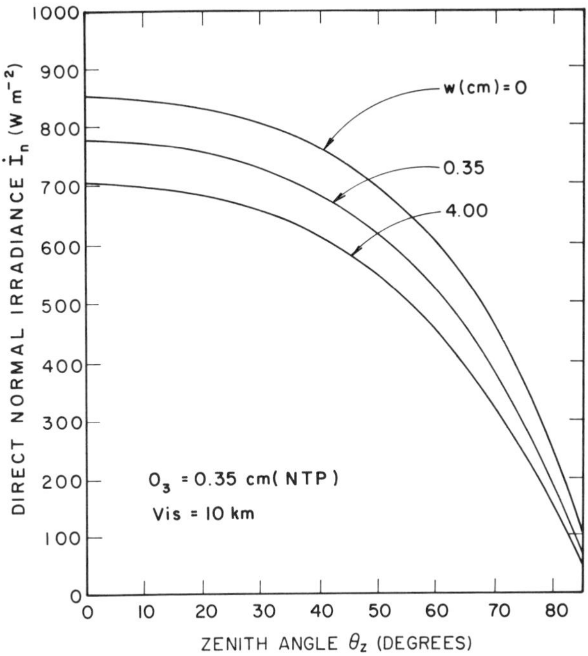

Figure 7.7.5 Direct normal irradiance as a function of precipitable water.

(7.2.2.). The results of such an exercise carried out at two visibilities, 10 and 23 km, are shown in Fig. 7.7.6. In these plots, $O_3 = 0.35$ cm(NTP) and $w = 2$ cm. The World Radiation Center (WRC) spectrum yields higher values. Irrespective of the visibility, there is a constant difference of slightly less than 2%. This is an interesting result in view of the fact that the difference between the two solar constants is only 1%. Therefore, it seems correct to recommend that the direct normal irradiances obtained from Models A and B should be increased by 2% to account for the current values of the extraterrestrial spectral radiation.

B. *Diffuse Irradiance on a Horizontal Surface*

Before we compare the diffuse irradiance resulting from the three models, we shall study the atmospheric albedo ρ_a'. The variation of ρ_a' with α and β is shown in Fig. 7.7.7. As expected, the atmospheric albedo increases with β, which represents the amount of aerosols present in the atmosphere. The atmospheric albedo also increases as the value of α increases. An increase in α represents a larger amount of small particles, and small particles scatter more radiation than do the large particles.

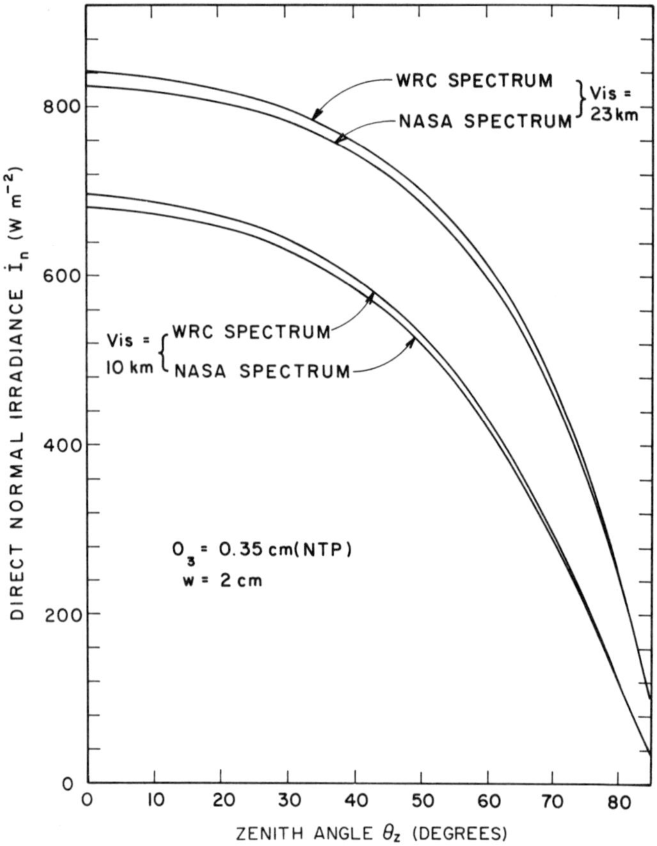

Figure 7.7.6 Comparison of the direct normal irradiance obtained from the WRC spectrum and the NASA spectrum.

Figure 7.7.8 represents a comparison of diffuse irradiance on a horizontal surface computed from the three models. Model C yields the lowest values, and the difference between this model and the other two is substantial. However, such differences depend on the parameters of atmospheric turbidity. As Model C is derived from rigorous studies [19, 20], it is believed to be more accurate than the other two models.

C. *Global Irradiance on a Horizontal Surface*

Figure 7.7.9 compares global irradiance on a horizontal surface. Model C yields values lower than those obtained from the other two models. This is

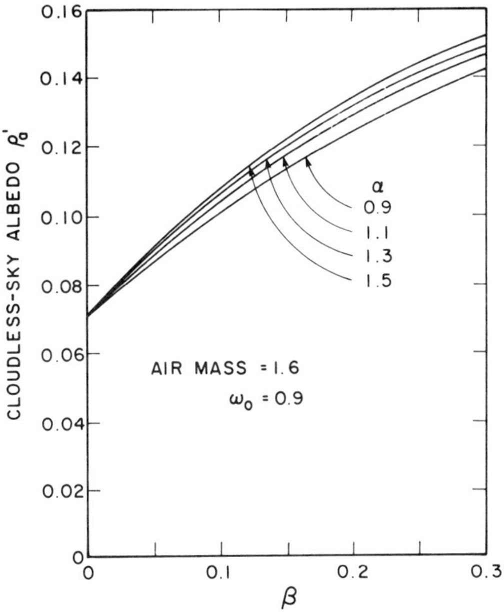

Figure 7.7.7 Variation of the cloudless-sky albedo as a function of the turbidity parameters α and β.

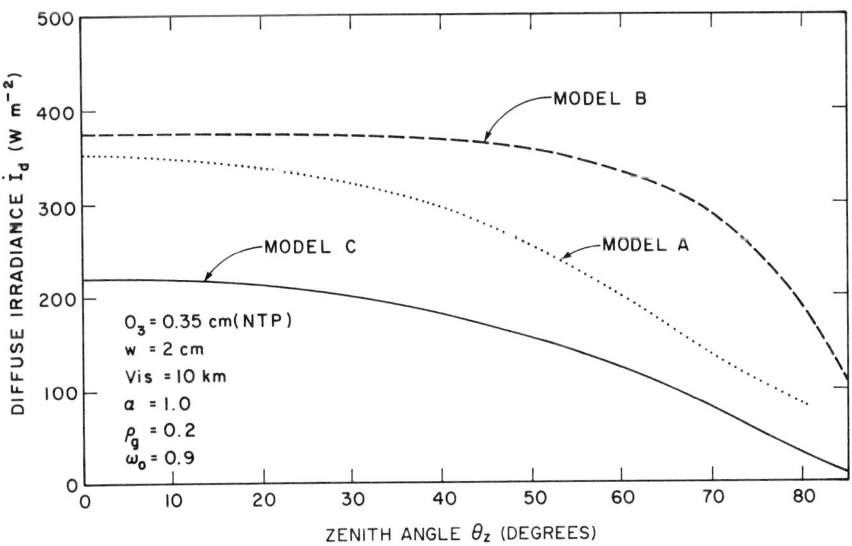

Figure 7.7.8 Comparison of the diffuse irradiance calculated from the three parameterization models.

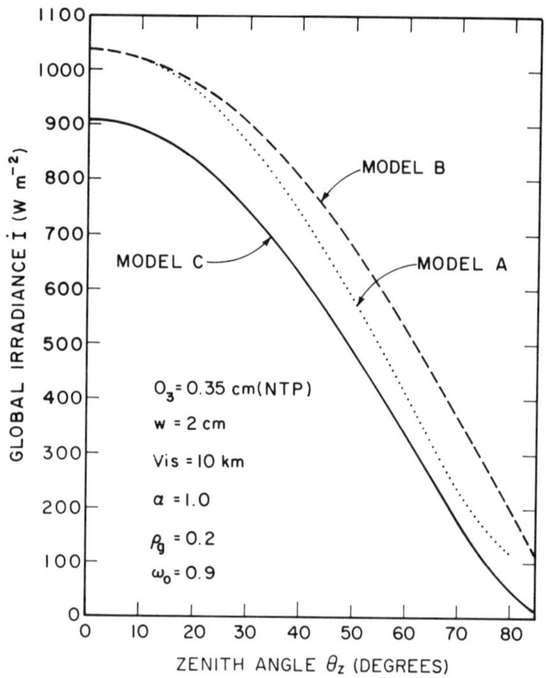

Figure 7.7.9 Comparison of the global irradiance calculated from the three parameterization models.

due mainly to the differences in diffuse components of the three models. Again, the degree to which these models differ depends on the parameters of atmospheric turbidity.

7.8 The ASHRAE Algorithm and Its Improvement

The past several sections discussed the spectral approach and the parameterization approach. In the former procedure, the spectral and the total insolation could be evaluated in sufficient detail and with high precision. The latter method is quick, but may be less accurate. In this section, a method much quicker and simpler than even the parameterization approach is presented. However, as happens so often in scientific formulations, simplicity may be gained at the expense of accuracy. The simple procedure referred to is the ASHRAE algorithm described in [21, 22] and very widely used by the engineering and architectural communities.

It is useful to describe briefly the historical development of the ASHRAE method in order to obtain an appreciation of its procedure. This method is based on four principal publications: Moon [23], Threlkeld and Jordan [24], Threlkeld [25], and Stephenson [26].

Moon modeled the extraterrestrial insolation spectrum using ground-based measurements obtained by the Smithsonian Institution over several years (1920–1934). He used the work of several researchers of attenuation coefficients for the various atmospheric constituents, and utilized the following values of the various transmittances, most of which were already presented in Chapter 6. For Rayleigh scattering:

$$\tau_{r\lambda} = \exp(-0.008\,85\lambda^{-4}m_a). \tag{7.8.1}$$

The above equation is practically identical to Eq. (6.4.1). For water-vapor scattering [Eq. (6.5.3)]:

$$\tau_{ws\lambda} = \exp(-0.008\,635\lambda^{-2}wm_r).$$

For dust scattering [Eq. (6.5.4)]:

$$\tau_{d\lambda} = \exp(-0.081\,28\lambda^{-0.75}(d/800)m_a).$$

For ozone absorption [Eq. (6.12.1)]:

$$\tau_{o\lambda} = \exp(-k_{o\lambda}lm_r).$$

For Ozone, Moon employed the attenuation coefficients of Wulf [27] and Läuchli [28]. The various symbols and meaning of the above equation have been discussed earlier.

For the transmittance due to water-vapor absorption, $\tau_{wa\lambda}$, Moon utilized Fowl's [29] data. The spectral transmittance due to the attenuators mentioned above can be written as follows:

$$\tau_\lambda = \tau_{r\lambda}\tau_{ws\lambda}\tau_{d\lambda}\tau_{o\lambda}\tau_{wa\lambda}. \tag{7.8.2}$$

Utilizing the extrapolated values of the solar spectrum outside the earth's atmosphere (resulting in a solar constant value of 1322 W m^{-2}) and taking $\sec\theta_z$ for optical mass throughout in Eq. (7.8.2), Moon calculated the spectral- as well as wavelength-integrated values of the cloudless-sky direct normal irradiance. Because the results were based on a certain dust concentration, precipitable water vapor, and ozone in Washington, DC, a procedure (and a simple one) was needed to account for the variability in the atmospheric constituents in other parts of the continental United States. Threlkeld and Jordan [24] undertook this task.

Threlkeld and Jordan used the technique developed by Moon to calculate the broadband normal-incidence direct radiation. They proposed a "basic atmosphere" containing 0.25 cm(NTP) ozone, 200 dust particles per cubic centimeter, and a variable water-vapor content (listed in Table 7.8.1). Evidently, this basic atmosphere represents typical nonindustrial midlatitude conditions, except for ozone thickness. Calculations were then made for hourly values of the direct normal radiation on the first day of each month at four north latitudes: 30°, 36°, 42°, and 48°. Corresponding diagrams were prepared for each of the four latitudes and were proposed as standard clear-day insolation curves. It was also proposed that these curves be multiplied by a clearness number to account for the local variation of clear-day precipitable water. The local clear-day water vapor was defined as 85% of its monthly average value. This clearness number (CN) may be defined as follows:

$$(\mathrm{CN}) = \frac{\dot{I}_{\mathrm{n}}(\text{calculated with local mean clear-day water vapor})}{\dot{I}_{\mathrm{n}}(\text{calculated with water vapor according to basic atmosphere})}. \tag{7.8.3}$$

A contour map of the clearness numbers for the United States was developed by Threlkeld and Jordan.

It would have been more useful had the clearness number been obtained as a ratio of the measured to the calculated direct normal irradiance. Such a clearness number would then have accounted for the total atmospheric clarity. In fact, Threlkeld and Jordan did attempt to obtain such a factor

Table 7.8.1

Equivalent Turbidity Parameter β of Ångström and Visibility against ASHRAE's "Basic Atmosphere" with 200 Dust Particles and Water Vapor[a]

Month	w (cm)[a]	β ($\alpha = 1.3$)	Equivalent visibility (km)
January	0.795	0.0259	177
February	0.855	0.0267	173
March	1.120	0.0302	155
April	1.778	0.0390	117
May	2.377	0.0465	92
June	2.760	0.0514	79
July	2.800	0.0520	78
August	2.620	0.0497	83
September	1.823	0.0394	116
October	1.253	0.0320	147
November	0.947	0.0280	166
December	0.795	0.0259	177

[a] Data from Threlkeld and Jordan [24, Fig. 27].

from four stations (Madison, Lincoln, Blue Hill, and Albuquerque) where measured values of the direct normal irradiance were available. However, as data from four stations were insufficient to develop contour maps for the whole country, (CN) was based on calculated values defined in Eq. (7.8.3).

The four standard clear-day insolation curves of Threlkeld and Jordan [Figs. 28–31 of [24]) corresponding to the north latitudes 30°, 36°, 42°, and 48° require tedious interpolation in order to estimate direct normal insolation at other latitudes. Stephenson [26] developed the following simple expression:

$$\dot{I}_{\mathrm{n}} = (\mathrm{CN}) A \exp(-B \sec \theta_{\mathrm{z}}), \tag{7.8.4}$$

where the coefficients A and B were obtained by fitting this equation to the proposed standard clear-day insolation curves of Threlkeld and Jordan. Monthly variations of the constants A and B are listed in Table 7.8.2. A is called the apparent extraterrestrial irradiance and takes into account the variations in sun–earth distance. Therefore, in fact,

$$A = \dot{I}_{\mathrm{sc}} E_{\mathrm{o}}(\mathrm{const}). \tag{7.8.5}$$

The value of the multiplying constant is close to 0.9. The variable B represents an overall broadband value of the atmospheric attenuation coefficient for the basic atmosphere of Threlkeld and Jordan, and (CN) is the clearness number.

Table 7.8.2

Constants A, B, and C for Calculation of Solar Irradiance According to ASHRAE [22] *and the Revised Recommended Values*

	ASHRAE			Revised recommended values		
Date	A (W m^{-2})	B (air mass)$^{-1}$	C (dimensionless)	A (W m^{-2})	B (air mass)$^{-1}$	C (dimensionless)
21 Jan	1230	0.142	0.058	1202	0.141	0.103
21 Feb	1215	0.144	0.060	1187	0.142	0.104
21 Mar	1186	0.156	0.071	1164	0.149	0.109
21 Apr	1136	0.180	0.097	1130	0.164	0.120
21 May	1104	0.196	0.121	1106	0.177	0.130
21 Jun	1088	0.205	0.134	1092	0.185	0.137
21 Jul	1085	0.207	0.136	1093	0.186	0.138
21 Aug	1107	0.201	0.122	1107	0.182	0.134
21 Sep	1151	0.177	0.092	1136	0.165	0.121
21 Oct	1192	0.160	0.073	1136	0.152	0.111
21 Nov	1221	0.149	0.063	1190	0.144	0.106
21 Dec	1233	0.142	0.057	1204	0.141	0.103

Equation (7.8.4) was developed for sea-level conditions. It can, however, be easily adapted for other atmospheric pressures by writing

$$\dot{I}_{\mathrm{n}} = (\mathrm{CN})A \exp[-B \sec\theta_z\,(p/p_0)], \tag{7.8.6}$$

where p is the actual station pressure and p_0 is the standard pressure.

Although the ASHRAE formulation for clear-sky direct normal irradiance is still used by engineers and architects, it is useful to summarize some of its shortcomings:

(i) the solar constant used (1322 W m^{-2}) and the extraterrestrial spectral irradiance employed by Moon are outdated;
(ii) the attenuation coefficients utilized by Moon were empirical functions and were based on site-specific data (Mount Wilson and Washington, DC);
(iii) the experimental data were based on the use of old instruments; and
(iv) when ln $\dot{I}_{\mathrm{n}}$ versus m_{a}, curves were plotted from Eqs. (7.8.6) or (7.8.4), the result was a straight line. The ordinate gives the apparent solar constant A, and the slope of the straight line is the overall broadband atmospheric attenuation coefficient B. This linear relationship is incorrect for broadband irradiance.

We attempt to explain the last point by rewriting Bouguer's Law for monochromatic irradiance [Eq. (6.2.1)]:

$$\dot{I}_{\mathrm{n}\lambda} = \dot{I}_{0\mathrm{n}\lambda} \exp(-k_\lambda m).$$

Plotting ln $\dot{I}_{\mathrm{n}\lambda}$ versus m, one obtains a straight line with slope k_λ. This is correct for monochromatic irradiance. However, this law fails for broadband insolation that includes wavelength-selective molecular absorption.

The ASHRAE formulation also includes a procedure for computing clear-sky diffuse irradiance on horizontal and vertical surfaces. According to this method, which is based on Threlkeld [25], the irradiance on a horizontal surface is given by

$$\dot{I}_{\mathrm{d}} = C\dot{I}_{\mathrm{n}}, \tag{7.8.7}$$

where C depends on the water content of the atmosphere and varies from summer to winter. The values of C are also listed in Table 7.8.2.

Once direct and diffuse irradiance are available, global irradiance on a horizontal surface is obtained through Eq. (7.2.5).

The ASHRAE algorithm for diffuse irradiance has two main drawbacks: (i) there is no provision to incorporate ground albedo and hence contribution by multiple reflections cannot be varied; and (ii) there is no separate provision to incorporate aerosol-generated diffuse irradiance.

From the foregoing it can be speculated that the ASHRAE clear-sky algorithm is not accurate. How inaccurate is it? We develop below a procedure to answer this question quantitatively.

In responding to the above question, the main difficulty lies in equating the scattering parameters in the ASHRAE procedure with Ångström's turbidity parameters. That is, the problem is to establish an equivalence between $\tau_{ws\lambda}$ plus $\tau_{d\lambda}$ on one side, and $\tau_{a\lambda}$ on the other. This difficulty can be overcome by spectrally integrating the two sides and equating; namely,

$$\sum_0^{\infty} \dot{I}_{0n\lambda}\{\exp(-0.008\,635\lambda^{-2}wm_r) \times \exp[-0.081\,28\lambda^{-0.75}(d/800)m_a]\}\,\Delta\lambda$$

$$= \sum_0^{\infty} \dot{I}_{0n\lambda}[\exp(-\beta\lambda^{-\alpha}m_a)]\,\Delta\lambda. \quad (7.8.8)$$

The monthly values of w are obtained from Table 7.8.1. For dust concentration we assume $d = 200$, the value employed in the "basic atmosphere" of ASHRAE. Equation (7.8.8) can now be readily solved, and a correspondence between the precipitable water vapor and Ångström's turbidity parameter β evaluated. Table 7.8.1 also lists the resulting equivalent values of the parameter β. The air mass appears to have minimal influence on this parameter. A further examination has shown that a slight variation in α, $0.7 < \alpha < 1.3$, also has negligible effect on β. The values in Table 7.8.1 are for $\alpha = 1.3$. We have also listed the corresponding values of horizontal visibility in this table. It is apparent that the basic atmosphere of ASHRAE represents very clear skies.

We are now in a position to employ parameterization Model C—that is, the model of Bird and Hulstrom—and to calculate the monthly direct normal irradiance. We keep ozone constant at 0.25 cm(NTP), water vapor and turbidity parameters as in Table 7.8.1. We may use Eq. (7.4.11) or (7.6.6) to compute τ_a as input to Model C. In this manner $\dot{I}_n$ can be calculated for the 21st day of each month. The resulting calculations, when compared with the ASHRAE algorithm, would answer our question regarding the accuracy of the latter model. In Fig. 7.8.1 we present a comparison for the two extreme cases. At $\theta_z = 0$ a maximum difference of not more than 6% is obtained between Model C and the ASHRAE algorithm. The maximum differences are during December and January. For all other months, differences are less than 6%. In May and June the correspondence is exact when $\theta_z = 0$. In general, during summer months (April–September) correspondence between the two models is very good when $\theta_z < 60°$. These results should lend confidence to the users of the ASHRAE clear-day direct-irradiance algorithm.

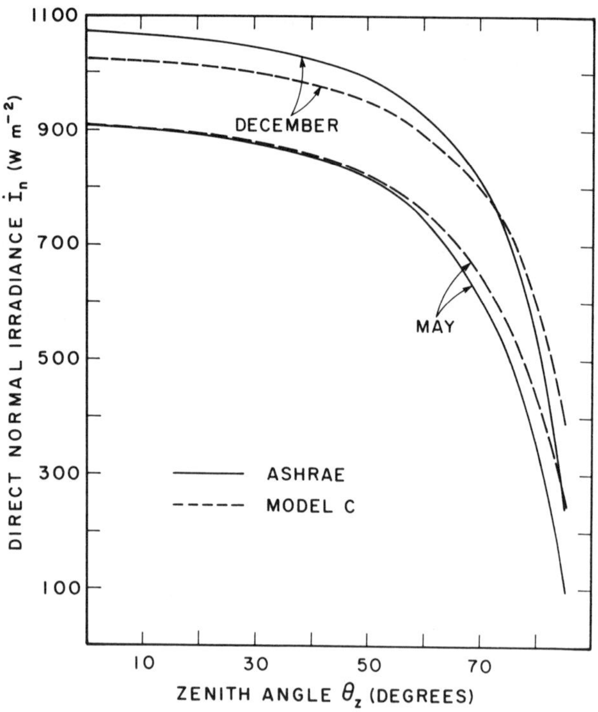

Figure 7.8.1 Comparison of the direct normal irradiance calculated from the ASHRAE model and Model C.

Because it appears that Model C represents state-of-the-art thought on this subject, the constants A and B of the ASHRAE algorithm may be slightly modified to conform to this model. The revised recommended values of A and B are listed in Table 7.8.2.

A reappraisal of the ASHRAE procedure to compute clear-sky diffuse irradiance has also been carried out [6]. Assuming ground albedo 0.2, the ASHRAE values have been compared with those obtained through Model C. For the winter months, the ASHRAE values are very low, with differences as large as 80% during December and January (Fig. 7.8.2). The differences diminish during summer, and the correspondence between the two procedures becomes exact in June and July. Accordingly we have revised the values of C, and they are also listed in Table 7.8.2.

In review, the ASHRAE direct-irradiance model results in a maximum error of 6% during winter months, and the diffuse-irradiance model seems to be substantially in error during winter months. Since diffuse radiation is only a small fraction of the direct radiation, an error in the former may not

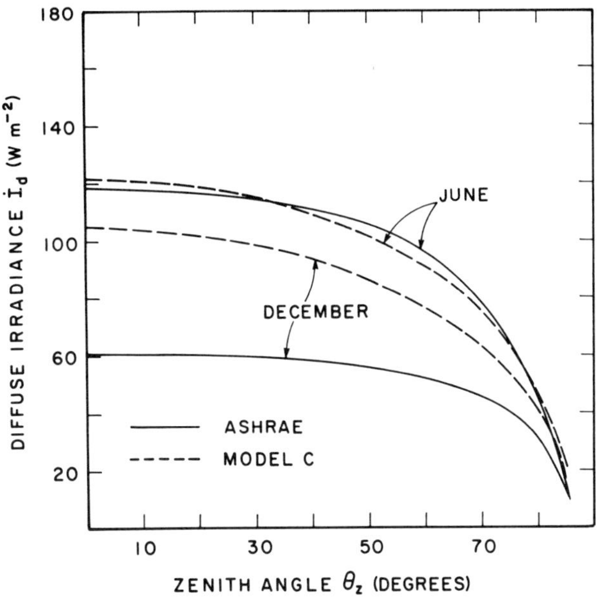

Figure 7.8.2 Comparison of the diffuse irradiance calculated from the ASHRAE model and Model C.

be too serious in many applications. It is fortunate that the ASHRAE algorithm is quite accurate for the summer months, the period for which the engineers use it to compute cooling loads. In any event, revised values of the constants *A*, *B*, and *C* should be used to obtain more accurate results. However, Model C should be preferred over the ASHRAE procedure with the revised constants—unless, of course, the atmospheric parameters conform to the basic atmosphere of ASHRAE.

☐ EXAMPLE 7.8.1. Example 7.4.1 repeated using the ASHRAE model.

Solution. From Example 7.4.1, $\cos\theta_z = 0.5075$.
ASHRAE model: because visibility is 160 km, assume (CN) = 1.

$$\dot{I}_n = A\exp[-B\sec\theta_z(p/p_o)], \qquad \dot{I}_d = C\dot{I}_n.$$

From Table 7.8.2, for January,

$$A = 1\,230\ \mathrm{W\,m^{-2}}, \qquad B = 0.142, \qquad C = 0.058.$$

Therefore

$$\dot{I}_n = 1\,230\exp[-(0.142/0.507\,5)(990/1013.25)] = 935.79\ \mathrm{W\,m^{-2}},$$

(a difference of not more than +4% from that obtained from Model C, Example 7.6.1)

$$\dot{I}_b = \dot{I}_n \cos\theta_z = 935.79(0.507\,5) = 474.91 \text{ W m}^{-2},$$

$$\dot{I}_d = 0.058(935.79) = 54.28 \text{ W m}^{-2}$$

(a difference of −34% from that obtained from Model C, Example 7.6.1)

$$\dot{I} = \dot{I}_b + \dot{I}_d = 474.91 + 54.28 = 529.2 \text{ W m}^{-2}$$

(almost identical to that obtained from Model C, Example 7.6.1). □

7.9 Further Reading

A very good summary of the clear-sky direct-insolation models is presented by Bird and Hulstrom [15]. This summary discusses and compares some elements of the various models presented in this chapter and a number of other models. The conclusion of this summary is that Model C, presented in this chapter, is the most accurate.

Schmetz and Raschke [30] presented a simple model somewhat similar to the ASHRAE formula. Rizzi *et al.* [31] developed a procedure to compute direct irradiance. This procedure is based on the spectral formulation. Hottel [32] developed a clear-sky direct-irradiance procedure which is limited to a specified set of atmospheric parameters. Following Moon [23] and Threlkeld and Jordan [24], Cole [33, 34] parameterized the ASHRAE formula in which the dust and the water-vapor concentration are independent variables.

Nomenclature[5]

A	Apparent extraterrestrial irradiance in the ASHRAE algorithm (W m^{-2})
B	Atmospheric attenuation coefficient in the ASHRAE algorithm [(air mass)$^{-1}$]
C	A dimensionless constant in the ASHRAE algorithm [see Eq. (7.8.7)]
(CN)	Clearness number [see Eq. (7.8.3)]
d	Number of particles per cm^3
E_o	Eccentricity correction factor of the orbit (dimensionless)
F_c	Fraction of forward scattering to total scattering (dimensionless)
$f(m_a)$	Rayleigh scattering function [see Eq. (7.5.5)]
$g(\beta)$	Aerosol scattering function [see Eqs. (7.5.7) and (7.5.8)]
$\dot{I}$	Total global (beam plus diffuse) irradiance on a horizontal surface (W m^{-2})
$\dot{I}_b$	Total beam irradiance on a horizontal surface (W m^{-2})
$\dot{I}_d$	Total diffuse irradiance on a horizontal surface (W m^{-2})

[5] Primes are used to indicate values obtained at a prescribed air mass.

$\dot{I}_{da}$ Diffuse radiation produced by aerosol scattering that reaches the ground after the first pass through the atmosphere (W m^{-2})

$\dot{I}_{dm}$ Diffuse irradiance produced by multiple reflections between the ground and the atmosphere (W m^{-2})

$\dot{I}_{dr}$ Diffuse radiation produced by molecular scattering that reaches the ground after the first pass through the atmosphere (W m^{-2})

$\dot{I}_n$ Direct normal irradiance (W m^{-2})

$\dot{I}_{0n\lambda}$ Extraterrestrial spectral irradiance at mean sun–earth distance (W m^{-2} μm^{-1})

$\dot{I}_{sc}$ Solar constant, 1 367 W m^{-2}

K Turbidity constant [see Eq. (7.4.9)]

k_a Aerosol optical thickness (dimensionless)

$k_{o\lambda}$ Spectral attenuation coefficient for ozone absorption (cm^{-1})

k_λ Spectral attenuation coefficient (dimensionless)

l Ozone layer thickness [cm(NTP)]

m Air mass (dimensionless)

m_a Air mass at actual pressure (dimensionless)

m_r Air mass at standard pressure, 1 013.25 mbars (dimensionless)

p Atmospheric pressure (mbars)

p_0 Standard atmospheric pressure, 1 013.25 mbars

Q See Eq. (7.5.13)

S_i Fraction of the incident energy scattered by an individual atmospheric constituent (dimensionless)

U_1 wm_r, total pressure-corrected relative optical path length for water vapor (cm)

U_3 lm_r, total optical path length for ozone [cm(NTP)]

Vis Horizontal visibility (km)

w Precipitable water-vapor thickness (cm)

α Wavelength exponent in Ångström's turbidity equation (dimensionless)

α_a α_4, fraction of the incident energy absorbed by aerosols (dimensionless)

α_i Fraction of the incident energy absorbed by a particular atmospheric constituent (dimensionless)

α_g α_2, fraction of the incident energy absorbed by carbon dioxide and oxygen (dimensionless)

α_o α_3, fraction of the incident energy absorbed by ozone (dimensionless)

α_w α_1, fraction of the incident energy absorbed by water vapor (dimensionless)

α_{wg} Fraction of the incident energy absorbed by water vapor plus carbon dioxide plus oxygen (dimensionless)

β Ångstrom's turbidity parameter (dimensionless)

θ Angle of incidence, the angle between normal to the surface and sun–earth line (deg)

θ_z Zenith angle, the angular position of the sun with respect to the local vertical (deg)

λ Wavelength (μm); as a subscript λ indicates monochromaticity

ρ'_a Albedo (fraction of incident energy reflected) of the cloudless-sky atmosphere (dimensionless)

ρ_g Albedo of ground (or ground cover) (dimensionless)

τ Transmittance, fraction of the incident energy transmitted by the atmosphere (dimensionless)

τ_a Fraction of the incident energy transmitted by aerosols (dimensionless)

τ_{aa} Fraction of the incident energy transmitted after absorption effects of aerosols (dimensionless

τ_{as} Fraction of the incident energy transmitted after scattering effects of aerosols (dimensionless)

τ_g Fraction of the incident energy transmitted after absorption by uniformly mixed gases (dimensionless)

τ_r Fraction of the incident energy transmitted after scattering by clean, dry air molecules, dimensionless

τ_o Fraction of the incident energy transmitted after absorption by ozone (dimensionless)

τ_w Fraction of the incident energy transmitted after absorption by water vapor (dimensionless)

ω_0 Single-scattering albedo, fraction of the incident energy scattered to total attenuation by aerosols (dimensionless)

References

1. G. W. Paltridge and C. M. R. Platt, "Radiative Processes in Meteorology and Climatology." Elsevier, New York, 1976.
2. J. A. Davies, W. Schertzer and M. Nunez, Estimating global solar radiation. *Boundary-Layer Meteorol.* **9,** 33–52 (1975).
3. P. W. Suckling and J. E. Hay, Modelling direct, diffuse and total solar radiation for cloudless days. *Atmosphere* **14,** 298–308 (1976).
4. J. A. Davies and J. E. Hay, Calculation of the solar radiation incident on a horizontal surface. Proc. *Canad. Sol. Radiat. Data Workshop, 1st, Toronto, Ontario, 1978.*
5. J. A. Davies, Models for estimating incoming solar irradiance. Report to A.E.S. Department of Geography, McMaster University, Hamilton, Ontario (1980).
6. M. Mächler, Parameterization of solar irradiation under clear skies. M.A.Sc. Thesis, University of British Columbia, Vancouver, Canada (1983).
7. A. A. Lacis and J. E. Hansen, A parameterization for the absorption of solar radiation in the earth's atmosphere. *J. Atmos. Sci.* **31,** 118–132 (1974).
8. G. Yamamoto, Direct absorption of solar radiation by atmospheric water vapor, carbon dioxide and molecular oxygen. *J. Atmos. Sci.* **19,** 182–188 (1962).
9. H. G. Houghton, On the annual heat balance of the northern hemisphere. *J. Meteorol.* **11** (1), 1–9 (1954).
10. T. Sasamori, J. London and D. V. Hoyt, Radiation budget of the Southern Hemisphere. *Am. Meteorol. Soc. Mon. Boston* **13,** 9–23, (1972).
11. D. V. Hoyt, A model for the calculation of solar global insolation. *Sol. Energy* **21** (1), 27–35 (1978).
12. Solmet Vol. 2, Hourly solar radiation—surface meteorological observations—Final Report TD-9724. National Climatic Center, Asheville, North Carolina (1979).
13. S. Manabe and R. F. Strickler, Thermal equilibrium of the atmosphere with a convective adjustment. *J. Atmos. Sci.* **21,** 361–385 (1964).
14. D. Labs and H. Neckel, The radiation of the solar photosphere from 2000 Å to 100 μ. *Z. Astrophys.* **69,** 1–73 (1968).
15. R. Bird and R. L. Hulstrom, Direct insolation models. *Trans. ASME J. Sol. Energy Eng.* **103,** 182–192 (1981). See also SERI/TR-335-344 Solar Energy Research Institute, Golden, Colorado (1980).
16. R. E. Bird and R. L. Hulstrom, A simplified clear sky model for direct and diffuse insolation on horizontal surfaces. SERI/TR-642-761, Solar Energy Research Institute, Golden, Colorado (1981).
17. J. E. A. Selby and R. A. McClatchey, Atmospheric transmittance from 0.25 to 28.5 microns. Computer Code Lowtran 3. Air Force Cambridge Research Labs., AFCRL-TR-75-0255, AD-A017734 (1975).

18. J. E. Selby, F. Y. Kneizys, J. H. Chetwind Jr., and R. A. McClatchey, Atmospheric Transmittance/Radiance: Computer Code Lowtran 4, Air Force Cambridge Research Labs., AFGL-TR-78-0053 (1978).
19. J. V. Dave, Extensive datasets of the diffuse radiation in realistic atmospheric models with aerosols and common absorbing gases. *Sol. Energy* **21** (5), 361–369 (1979).
20. R. E. Bird and R. L. Hulstrom, Application of Monte Carlo Technique to insolation characterization and prediction. SERI/RR-36-306, Solar Energy Research Institute, Golden, Colorado (1979).
21. E. A. Farber and C. A. Morrison, Clear Day Design Values. ASHRAE GRP. 170, ASHRAE, New York (1977).
22. ASHRAE Applications Handbook, Chapter 58, Solar Energy Utilization for Heating and Cooling. ASHRAE, New York (1978).
23. P. Moon, Proposed standard solar radiation curves for engineering use. *J. Franklin Inst.* **230,** 583–617 (1940).
24. J. L. Threlkeld and R. C. Jordan, Direct solar radiation available on clear days. *ASHRAE Trans.* **64,** 45–68 (1958).
25. J. L. Threlkeld, Solar irradiation of surfaces on clear days. *ASHRAE Trans.* **69,** 24–36 (1963).
26. D. G. Stephenson, Tables of solar altitude and azimuth, intensity and solar heat gain tables. Tech. Paper No. 243 National Research Council of Canada, Ottawa (1967).
27. O. R. Wulf, The determination of ozone by spectrobalometric measurements. *Smithsonian Misc. Coll.* **85** (9), 1–12 (1931).
28. A. Von Läuchli, Zur absorption der ultravioletten Strahlung in Ozon. *Z. Phys.* **53,** 92–94 (1929).
29. F. E. Fowle, The transparency of aqueous vapor. *J. Astrophys.* **42,** 394–411 (1915).
30. J. Schmetz and E. Raschke, A method to parameterize the downward solar radiation at ground. *Arch. Meteorol. Geophys. Bioklimatol. Ser. B* **26,** 143–151 (1978).
31. R. Rizzi, C. Serio, R. Guzzi, and M. Francesca, Solar direct irradiance at the ground: A parametric approach. *Sol. Energy* **25** (1), 15–20 (1980).
32. H. C. Hottel, A simple model for estimating the transmittance of direct solar radiation through clear atmospheres. *Sol. Energy* **18** (3), 129–134 (1976).
33. R. J. Cole, Direct solar radiation data as inputs into mathematical models describing the thermal performance of buildings–I, A review of existing relationships which predict the direct component of solar radiation. *Build. Environ.* **11,** 173–179 (1976).
34. R. J. Cole, Direct solar radiation data as inputs into mathematical models describing the thermal performance of buildings–II, Development of relationships. *Build. Environ.* **11,** 181–186 (1976).

Chapter 8

SOLAR RADIATION UNDER CLOUDY SKIES

8.1 Introduction

In the previous two chapters, direct and diffuse radiation under cloudless skies was treated. Direct radiation on a horizontal surface was seen to be several times greater than diffuse radiation, except, of course, at low solar altitudes. This chapter can be read without this background, but the previous discussion serves to introduce the reader to the various elements of the cloudless atmosphere that absorb and scatter solar radiation.

For a brief introduction to the radiation arriving on a horizontal surface under cloudy skies, consider Fig. 8.1.1. The beam radiation (radiation coming directly from the solar disk) is attenuated by the presence of clouds in its path, as well as by the various elements of the cloudless atmosphere already studied. The depletion of the direct beam by the clouds depends on the type of clouds, their thickness, and the number of layers.

The downward radiation received on a horizontal surface from a solid angle of 2π, with the exception of the solid angle subtended by the sun, is called diffuse radiation. The diffuse component consists of several parts. Scattering by air molecules and aerosols remains the same as described in the previous chapter. In addition, there is the interaction of direct solar

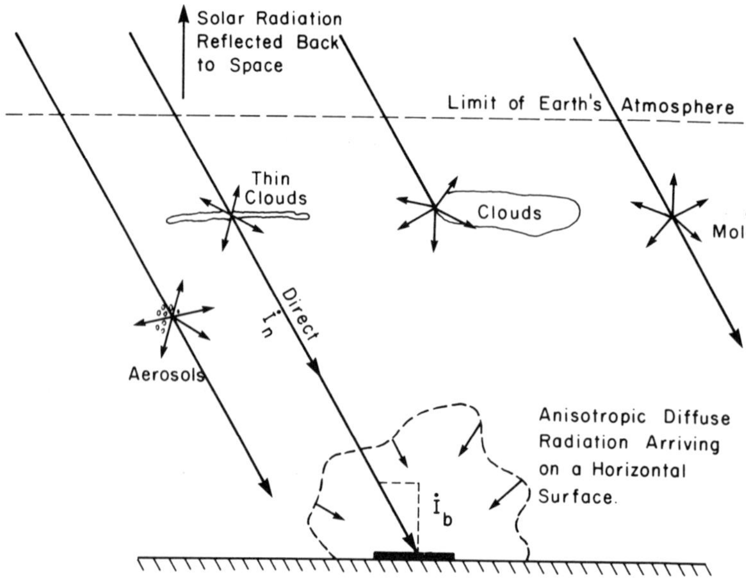

Figure 8.1.1 Radiation arriving on the ground under cloudy skies.

radiation with the clouds (and cloud layers if the clouds are in layers), resulting in reflected diffuse radiation. Further, a portion of the direct and the diffuse radiation reaching the earth after the first pass through the atmosphere is reflected back to the sky, which contributes to the multiply reflected irradiance, as we have seen earlier. This multiply reflected irradiance (not shown in Fig. 8.1.1) will depend strongly on the reflectance properties of the cloud system. It is quite obvious that the directional intensity of the diffuse irradiance will depend directly on the position of the clouds. When the directional intensity of the diffuse irradiance is not uniform over the sky hemisphere, it is called anisotropic diffuse radiation. When the sky is completely overcast by clouds, the diffuse irradiance is almost isotropic.

Theoretical determination of direct, diffuse, and directional intensity of diffuse irradiance is quite difficult. Such a task requires data on the type and optical properties of clouds, cloud amount, thickness, position, and the number of layers. Such data are rarely collected on a routine basis. However, sunshine and total cloud-cover data (fraction of sky covered by clouds) are widely available and are easily available to most users. Therefore, we shall develop this subject to the extent that it requires only simple and readily available data such as sunshine hours, total cloud cover, cloud albedo, and ground albedo.

The sum total of direct and diffuse radiation is called global radiation. In North America, solar engineering literature often uses the word "total" or "hemispherical" instead of "global." The term "global," however, is more commonly used internationally and will be retained in this text.

A common approach and also the one presented in this chapter is to estimate the diffuse component from the measured or estimated global radiation. In this chapter, the two main objectives are to present (i) simple methods of predicting global radiation and (ii) methods of estimating diffuse radiation from the measured or predicted global radiation.

In order to present this matter in sequence, the definitions of elementary radiation quantities, their time scale, and recording of their observed values are discussed first. In the latter sections, we use the measured data to demonstrate the accuracy of the predictive methods.

Unless otherwise stated, in this chapter the term "radiation" means the downward solar radiation on a horizontal surface. Further, all radiation terms used here represent the spectrally integrated totals. Almost all predictive correlations in this chapter developed by the various authors are based on the NASA design solar constant, 1 353 W m^{-2}. However, the results remain substantially unaffected if the new value of 1 367 W m^{-2} is used. We shall use the new value.

We begin with the definitions of the hourly and daily radiation quantities.

8.2 Hourly Global I, Diffuse I_d, and Beam I_b Radiation on a Horizontal Surface

The radiation instruments and measurements are discussed in Chapter 12. In this section we are interested primarily in identifying the measured quantities. However, some reference to the various instruments used for this purpose will be beneficial.

Global radiation is measured by an instrument called a pyranometer. It measures all radiation incident on it within a solid angle of 2π. Let us consider an imaginary trace of the instantaneous radiative flux recorded by such an instrument in Fig. 8.2.1. In this sketch, (a) represents a typical profile of the flux on a particularly clear day and (b) represents the same on a partly cloudy day. The global radiation I is the quantity collected during an hour and is represented by the shaded areas. Data are usually recorded on an hourly basis. Radiation for a period shorter than an hour is rarely measured on a routine basis.

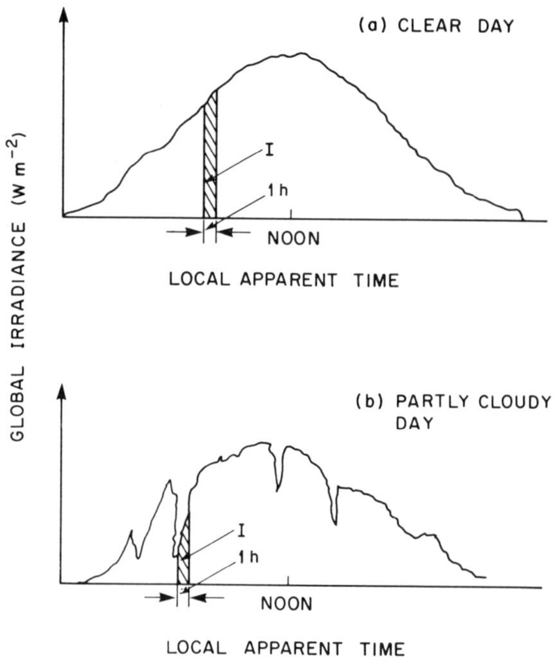

Figure 8.2.1 Diurnal variation of the radiative flux during (a) clear day, (b) partly cloudy day.

Diffuse radiation is measured by a pyranometer fitted with a device that occults[1] the direct radiation from the solar disk. There are two types of occulting devices: (i) a shadow band mounted on an east–west axis and (ii) a circular disk 10 cm in diameter placed about a meter away from the sensor. The disk shades the sensor from the direct solar beam. However, for this mechanism, an equatorial mount is required to track the sun continuously. The reading recorded by a shadow-band instrument requires a slight adjustment to account for the loss in irradiance obstructed by the ring [1]. However, no such correction is required for a disk-mounted instrument. The total amount of (duly corrected) diffuse radiation received during a period of an hour is denoted by I_d.

Beam radiation can be measured by either of the two procedures: (i) direct or (ii) indirect. The direct method utilizes a pyrheliometer, an instrument with a small conical aperture slightly wider than the solid angle subtended by the solar disk. This instrument requires an equatorial mount to follow the sun. It yields measurement of the direct normal irradiance; unavoidably, it contains some portion of the circumsolar radiation, the radi-

[1] In meteorology and astronomy, it means a device for hiding a planet or a star from our eyesight.

ation coming from the immediate vicinity of the sun. From the direct normal irradiance recorded by a pyrheliometer, the beam irradiance on a horizontal surface can be written

$$\dot{I}_b = \dot{I}_n \cos \theta_z, \tag{8.2.1}$$

where $\dot{I}_n$ is the pyrheliometer reading. Integrating the above quantity over a period of 1 h yields the hourly beam radiation:

$$I_b = \int^{1\text{ h}} \dot{I}_n \cos \theta_z \, dt. \tag{8.2.2}$$

The indirect method utilizes readings of two pyranometers, one with and the other without an occulting device. The hourly beam radiation on a horizontal surface is deduced from the difference between the two readings. Thus

$$I_b = I - I_d. \tag{8.2.3}$$

In the event diffuse radiation is not measured but the global and beam radiation are measured, it is quite obvious that the diffuse radiation can be obtained from the following:

$$I_d = I - I_b. \tag{8.2.4}$$

In the remainder of this chapter we shall assume that direct measurements of I_b are not available; this is, in fact, the case in most parts of the world. However, it is important to point out that in the U.S. network, beam radiation is measured on a regular basis and the diffuse radiation is derived through Eq. (8.2.4).

8.3 Daily Global H, Diffuse H_d, and Beam H_b Radiation on a Horizontal Surface

Each of the above quantities is obtained from an integration over a day of the corresponding hourly global I, diffuse I_d, and beam I_b radiation. Thus H is the daily global radiation on a horizontal surface.

$$H = \int^{\text{day}} \dot{I} \, dt \tag{8.3.1}$$

and is equivalent to the area under the curves in Fig. 8.2.1. H_d is the daily diffuse radiation on a horizontal surface, and

$$H_d = \int^{\text{day}} \dot{I}_d \, dt. \tag{8.3.2}$$

Sometimes the term daily beam radiation is also utilized. It is denoted H_b and is defined as the daily beam radiation on a horizontal surface.

$$H_b = H - H_d. \tag{8.3.3}$$

In the literature, another term, *daily direct normal radiation*, has also been employed. It represents the integral of I_n over a period of a day. However, in this text we do not need to use this term.

8.4 The Availability of Data

Many countries maintain a network of radiation measuring stations. Among the industrialized countries, it is usual to measure regularly the hourly values on a horizontal surface. In other countries, only daily values are recorded. Although records of the hourly and daily global radiation are quite common, the measurement of diffuse radiation is very rare. For example, in Canada there are at present over 50 stations regularly measuring hourly global radiation; however, only about seven of them measure the diffuse component as well. Figure 8.4.1 shows a map of the Canadian network. In 1977, the United States established a new solar radiation monitoring network operated by NOAA (National Oceanic and Atmospheric Administration). It consists of 38 stations equipped with pyranometers and pyrheliometers. Ten stations also have shadow-band pyranometers for measurement of diffuse horizontal radiation. Stations belonging to this network, with their names underlined, are shown in Fig. 8.4.2. Locations without their names underlined belong to the older network, and most of these locations have only derived data.

Tables 8.4.1 and 8.4.2 illustrate the manner in which published data are available in Canada [2].[2] These tables (for June 1976) are for global and diffuse radiation, respectively. In other countries, data may be tabulated in a slightly different manner. The amount of radiant energy collected during one full hour is listed under the hour column. In Table 8.4.2, the diffuse-radiation values have already been compensated for the shadow-band effects. The last column contains daily values; the E after some of these values means that one or more of the hours during the day are interpolated. The last row in these tables contains the hourly and daily averages for that particular month of 1976. These two tables will be used as a basis for the verification of some correlations presented later in this chapter, and also for the development of mean radiation data.

[2] As of 1978, the format has slightly changed and the units are in SI units. However, the older format is more appropriate for use in this text.

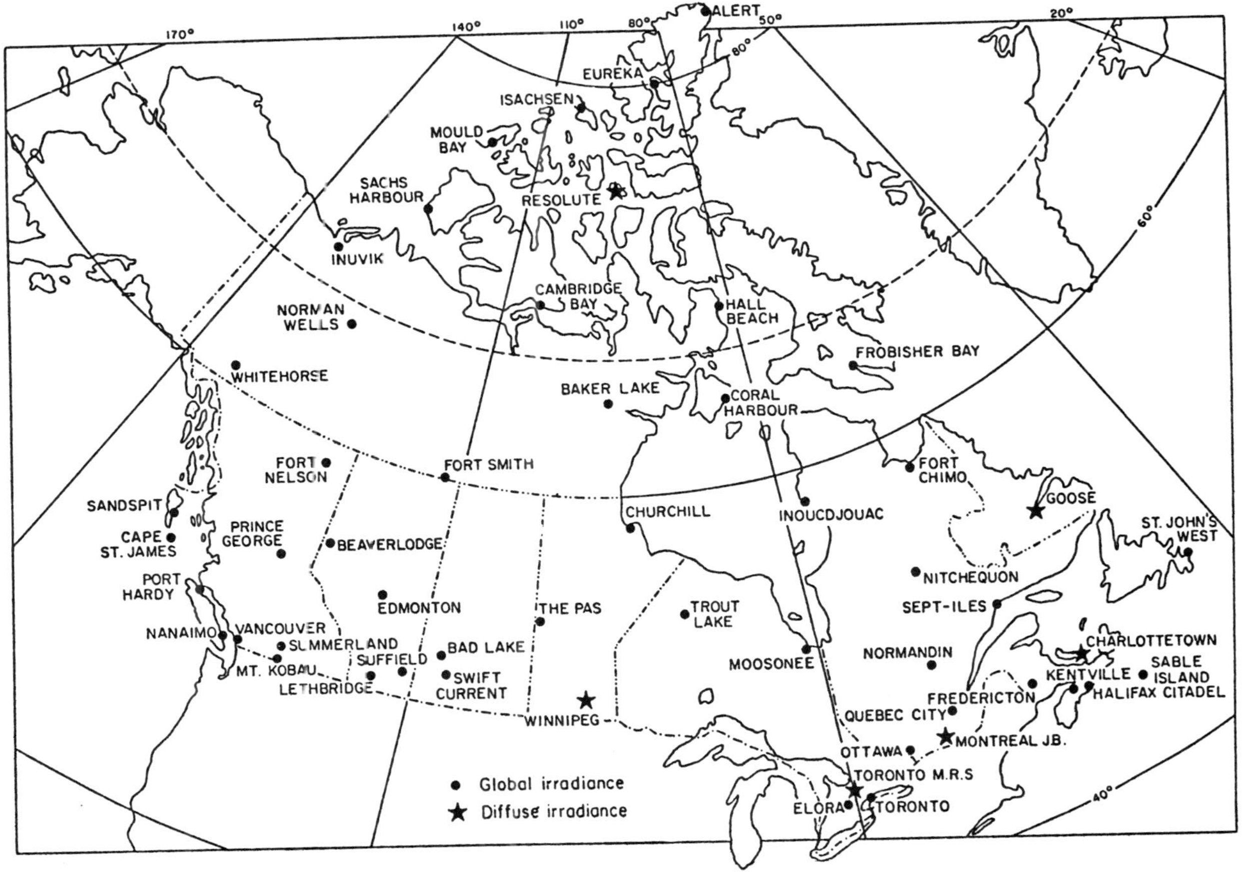

Figure 8.4.1 Canadian solar radiation network.

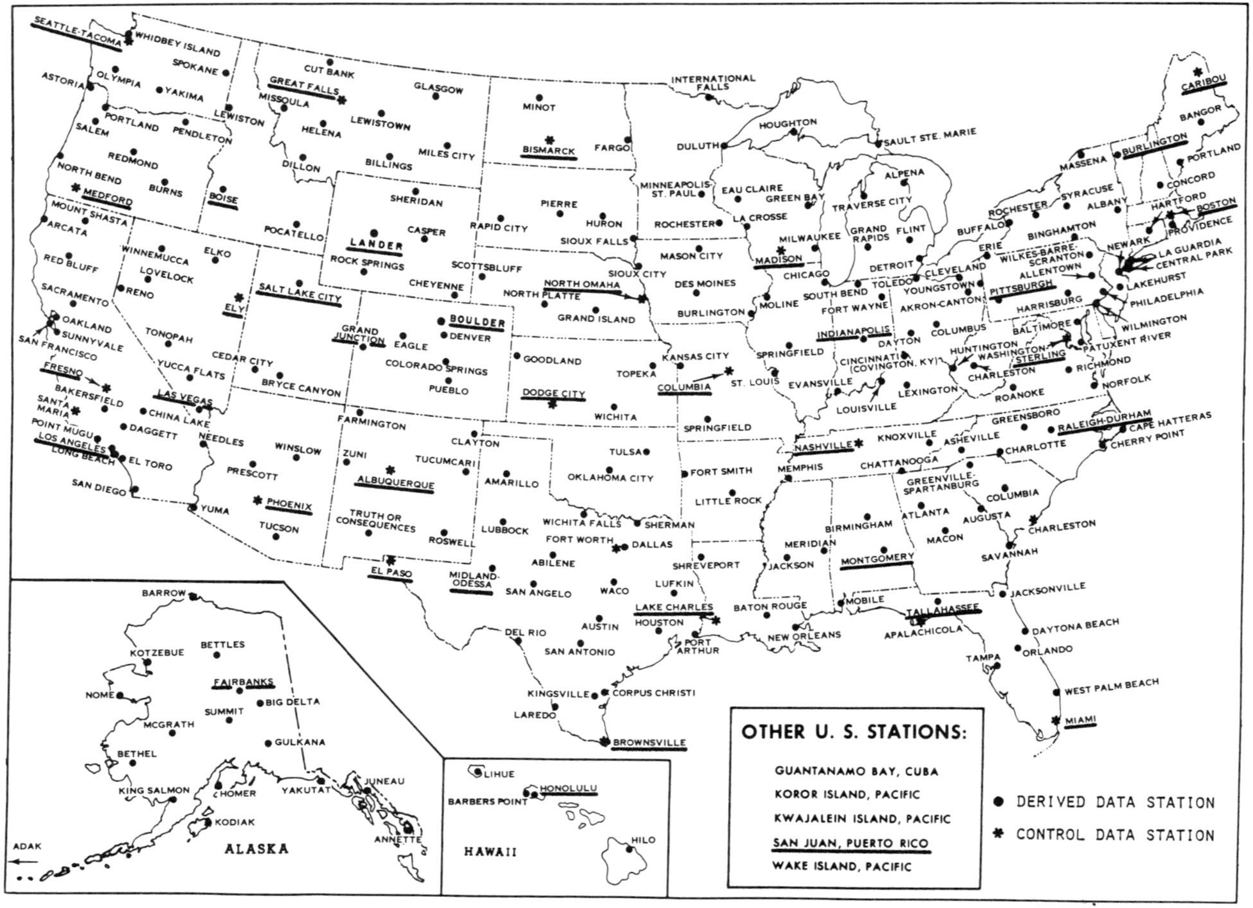

Figure 8.4.2 Stations underlined belong to the NOAA network established in 1977 in which global and direct normal radiation is measured. Most of these stations are at the same location as the older network of 26 "Control Data Stations."

Many solar engineering problems are solved with the use of long-term averages of radiation data, called mean values. These are, in fact, long-term averages of the quantities in the last rows of Tables 8.4.1 and 8.4.2. The radiation quantities based on long-term averages are defined below.

$\bar{I}$ is the monthly average hourly global radiation received on a horizontal surface.

$$\bar{I} = \left(\sum_{1}^{n_y} \sum_{1}^{n_d} I \right) \Big/ n_y n_d \tag{8.4.1}$$

where n_d is the number of days in the month under consideration and n_y is the number of years of data utilized. Similarly, $\bar{I}_d$ is the monthly average hourly diffuse radiation received on a horizontal surface.

$$\bar{I}_d = \left(\sum_{1}^{n_y} \sum_{1}^{n_d} I_d \right) \Big/ n_y n_d . \tag{8.4.2}$$

From (8.4.1) and (8.4.2), $\bar{I}_b$, the monthly average hourly beam radiation on a horizontal surface, can be deduced as follows:

$$\bar{I}_b = \bar{I} - \bar{I}_d . \tag{8.4.3}$$

The corresponding daily values can be written as follows. $\bar{H}$ is the monthly average daily global radiation received on a horizontal surface.

$$\bar{H} = \sum^{\text{day}} \bar{I}. \tag{8.4.4}$$

It should be noted that the measuring instruments generally record some predawn and postsunset radiation. The instrument-recorded day length is sometimes about two hours longer than the theoretical day length [Eq. (1.5.5)]. In a similar fashion as before, $\bar{H}_d$ is the monthly average daily diffuse radiation received on a horizontal surface.

$$\bar{H}_d = \sum^{\text{day}} \bar{I}_d . \tag{8.4.5}$$

Table 8.4.3 contains all four quantities for Montreal, $\bar{I}$, $\bar{I}_d$, $\bar{H}$, and $\bar{H}_d$ (Montreal is one of the few Canadian stations where both global and diffuse radiation are measured). Table 8.4.4. lists $\bar{H}$ and $\bar{H}_d$ values for five locations with latitudes ranging from 44° to 74° N.

We provide below a very short list of sources of radiation data.

(1) *The World Survey of Climatology* [3] contains radiation data and other useful climatological data for most parts of the world. World radiation data, in the form of contour maps, have been published by Löf *et al.* [4] and de Jong [5].

Table 8.4.1

Hourly Global Radiation Incident on a Horizontal Surface, I (cal cm^{-2} h^{-1})[a]

Day	Global radiation for each hour ending at (local apparent time)																							Daily global
	1	2	3	4	5	6	7	8	9	10	11	12	13	14	15	16	17	18	19	20	21	22	23	24	*H*
01					1	10	23	41	57	68	77	66	74	71	65	53	43	29	13	2					693 E
02					2	14	29	43	59	68	77	82	79	58	56	54	42	23	10	1					696
03					1	4	11	24	52	60	78	63	42	51	39	7	4	15	10	2					460
04					2	12	26	41	55	67	76	80	79	75	67	55	41	25	10	1					710
05					2	12	26	41	55	67	75	79	76	75	67	54	43	27	11	2					711
06					2	11	26	37	54	66	73	73	78	75	55	40	13	9	6	1					617
07					2	13	26	28	38	22	30	51	44	37	46	42	22	8	7	2					418
08					+	4	20	29	49	62	70	75	75	72	60	50	27	12	3	+					607
09					2	12	26	41	55	67	76	80	79	73	65	52	33	20	7	2					689
10					1	9	25	40	52	62	71	76	75	70	64	52	39	24	11	2					668
11					1	6	8	5	6	7	14	19	40	58	39	1	5	9	4	+					223
12					1	11	29	45	57	71	79	80	79	71	62	45	38	22	14	2					706
13					3	16	30	39	57	65	77	61	47	30	22	33	15	5	2	1					504

14	2	11	12	26	29	29	56	68	68	59	53	40	27	13	4	1	497
15	2	11	25	40	53	65	73	77	78	73	66	54	39	24	10	2	690
16	2	6	12	18	14	19	14	14	36	25	7	2	3	4	7	1	181
17	1	4	7	19	41	36	69	64	70	26	30	44	37	18	8	1	474
18	2	10	26	41	56	67	76	79	78	73	65	47	38	24	11	1	692
19	1	7	24	33	50	62	55	53	66	67	55	43	36	17	2	+	569
20	+	1	7	6	4	6	8	16	19	12	11	11	8	10	8	2	130
21	2	4	8	13	20	23	22	21	31	26	38	37	29	12	5	1	291
22	+	4	7	9	18	46	31	43	39	38	26	34	21	24	11	2	352
23	1	5	9	17	28	56	72	52	63	58	59	52	40	25	11	2	547
24	1	6	12	38	45	39	50	32	18	35	32	19	16	11	4	1	358
25	1	2	3	6	4	9	31	41	50	60	33	8	28	15	1	1	292
26	1	3	10	24	37	44	56	67	69	49	68	45	40	22	10	1	545
27	1	3	8	19	14	21	54	32	45	46	13	39	40	25	10	2	370
28	2	3	2	33	52	36	47	49	69	75	64	55	32	26	7	2	554
29	2	6	11	8	9	9	13	13	14	17	68	54	28	9	2	1	262
30	1	6	16	42	54	57	57	55	21	15	26	10	5	3	1	+	369
Mean	1	8	17	28	39	46	55	55	57	52	47	38	28	17	7	1	496

[a] Montreal, Quebec (45°30′ N, 73°37′ W), using a Kipp CM 6 pyranometer in June 1976.

Table 8.4.2

Hourly Diffuse Radiation Incident on a Horizontal Surface, I_d (cal cm^{-2} h^{-1})[a]

Day	Diffuse radiation for each hour ending at (local apparent time)																								Daily diffuse
	1	2	3	4	5	6	7	8	9	10	11	12	13	14	15	16	17	18	19	20	21	22	23	24	H_d
01					1	6	11	8	8	8	8	9	19	19	18	15	9	5	4	1					148 E
02					1	4	6	7	8	9	9	10	12	18	21	13	9	6	4	1					137
03					1	4	10	13	19	23	12	37	28	23	25	6	4	13	5	1					225
04					1	4	6	8	8	8	9	9	10	11	11	9	8	7	4	1					113
05					1	4	6	8	8	9	10	11	13	10	10	11	9	7	5	1					123 E
06					1	7	10	12	11	12	14	24	21	17	26	29	13	9	6	1					211
07					2	9	17	25	34	22	30	45	38	33	34	25	15	7	6	1					342
08					+	4	12	17	20	22	24	23	22	21	26	22	18	10	3	0					245
09					1	6	8	10	11	10	10	10	10	12	15	19	17	13	6	2					158
10					1	6	11	14	18	23	25	22	24	25	19	16	14	12	7	2					237 E
11					1	6	8	5	6	7	13	19	28	35	29	1	5	9	4	+					177
12					1	5	6	7	7	7	13	15	20	18	26	22	16	11	9	2					184
13					2	7	13	15	12	13	12	26	36	27	21	26	14	5	2	1					231

14	2	10	11	20	25	22	35	38	36	38	34	29	19	12	4	1	334
15	1	6	11	14	17	18	18	19	20	18	15	13	13	11	6	1	201
16	2	6	11	17	14	18	14	13	34	25	7	2	3	4	7	+	176
17	1	4	7	18	21	22	21	25	35	26	29	28	14	10	6	1	267
18	1	8	13	10	11	13	12	13	20	19	20	19	15	10	6	1	191
19	1	6	17	22	27	29	39	37	35	35	31	25	20	11	2	+	336
20	+	1	7	6	4	6	8	15	19	12	10	11	8	10	7	2	125
21	2	4	7	13	20	23	22	21	31	26	36	33	25	12	5	1	279
22	+	4	7	8	17	35	31	40	34	35	25	28	17	17	9	2	309
23	1	5	9	17	27	41	40	39	32	29	26	17	13	10	7	2	313
24	1	6	11	23	36	38	44	31	17	33	30	19	16	10	4	1	319
25	1	2	3	6	4	9	30	37	36	35	28	8	19	12	1	1	231
26	+	3	10	23	30	31	28	22	32	32	13	15	13	13	7	1	273
27	1	3	8	13	13	20	45	31	40	39	13	24	12	7	5	2	281
28	2	3	2	11	19	30	38	34	34	20	16	14	20	16	6	2	267
29	2	5	11	8	9	8	12	13	13	16	25	15	15	8	2	1	163
30	1	6	13	15	20	35	34	43	21	14	25	10	5	3	1	+	246
Mean	1	5	9	13	15	19	22	24	26	24	22	17	13	10	5	1	228

[c] Montreal, Quebec (45°30′ N, 73°37′ W), using a Kipp CM 6 pyranometer in June 1976.

Table 8.4.3

Monthly Average Hourly Horizontal Diffuse $\bar{I}_d$ and Global $\bar{I}$ Radiation at Montreal, 45°30′ N (October 1964 to December 1975)

	Solar radiation (kJ m^{-2} h^{-1}), solar time ending at																Daily values (MJ m^{-2} day^{-1})
Month	05	06	07	08	09	10	11	12	13	14	15	16	17	18	19	20	
Jan																	
diffuse				21	155	318	444	503	494	436	323	168	29				2.89
global				29	239	545	817	964	964	825	582	264	38				5.27
Feb																	
diffuse			8	113	310	494	620	675	666	603	465	302	122	8			4.39
global			8	176	540	943	1249	1412	1408	1249	968	574	193	13			8.73
Mar																	
diffuse		4	80	264	457	624	737	825	813	729	582	436	247	75	4		5.88
global		4	113	478	922	1345	1621	1781	1785	1605	1320	918	482	121	4		12.50
Apr																	
diffuse		42	193	356	532	649	733	792	788	729	662	528	360	197	46		6.61
global		63	373	809	1265	1638	1890	2015	1990	1827	1554	1190	783	356	59		15.81
May																	
diffuse	17	134	293	453	599	767	846	905	922	859	767	654	499	314	143	17	8.19
global	21	214	587	1052	1500	1860	2082	2196	2128	1957	1714	1383	1018	591	218	21	18.54

Jun																	
diffuse	38	184	356	532	712	855	960	981	981	922	817	704	541	369	197	46	9.19
global	50	310	700	1144	1601	1994	2250	2380	2359	2200	1894	1571	1156	721	318	54	20.70
Jul																	
diffuse	25	172	348	528	696	834	943	1031	1018	960	872	721	545	360	176	29	9.26
global	33	281	708	1186	1668	2053	2321	2480	2443	2258	1982	1609	1148	700	281	38	21.19
Aug																	
diffuse	4	80	260	440	529	771	884	930	943	888	771	620	444	260	84	4	8.01
global	4	126	478	922	1349	1709	1948	2087	2078	1965	1693	1332	897	469	121	4	17.18
Sep																	
diffuse		13	117	285	444	591	687	746	767	717	624	474	289	122	8		5.88
global		17	201	587	1018	1425	1672	1793	1760	1617	1362	997	566	188	13		13.22
Oct																	
diffuse			21	143	289	427	545	582	574	515	411	298	138	21			3.96
global			29	239	574	905	1152	1270	1261	1127	871	574	235	25			8.26
Nov																	
diffuse				34	147	277	381	436	436	377	281	155	34				2.56
global				50	243	486	683	792	779	666	473	247	46				4.47
Dec																	
diffuse				8	105	247	369	427	406	339	247	113	13				2.28
global				13	147	385	607	733	716	595	411	163	13				3.78

Table 8.4.4

Monthly Average Daily Global $\overline{H}$ and Diffuse Radiation $\overline{H}_d$ on Horizontal Surfaces
($MJ\ m^{-2}\ day^{-1}$)

Station	$\overline{H}$ or $\overline{H}_d$	Jan	Feb	Mar	Apr	May	Jun	Jul	Aug	Sep	Oct	Nov	Dec
Resolute													
(Canada)	$\overline{H}$	0.00	0.62	5.16	14.82	23.39	25.14	18.45	11.13	5.22	1.26	0.01	0.00
74°43′ N	$\overline{H}_d$	0.00	0.43	2.90	7.59	14.51	15.26	10.38	7.09	3.79	0.97	0.01	0.00
94°59′ W	$\overline{K}_T = \overline{H}/\overline{H}_0$	—	—	0.68	0.72	0.66	0.59	0.47	0.43	0.45	0.61	—	—
Bergen													
(Norway)	$\overline{H}$	0.69	2.57	6.01	11.53	14.60	17.04	14.39	12.14	6.74	3.32	1.07	0.41
60°24′ N	$\overline{H}_d$	0.62	1.71	3.59	6.00	7.96	8.56	8.09	6.79	4.12	2.08	0.86	0.41
05°19′ E	$\overline{K}_T = \overline{H}/\overline{H}_0$	0.21	0.32	0.37	0.43	0.40	0.42	0.38	0.40	0.34	0.32	0.25	0.20
Trappes													
(France)	$\overline{H}$	3.09	5.60	9.47	14.50	17.50	19.60	20.01	15.65	13.04	7.58	3.80	2.17
48°46′ N	$\overline{H}_d$	2.04	3.31	5.18	7.56	8.38	9.37	9.08	7.71	6.04	3.88	2.43	1.48
02°01′ E	$\overline{K}_T = \overline{H}/\overline{H}_0$	0.32	0.37	0.41	0.46	0.46	0.48	0.50	0.46	0.50	0.44	0.35	0.26
Montreal													
(Canada)	$\overline{H}$	5.27	8.73	12.50	15.81	18.54	20.70	21.19	17.18	13.22	8.26	4.47	3.78
45°30′ N	$\overline{H}_d$	2.89	4.39	5.88	6.61	8.19	9.19	9.26	8.01	5.88	3.96	2.56	2.28
73°37′ W	$\overline{K}_T = \overline{H}/\overline{H}_0$	0.45	0.52	0.51	0.48	0.48	0.51	0.53	0.49	0.48	0.43	0.35	0.37
Carpentras													
(France)	$\overline{H}$	5.7	10.58	13.46	18.65	22.27	25.23	26.67	22.04	16.79	11.40	6.82	5.11
44°05′ N	$\overline{H}_d$	2.62	4.65	5.82	6.33	7.34	9.02	8.69	6.95	5.54	3.86	2.91	2.53
05°03′ E	$\overline{K}_T = \overline{H}/\overline{H}_0$	0.45	0.60	0.53	0.56	0.57	0.61	0.67	0.63	0.60	0.57	0.50	0.46

(2) On a national level, contour maps of solar radiation for the United States and Canada are available in [6] and [7], respectively. Tabulated values of the U.S. and the Canadian data have been prepared by Cinquemani *et al.* [8] and by Phillips and Aston [7]. The U.S. publication lists mean daily global horizontal data for 248 locations shown in Fig. 8.4.2 (with a small exception in this map: Boulder, Colorado, and Lander, Wyoming, do not belong to this list). Furthermore, among these 248 locations, only 26 stations have corrected (also called rehabilitated!) data, the remaining 222 having derived values. The Canadian publication gives mean daily and hourly global horizontal values for 57 stations, and the corresponding diffuse-radiation data for 5 stations.

In many countries national meteorological offices publish local data in more or less the same manner as those published by the United States and Canada. Such offices should be contacted for more detail.

8.5 Correlation of Average Daily Global Radiation with Hours of Sunshine

It seems obvious that more sunshine results in more insolation and vice versa. Therefore, there should be a direct relationship between the number of sunshine hours during a day and insolation.

On a long-term basis, the monthly average daily global radiation on a horizontal surface can be estimated through the number of bright[3] sunshine hours. Bright sunshine hours are recorded at many locations around the world, in fact at more locations than those measuring insolation. Records of sunshine for more than thirty years also exist at many locations. Kimball [9] studied the relation between duration of sunshine and insolation. Ångström [10] proposed the following relationship to predict insolation:

$$\bar{H}/H_c = a_1 + (1 - a_1)S = a_1 + b_1 S, \tag{8.5.1}$$

where H_c is the perfectly clear day horizontal insolation and S the monthly mean daily fraction of possible sunshine obtained from

$$S = \bar{n}/\bar{N}_d. \tag{8.5.2}$$

[3] The word *bright* is used to emphasize the fact that most sunshine recorders are activated only when the direct radiation is above a certain threshold of brightness, approximately 210 W m^{-2}, and this figure varies with humidity. Also, the sun is considered shining when an object under its direct light casts a well-defined shadow.

The symbol $\bar{n}$ is the monthly average number of instrument-recorded bright sunshine hours per day, and $\overline{N}_d$ is the average day length. For a given month, the average day length can be calculated from

$$\overline{N}_{\mathrm{d}} = \frac{1}{n_2 - n_1} \sum_{n=n_1}^{n=n_2} \left[\frac{2}{15} \cos^{-1}(-\tan\phi \tan\delta) \right], \tag{8.5.3}$$

where n_1 and n_2 are day numbers at the beginning and end of a month. $\overline{N}_d$ can also be obtained from Eq. (1.5.5) by letting $\delta = \delta_c$.

In Eq. (8.5.1), a_1 and b_1 are the empirical coefficients obtained from regression analysis using measured values of $\overline{H}$. When a_1 and b_1 are known, Eq. (8.5.1) can be applied to other locations with climatological characteristics similar to those from where the original data are obtained. Ångström recommended values of 0.25 and 0.75, respectively, for these coefficients based on data from Stockholm. From Eq. (8.5.1) it is obvious that $a_1 + b_1 = 1$, because on clear days S is supposed to be equal to 1. However, because of problems inherent in sunshine recorders, measurements of S never equal 1. Consequently, for clear-sky conditions, Eq. (8.5.1), with Ångström's coefficients, underestimates insolation. The same is true for overcast days.

From the above, it can be concluded that the sum of the coefficients a_1 and b_1 should not be equal to unity. Fritz and MacDonald [11], using data from 11 stations in the continental United States, obtained values of 0.35 and 0.61 for a_1 and b_1, respectively.

A poor correspondence between the coefficients and the data is sometimes attributed to the type of sunshine recorder used and local practice of obtaining the data. Characteristics of the various sunshine recorders are quite different. For example, the Campbell–Stokes sunshine recorder generally does not record bright sunshine when the sun is less than about 5° above the horizon. At some locations, and during certain periods, as in winter in Canada, the period during which the sun is less than 5° above the horizon may be a considerable portion of the day. Taking this factor into account, Mateer [12] developed the following correlation based on Canadian data:

$$\overline{H}/H_c = 0.43 + 0.58S(1 + 5.54j^2), \tag{8.5.4}$$

where j is the fraction of the total daylight period during which the zenith angle is greater than 85° (that is, the sun is less than 5° above the horizon).

Apart from the difficulty arising from the use of a particular type of sunshine recording instrument, these correlations suffer from another problem: the lack of a good definition of a perfectly clear day from which to calculate H_c. As seen in the preceding chapters, even if such a definition existed, the calculation of H_c would not be a straightforward matter. Al-

Table 8.5.1

Monthly Average Bright Sunshine Hours [$\bar{n}$ *(number of days in the month)*]

Station	Jan	Feb	Mar	Apr	May	Jun	Jul	Aug	Sep	Oct	Nov	Dec	Annual
Resolute	0	17	145	267	274	244	276	159	56	21	0	0	1 459
Bergen	16	51	98	164	182	205	161	169	91	66	27	10	1 243
Trappes	55	80	135	174	210	225	230	200	170	125	60	43	1 707
Montreal	96	122	171	188	241	248	261	234	196	150	75	79	2 061
Carpentras	140	158	205	250	305	315	365	315	250	205	140	125	2 773

though it appears very difficult to alleviate the problems associated with the sunshine data, Prescott [13] and others have modified the insolation–sunshine correlation to base it on the extraterrestrial irradiation, a quantity that can be readily computed. Such an equation can be written as follows:

$$\bar{H}/\bar{H}_0 = a + b(\bar{n}/\bar{N}_d). \tag{8.5.5}$$

Löf *et al.* [4] and many other researchers have presented values of the coefficients a and b for many parts of the world. Such coefficients are then applied to compute $\bar{H}$ for nearby locations with similar climatic conditions. It is a common practice to present these coefficients as yearly averages. However, for some parts of the world, it is sometimes necessary to develop seasonal values when there are major climatological changes, as in the 12-month period in Canada and the monsoon period in south and southeast Asia.

Monthly mean numbers of bright sunshine hours for a number of U.S. and Canadian stations are available in Refs. [6] and [14], respectively. Table 8.5.1 lists the sunshine data of stations mentioned in Table 8.4.4.

Table 8.5.2

Coefficients a and b in $\bar{H}/\bar{H}_0 = a + b(\bar{n}/\bar{N}_d)$

Station	a	b
Resolute	0.319	0.683
Bergen	0.175	0.633
Trappes	0.219	0.468
Montreal	0.295	0.371
Carpentras	0.313	0.386

From Eq. (8.5.5), the constants a and b have been developed for the five stations and are given in Table 8.5.2.

Rietveld [15] examined several published values of a and b and noted that a is related linearly and b hyperbolically to the appropriate mean value of $\bar{n}/\bar{N}_d$ such that

$$a = 0.1 + 0.24(\bar{n}/\bar{N}_d) \tag{8.5.6}$$

and

$$b = 0.38 + 0.08(\bar{N}_d/\bar{n}). \tag{8.5.7}$$

Substituting (8.5.6) and (8.5.7) into (8.5.5), we have

$$\bar{H}/\bar{H}_0 = 0.18 + 0.62(\bar{n}/\bar{N}_d). \tag{8.5.8}$$

The above is believed to be applicable anywhere in the world and yields particularly superior results for cloudy conditions, $\bar{n}/\bar{N}_d < 0.4$.

Some authors, attempting to improve on the insolation–sunshine correlation, have included geographic factors such as latitude and elevation. Locations at higher latitudes on an average have higher air mass than locations at lower latitudes. Glover and McCulloch [16] included latitude effects and presented the following correlation:

$$\bar{H}/\bar{H}_0 = 0.29 \cos\phi + 0.52(\bar{n}/\bar{N}_d), \qquad \phi < 60^\circ. \tag{8.5.9}$$

The effect of elevation has been treated by Bennett [17].

The foregoing insolation–sunshine correlations are valid only for long-term averages. Monthly averages are most common, and 10-day averages have also been attempted. However, these correlations are not valid when global radiation for an individual day is required. An exception to this may be at a location where the skies are always very clear.

At this stage, it is necessary to list several points concerning the use of the insolation–sunshine correlations.

(i) In the past, different researchers have used different values of I_{sc}, δ, and E_0 in computing $\bar{H}_0$.

(ii) The radiation data from the stations were not all derived from the same periods, type of instrument, instrument calibrations, or pyrheliometer scales.

(iii) The sunshine data from different instruments working under different conditions have often been mixed in order to obtain the coefficients a and b.

The coefficients a and b in general are site dependent. They are affected by the optical properties of the cloud cover, ground reflectivity, and average air mass. Hay [18] developed a generalized procedure that takes these factors into account. His site-independent correlation,

$$\frac{\bar{H}}{\bar{H}_0} = \frac{0.157\,2 + 0.556\,6(\bar{n}/\bar{N}_j)}{1 - \rho[\rho_a(\bar{n}/\bar{N}_j) + \rho_c(1 - \bar{n}/\bar{N}_j)]}, \tag{8.5.10}$$

incorporates the ground albedo ρ, cloudless-sky albedo ρ_a, and cloud albedo ρ_c. The numerical constants in this equation are obtained assuming $\rho_a = 0.25$ and $\rho_c = 0.6$. (Note: $\rho_a = 0.25$ is rather too high when compared with the values in Fig. 7.7.7.) $\bar{N}_j$ is the modified day length and excludes the fraction during which the solar zenith angle is greater than 85°. The modified day length is obtained from

$$\bar{N}_j = \frac{1}{7.5} \cos^{-1}\left(\frac{\cos 85 - \sin\phi \sin\delta_c}{\cos\phi \cos\delta_c}\right), \tag{8.5.11}$$

where δ_c is the characteristic declination (Table 4.2.1).

In Eq. (8.5.10), although constant values of the cloud albedo and clear-sky albedo have been employed, monthly values of ground albedo ρ should be utilized. Tables 9.4.3 and 9.4.4. in Chapter 9 list the monthly variation of albedo for some U.S. and Canadian locations. These albedos are for large geographic areas.

We illustrate the use of insolation–sunshine correlations through the following example. Those readers not familar with elementary statistical tests are advised to study Appendix D before reading this example.

☐ EXAMPLE 8.5.1. Prediction of the annual variation of mean daily global radiation incident on a horizontal surface in Montreal; the results are compared with measured values, presented in tabulated form, and illustrated graphically.

Solution. In this example we make use of Eqs. (8.5.8)–(8.5.10). We also employ the values of a and b from Table 8.5.2. The sunshine data are obtained from Table 8.5.1. For Eq. (8.5.10) ground albedo is also required. It is obtained from Table 9.4.4.

In Table 8.5.3 we present the predicted values of $\bar{H}$ derived from the various correlations. In Table 8.5.4 we list the percentage difference between the predicted values and the measured values. Figure 8.5.1 shows the plot of the predicted values. It is evident that a visual comparison of these results is not sufficient to judge the superiority of one model over another. A comparison of the percent difference is also inadequate. It is necessary to compute at least the root-mean-square error (RMSE) and the mean bias error (MBE). Results of these tests are listed in the last two rows of Table 8.5.4.

The result with coefficients $a = 0.295$ and $b = 0.371$ is the best. However, this comparison is self-defeating in this particular case because these coefficients were obtained from Montreal data. The correlation of Glover and

Table 8.5.3

Predicted Values of the Monthly Average Daily Global Radiation on a Horizontal Surface in Montreal[a]

Month	$\bar{H}_0$ (MJ m^{-2} day^{-1})	$\frac{\bar{n}^b}{\bar{N}_d}$	Predicted $\bar{H}$ (MJ m^{-2} day^{-1}) Eq. (8.5.8)	Eq. (8.5.9)	Eq. (8.5.10)	With $a = 0.295$ $b = 0.371$	Measured
Jan	10.78	0.35	4.24	4.11	4.76	4.5	5.27
Feb	16.52	0.43	7.31	6.98	8.08	7.44	8.73
Mar	25.57	0.47	11.94	11.34	12.51	11.88	12.50
Apr	32.77	0.47	15.30	14.53	15.72	15.23	15.81
May	40.47	0.53	20.38	19.19	20.68	19.70	18.54
Jun	41.78	0.54	21.30	20.03	21.62	20.49	20.70
Jul	42.97	0.56	22.43	21.04	22.74	21.39	21.19
Aug	35.69	0.54	18.19	17.11	18.46	17.51	17.18
Sep	29.04	0.53	14.62	13.77	14.91	14.13	13.22
Oct	20.49	0.45	9.31	8.87	9.66	9.37	8.26
Nov	12.87	0.27	4.43	4.38	4.71	5.03	4.47
Dec	10.44	0.30	3.78	3.72	4.19	4.20	3.78

[a] See Example 8.5.1.
[b] For Eq. (8.5.10), $\bar{n}/\bar{N}_j$ was employed.

Table 8.5.4

Percentage Difference between the Measured and the Predicted Results and a Comparison of the Root-Mean-Square Error and Mean Bias Error[a]

Month	Eq. (8.5.8)	Eq. (8.5.9)	Eq. (8.5.10)	$a = 0.295$ $b = 0.371$
Jan	19.62	21.99	9.66	13.98
Feb	16.31	20.00	7.41	14.82
Mar	4.51	9.32	0.11	4.92
Apr	3.25	8.11	0.56	3.66
May	9.92	3.50	11.56	6.25
Jun	2.89	3.25	4.44	1.00
Jul	5.84	0.73	7.31	0.93
Aug	5.90	0.42	7.47	1.89
Sep	10.61	4.15	12.75	6.92
Oct	12.75	7.42	16.97	13.47
Nov	0.99	2.04	5.41	12.63
Dec	0.12	1.72	10.97	11.14
RMSE	0.809	0.806	0.913	0.766
MBE	0.245	−0.427	0.652	0.105

[a] See Example 8.5.1.

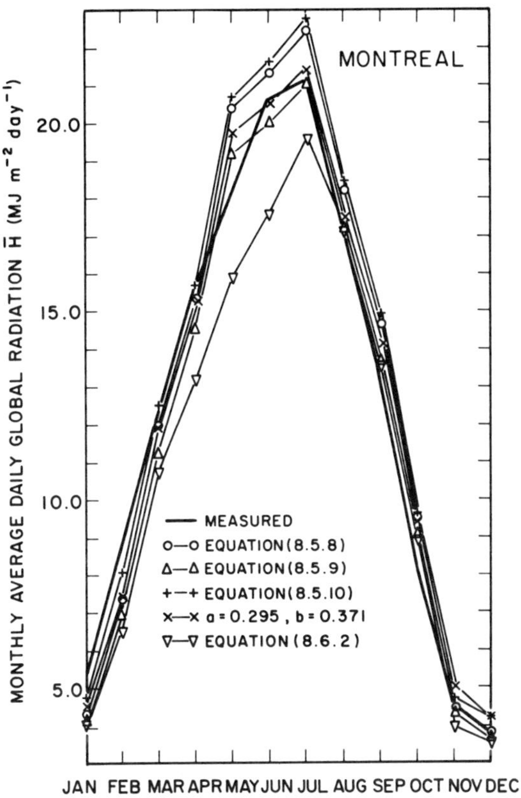

Figure 8.5.1 Comparisons of the various correlations to predict monthly average daily global radiation.

McCulloch [Eq. (8.5.9)] has the lowest root-mean-square error whereas the correlation of Rietveld [Eq. (8.5.8)] has the lowest mean bias error. The correlation by Hay [Eq. (8.5.10)] does not perform as well as the other two correlations. It seems to indicate that use of a modified day length results only in influencing the regression coefficients of a correlation.

This example is a study of one station only. It is necessary to carry out similar comparisons with data from many other stations to establish superiority of one correlation over the others. Such comparisons carried out here at the University of British Columbia have shown that, in general, Rietveld's correlation is superior to the other correlations.[4] □

[4] See C. C. Y. Ma and M. Iqbal, Statistical comparison of solar radiation correlations, monthly average global and diffuse radiation on horizontal surfaces. *Proc. ASES Ann. Meeting Minneapolis/St. Paul, Minnesota, 1–3 June, 1983*, to be published in *Sol. Energy*.

8.6 Correlation of Average Daily Global Radiation with Cloud Cover

Although sunshine measuring stations are more numerous than those measuring insolation, there are still too few of them to provide good geographic coverage. Therefore, investigators in the field of insolation climatology have looked at other estimators of insolation. It appears that the total cloud-cover (also called sky-cover) data are more numerous and, in some instances, geographically better-distributed than sunshine data. For example, in some regions of the high north latitudes, cloud-cover data are available but not sunshine data.

It has long been known that a relationship exists between insolation and the amount of sky covered by clouds [9]. The linear correlation usually proposed is

$$\overline{H}/\overline{H}_0 = a_2 - b_2 C, \tag{8.6.1}$$

where C is the monthly average fraction of the *daytime sky* obscured by clouds. Black [19], using data from many parts of the world, proposed the following quadratic equation:

$$\overline{H}/\overline{H}_0 = 0.803 - 0.340C - 0.458C^2 \qquad \text{with} \quad C \leq 0.8. \tag{8.6.2}$$

Black also developed world radiation contour maps.

Table 8.6.1 contains the monthly mean cloud-cover data (also called cloud normals) for a number of stations. Meteorologists publish the cloud-cover data in percentages, tenths, or eighths of sky covered by clouds. The data are based either on 24-h or daytime observations. The value of C in the above equation should be some fraction of unity of the daytime sky covered by clouds.

The above equation and a number of other correlations based on the

Table 8.6.1

Monthly Mean Cloud Cover, also called Cloud Normals[a]

Station	Jan	Feb	Mar	Apr	May	Jun	Jul	Aug	Sep	Oct	Nov	Dec	Annual
Resolute	3.9	4.1	4.0	4.7	6.9	7.2	7.5	7.9	8.2	7.5	4.6	4.0	5.9
Bergen	7.3	7.3	7.0	6.9	6.5	6.7	7.3	7.3	7.2	7.3	7.3	7.3	7.1
Trappes	8.0	7.6	7.3	6.8	7.2	6.6	6.5	5.7	6.1	7.2	7.5	7.6	7.0
Montreal	6.7	6.4	6.1	6.3	6.4	6.1	5.7	5.4	5.6	5.9	7.3	7.0	6.2
Carpentras	5.5	5.7	5.6	5.2	5.6	4.0	3.1	3.4	4.1	4.9	4.9	5.3	4.8

[a] Values in this table are in tenths of sky covered by clouds from sunrise to sunset; for use in Eq. (8.6.2) divide by ten to obtain C.

cloudiness index are known to be less reliable [20] than the corresponding insolation–sunshine correlations. A number of factors lead to this conclusion.

(i) Insolation and sunshine measurements are integrated over the entire day whereas the cloud-cover averages are simply averages of observations taken at definite times from sunrise to sunset, quite often every three hours.
(ii) The index C in itself does not directly give information as to which part of the sky is covered by the clouds. In an extreme case, one small cloud could keep the sun obscured by slowly traversing the sky. On the other hand, it is also possible that a small hole in the clouds could remain open to the sun for a long period.
(iii) In some instances, the reflection of solar radiation from the edges and sides of clouds can increase the insolation to even more than that received above the atmosphere.

It appears difficult to circumvent these disadvantages with respect to the use of cloudiness index as an estimator of insolation. However, Bennett [21] has shown that it is possible to improve on the above by introducing a new variable called *opaque sky cover*. It differs from the total cloud cover in that it includes only the amount of sky visually obscured by clouds. The opaque sky cover is observed at a number of stations. Bennett has shown that this parameter gives results very close to those obtained from the insolation–sunshine correlations.

The cloudiness index has been used in developing solar radiation data and regression equations for a large number of stations in the United States [22, 23]. In the tables prepared by Cinquemani *et al.* [8], radiation values for many of the so-called derived data stations were developed by such regression equations.

□ EXAMPLE 8.6.1. Prediction of the annual variation of mean daily global radiation on a horizontal surface in Montreal using cloud-cover data, comparison of the results with those obtained from Example 8.5.1.

Solution. The cloudiness index (fraction of sky covered by clouds) for Montreal is given in Table 8.6.1. Employing Eq. (8.6.2), we list below the predicted values of $\bar{H}$ in MJ m^{-2} day^{-1}.

Month	Jan	Feb	Mar	Apr	May	Jun	Jul	Aug	Sep	Oct	Nov	Dec
$\bar{H}$ Eq. (8.6.2)]	3.94	6.51	10.77	13.21	15.94	17.59	19.59	17.17	13.48	8.99	3.96	3.52
$\bar{H}$ (measured)	5.27	8.73	12.50	15.81	18.54	20.70	21.19	17.18	13.22	8.26	4.47	3.78
Percentage difference	25.2	25.2	13.9	16.5	14.0	15.0	7.57	0.06	1.99	8.81	11.4	6.84

The predicted values are also plotted in Fig. 8.5.1 and show that in general this method does not give a good correspondence with the measured values. The root-mean-square error is 1.742, and the mean bias error is -1.249. Both these tests indicate that this correlation does not perform as well as the correlations with sunshine data. □

In this and the preceding sections, we have studied methods of predicting the monthly average daily global radiation on a horizontal surface. In the next section we discuss a procedure to estimate its hourly component.

8.7 Estimation of the Monthly Average Hourly Global Radiation on a Horizontal Surface

Long-term averages of hourly global and hourly diffuse radiation are required for many design purposes. From the measured or estimated daily global radiation $\bar{H}$, the two hourly quantities can be predicted. In this section we treat only the hourly global radiation.

Let us reexamine Fig. 8.2.1. For either clear days or cloudy days, long-term averages of the shaded areas give the quantity $\bar{I}$. As an illustration, in Fig. 8.7.1 measured values of $\bar{I}$ for a specific location [Montreal data (Table 8.4.3)] are plotted for two months. The area under each curve gives the corresponding daily quantity $\bar{H}$. A plot of data from any other station will closely resemble this figure. Now, consider the corresponding plots of the extraterrestrial hourly radiation $\bar{I}_0$ for Montreal (Fig. 8.7.2). The area under these curves is $\bar{H}_0$. There is a very close resemblance between these two diagrams: it appears that the ratios $\bar{I}/\bar{H}$ and $\bar{I}_0/\bar{H}_0$ may well be identical. Whillier [24], while investigating this identity, developed first a relationship between hourly and daily beam radiation $\bar{I}_b/\bar{H}_b$, which can be written as follows:

$$\frac{\bar{I}_b}{\bar{H}_b} = \frac{\int^{1\text{ h}} \bar{\tau}_b(\omega)\dot{I}_{sc}E_0 \cos\theta_z \, d\omega}{\int^{1\text{ day}} \bar{\tau}_b(\omega)\dot{I}_{sc}E_0 \cos\theta_z \, d\omega}, \tag{8.7.1}$$

where $\bar{\tau}_b(\omega) = \bar{I}_b/\bar{I}_0$ is the atmospheric transmittance for beam radiation. The expression $\bar{\tau}_b(\omega)$ is identical to τ in Chapter 7. The bar over τ indicates that we are dealing with average values. By assuming the atmospheric transmittance for beam radiation $\bar{\tau}_b(\omega)$ to be constant throughout the day, the above reduces to the following:

$$\frac{\bar{I}_b}{\bar{H}_b} = \frac{\bar{I}_0}{\bar{H}_0} = \frac{\pi}{24} \frac{(24/\pi)\sin(\pi/24)\cos\omega_i - \cos\omega_s}{\sin\omega_s - (\pi/180)\omega_s\cos\omega_s}. \tag{8.7.2}$$

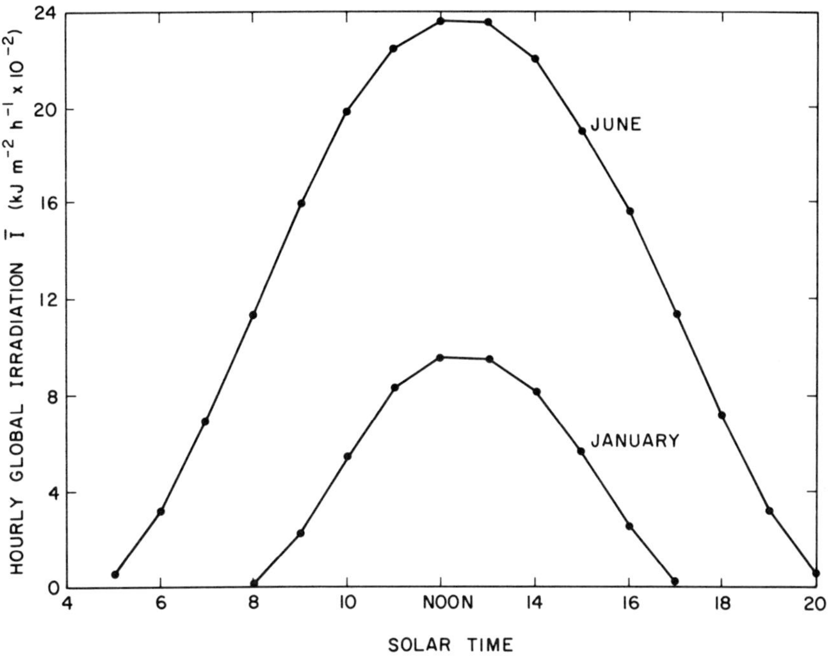

Figure 8.7.1 Diurnal variation of the global radiation at Montreal (45°30′ N).

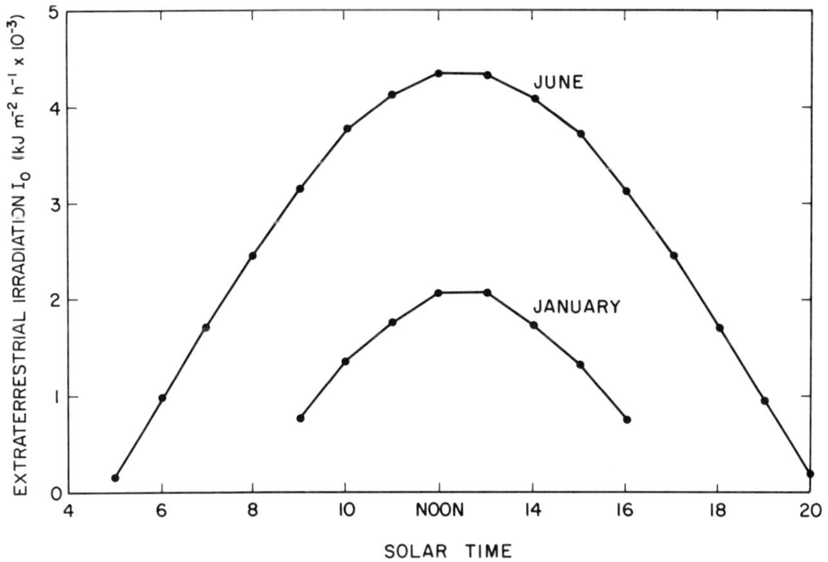

Figure 8.7.2 Diurnal variation of the extraterrestrial radiation above Montreal (45°30′ N).

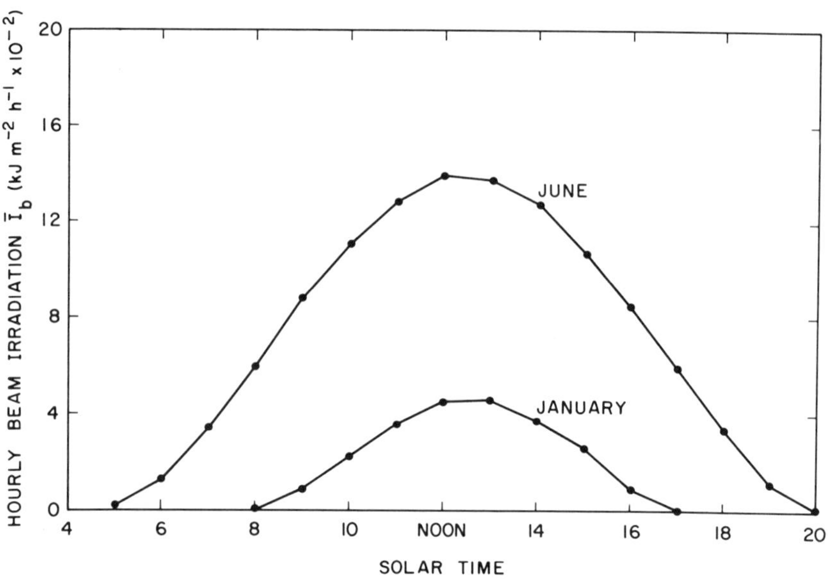

Figure 8.7.3 Diurnal variation of the beam radiation at Montreal (45°30′ N).

The right-hand side of Eq. (8.7.2) is simply a ratio of (4.2.8) to (4.2.20). In most instances $\overline{H}_b$ is not known and $\bar{I}_b$, therefore, cannot be determined. However, the diurnal variation of $\bar{I}_b$ is similar to that of $\bar{I}$, and is shown in Fig. 8.7.3. From an examination of data of several stations, Whillier [24, 25] noted that the correspondence between the measured $\bar{I}/\overline{H}$ values and the right-hand side of (8.7.2), although not perfect, is quite close. Consequently, he modified the theoretical relationship to fit the data. The slight discrepancy between Whillier's theoretical approach and the data occurs because, contrary to his assumption, beam transmittance through a day is generally not constant. For example, for Montreal, the daily variation in beam transmittance is substantial; this is demonstrated in Fig. 8.7.4. Nevertheless, Whillier's assumption points out a fact that does not appear to be obvious: the atmospheric transmittance is not a controlling factor in (8.7.1).

Liu and Jordan [26] extended the day length range of Whillier's graph; this is shown in Fig. 8.7.5. The solid lines in this diagram are a slight modification of Eq. (8.7.2). This graph is valid for the mean values of hour pairs around solar noon. It has been amply verified [27] and is applicable everywhere. For example, for the mean values of hour pairs around solar noon, the Vancouver data correspond very well with the solid lines in Fig. 8.7.5. However, there can be substantial differences between morning and afternoon radiation (Fig. 8.7.6); admittedly, Vancouver is an extreme case.

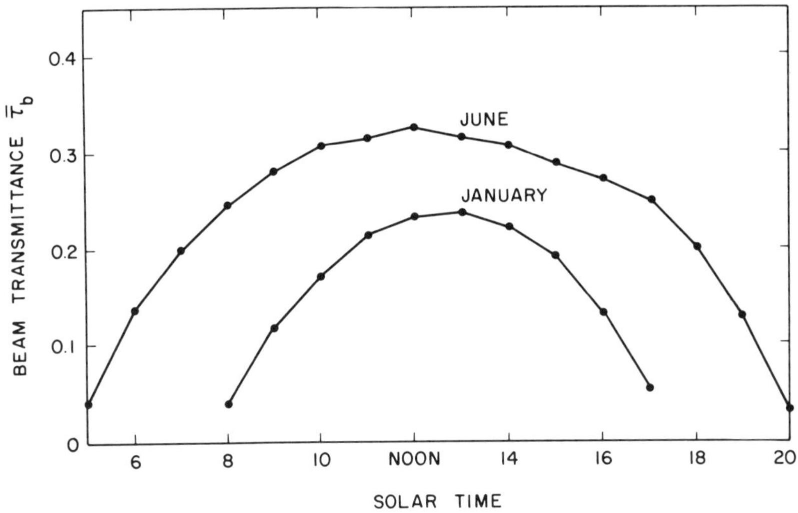

Figure 8.7.4 Diurnal variation of the beam transmittance at Montreal (45°30′ N).

The plots in Fig. 8.7.5 are valid for long-term averages of cloudy weather and for individual days of clear skies. They are plotted in terms of $\frac{1}{2}$, $1\frac{1}{2}$, etc., hours around solar noon. The time is taken at midhour because the measured data are recorded as hour-end values (see Table 8.4.1). Consequently, the plots represent the midhour values.

Collares-Pereira and Rabl [28] developed a mathematical expression for the plots in Fig. 8.7.5. This expression is

$$\bar{r}_t = \bar{I}/\bar{H} = (\bar{I}_0/\bar{H}_0)(a_3 + b_3 \cos \omega_i), \tag{8.7.3}$$

where

$$a_3 = 0.409 + 0.501\,6 \sin(\omega_s - 60°) \tag{8.7.3'}$$

and

$$b_3 = 0.660\,9 - 0.476\,7 \sin(\omega_s - 60°). \tag{8.7.3''}$$

In addition to Fig. 8.7.5, Whillier [29] presented two more diagrams that revealed some interesting quantitative features of solar radiation at any location. These graphs are reproduced in Figs. 8.7.7 and 8.7.8 and can be applied to latitudes ranging from the equator to 50° north and south. The graphs aid in answering such questions as:

How much ratiation would be received during a certain hour?
What is the average solar irradiation at noon?
What is the average solar irradiation at any instant?
How much radiation would be received between certain hours equally spaced around solar noon?

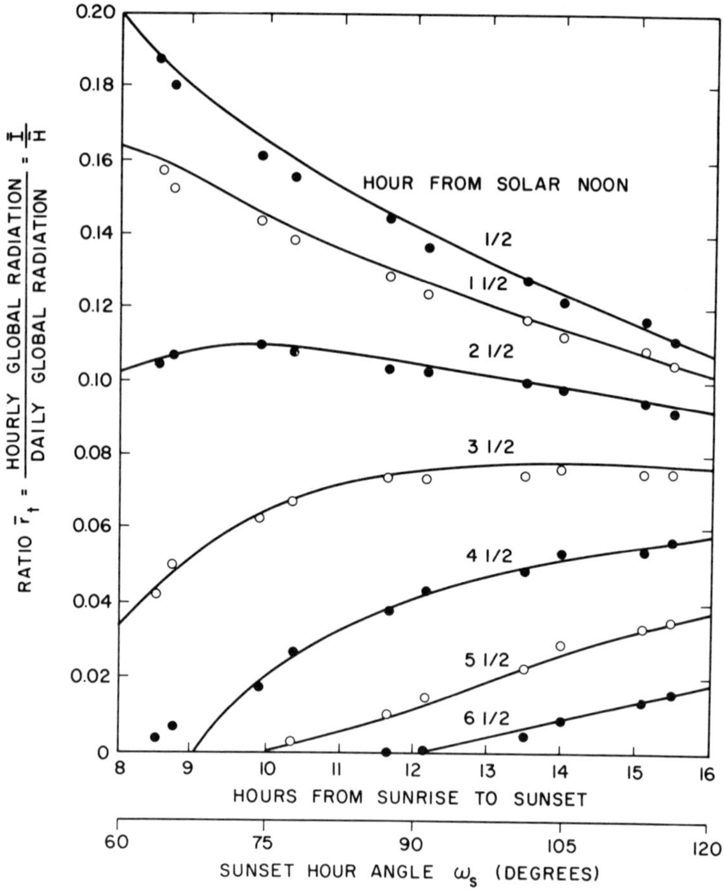

Figure 8.7.5 Distribution of the monthly average hourly global radiation at Vancourver (49°15′ N) for 1959–1975. Adapted with permission from Iqbal [27], copyright 1979, Pergamon Press, Ltd. (Note: these plots are valid all over the world.)

From Whillier's diagrams, it can be demonstrated that, at any time of year and at any location, 90% of the day's radiation exceeds 40% of the noontime radiation. It can also be demonstrated that 90% of the day's radiation is received during the middle two-thirds of the day. This means that for a fixed-position solar energy device, practically all the useful heat gain comes from the middle two-thirds of the day. This observation indicates that the early morning or the late afternoon climatological conditions are of little practical consequence in such applications. The following example illustrates the use of these diagrams.[5]

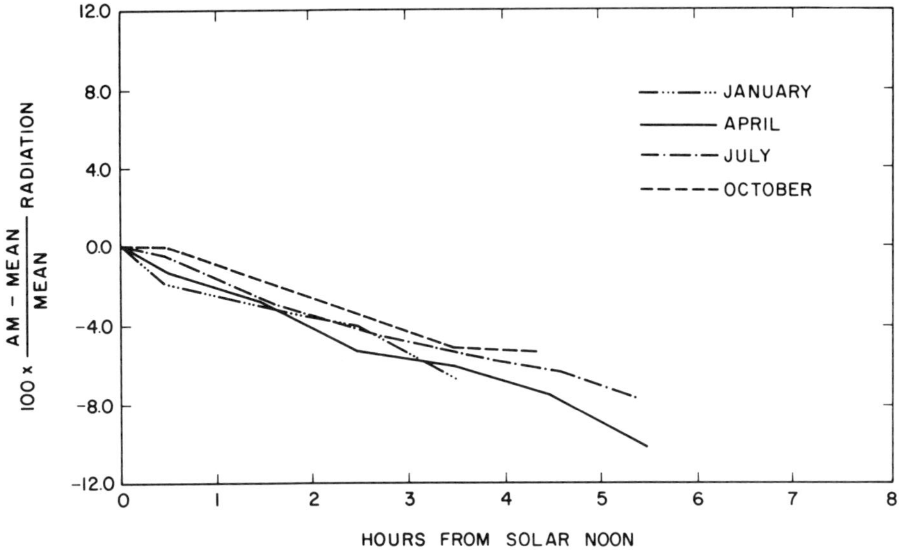

Figure 8.7.6 Hourly variation of the global radiation from its mean value at Vancouver (49°15′ N). Adapted with permission from Iqbal [27], copyright 1979, Pergamon Press, Ltd.

□ EXAMPLE 8.7.1. Prediction of (a) the hourly global radiation received between 10:00 and 11:00, (b) the average global radiation per minute at noon, (c) the average global radiation per minute at 16:15, and (d) the global radiation received between 8:00 and 16:00, during March in Montreal, when the monthly mean daily global radiation on a horizontal surface is 12.5 MJ m^{-2} day^{-1}.

Solution. (a) From Table 4.2.1, for March, $\delta_c = -2.4°$. From Eq. (8.5.3),

$$N_d = \tfrac{2}{15}\cos^{-1}(-\tan\phi\tan\delta) = 11.67 \text{ h}.$$

The global radiation received between 10:00 and 11:00 is equivalent to the hourly radiation at $1\frac{1}{2}$ h from solar noon. From Fig. 8.7.5 we have

$$\bar{r}_t = \bar{I}/\bar{H} = 0.13$$

or

$$\bar{I} = (0.13)12.5 \times 10^3 = 1\,625 \text{ kJ m}^{-2}\text{ h}^{-1}.$$

[5] This example shows a remarkable accuracy of only one of the many pioneering studies conducted by the late Dr. Austin Whillier, an outstanding contributor to the development of solar energy technology. These words are meant to pay homage to my first doctoral thesis advisor.

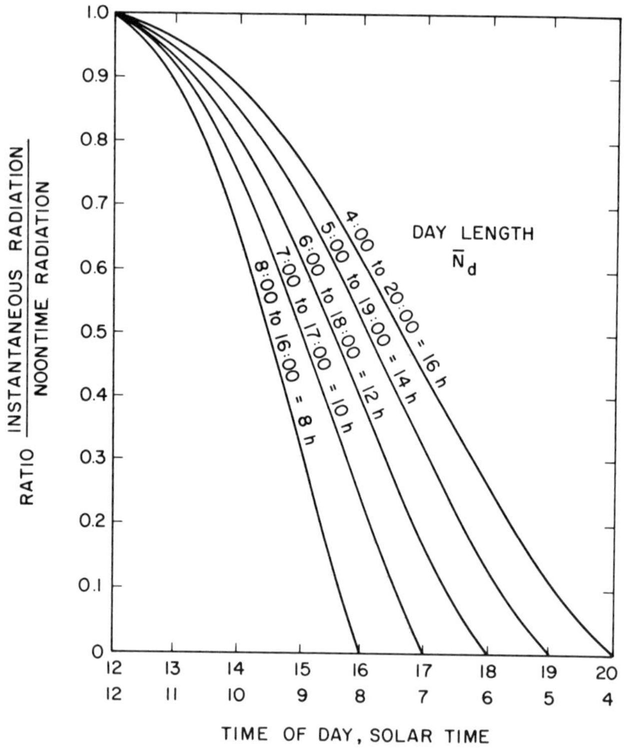

Figure 8.7.7 Variation of the instantaneous radiation as a function of the day length. Adapted with permission from Whillier [29], copyright 1965, Pergamon Press, Ltd.

The hourly global radiation can also be computed from Eq. (8.7.3). For this equation,

$$a_3 = 0.409 + 0.501\,6 \sin(87.56^\circ - 60^\circ) = 0.641\,1$$

and

$$b_3 = 0.660\,9 - 0.476\,7 \sin(87.56^\circ - 60^\circ) = 0.440\,3.$$

$\bar{I}_0$ can be computed from Eq. (4.2.11),

$$\bar{I}_0 = I_{sc}E_0 \cos \delta_c \cos \phi(\cos \omega_i - \cos \omega_s) \quad (4.2.11)$$

$$= 4\,921(1.010\,8) \cos(-2.4) \cos(45.5)[\cos(22.5) - \cos(87.56)]$$

$$= 3\,070 \text{ kJ m}^{-2} \text{ h}^{-1}.$$

$\bar{H}_0$ can be computed from Eq. (4.2.20),

$$\bar{H}_0 = (24/\pi)I_{sc}E_0 \cos \phi \cos \delta_c\{\sin \omega_s - [(\pi/180)\omega_s] \cos \omega_s\} \quad (4.2.20)$$

$$= (24/\pi)(4.921)(1.010\,8) \cos(45.5^\circ) \cos(-2.4^\circ)$$

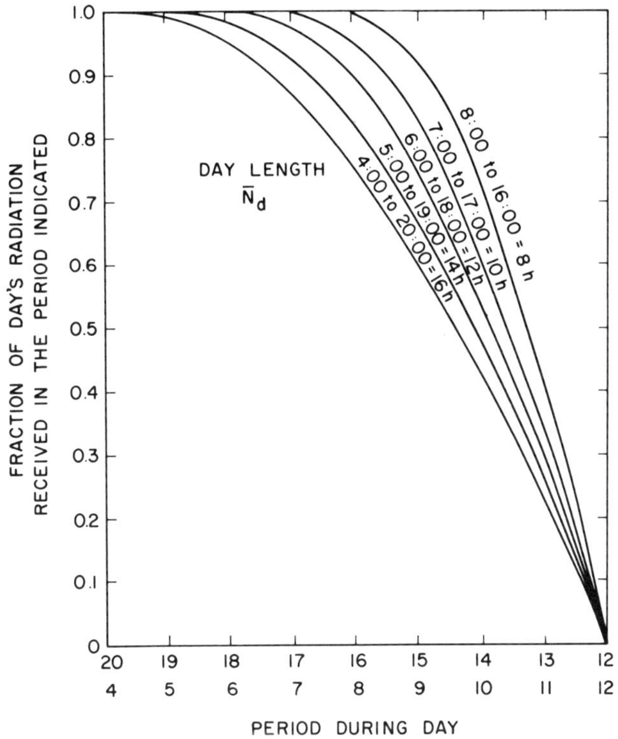

Figure 8.7.8 Variation of the fractional radiation as a function of the day length. Adapted with permission from Whillier [29], copyright 1965, Pergamon Press, Ltd.

$$\times \left[\sin(87.56^\circ) - (\pi/180)(87.56^\circ)\cos(87.56^\circ)\right]$$

$$= 25.6 \text{ MJ m}^{-2}\text{ day}^{-1}.$$

Consequently, we have

$$\begin{aligned}\bar{I} &= (\bar{H}/\bar{H}_0)\bar{I}_0(a_3 + b_3 \cos \omega_i)\\ &= (12.5/25.6)(3\,070)[0.641\,1 + 0.440\,3\cos(22.5^\circ)]\\ &= 1\,635 \text{ kJ m}^{-2}\text{ h}^{-1}.\end{aligned}$$

From Table 8.4.3, the measured value is $\bar{I} = 1\,621$ kJ m^{-2} h^{-1}. It is evident that the values predicted from Fig. 8.7.5 or Eq. (8.7.3) are very close to the measured value.

(b) From Fig. 8.7.7, against time of day 10:30 and $\bar{N}_d = 11.67$ h, we read the following:

$$\frac{\text{Instantaneous radiation at 10:30}}{\text{Noontime radiation}} = 0.87.$$

From (a) the average radiation per minute is

$$(1\,635 \text{ kJ m}^{-2}\text{ h}^{-1})/(60 \text{ min}) = 27.24 \text{ kJ m}^{-2}\text{ min}^{-1}.$$

Therefore the average noontime radiation is

$$27.24/0.87 = 31.3 \text{ kJ m}^{-2}\text{ min}^{-1}$$

(c) Again from Fig. 8.7.7, against time of day 16:15 and $\overline{N}_d = 11.67$ h, we read the following:

$$\frac{\text{Instantaneous radiation at 16:15}}{\text{Noontime radiation}} = 0.3.$$

Therefore the instantaneous radiation at 16:15 hour is

$$0.3(31.3) = 9.4 \text{ kJ m}^{-2}\text{ min}^{-1}.$$

(d) The fraction of radiation received between 8:00 and 16:00 is obtained from Fig. 8.7.8. Against 8-h period and $\overline{N}_d = 11.67$ h we read

$$\text{fraction of day's radiation received} = 0.905.$$

Therefore, the radiation received between 8:00 and 16:00 is

$$0.905(12.5) \text{ MJ m}^{-2} = 11.31 \text{ MJ m}^{-2}.$$

The above can be confirmed from Table 8.4.3 by summing all hour-end values from 9 to 16 h as follows:

$$(922 + 1\,345 + 1\,621 + 1\,781 + 1\,605 + 1\,320 + 918)$$
$$\times 10^{-3} = 11.30 \text{ MJ m}^{-2}.$$

□

8.8 Statistical Distribution of the Daily and Hourly Global Radiation on a Horizontal Surface

For certain applications, it is necessary to determine the statistical distribution of the daily (or hourly) global radiation for a particular location. For example, the frequency of radiation above a certain threshold level can be helpful in sizing thermal storage systems. Liu and Jordan [26] have shown that solar climate at a particular location can be characterized by the clear-

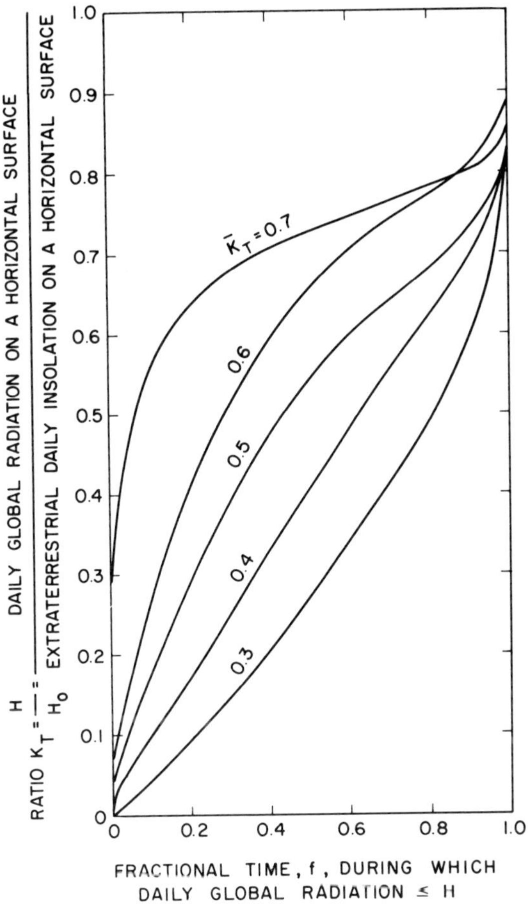

Figure 8.8.1 Frequency distribution of the daily global radiation as a function of the clearness index. Adapted with permission from Liu and Jordan [26], copyright 1960, Pergamon Press, Ltd.

ness parameter $\overline{K}_T$ (which they called the cloudiness index), where

$$\overline{K}_T = \overline{H}/\overline{H}_0. \tag{8.8.1}$$

They have demonstrated that, irrespective of latitude, the fractional time f during which daily global radiation is equal to or less than a certain value is directly dependent on the parameter $\overline{K}_T$. Their generalized curves, based on data from many widely separate locations, are shown in Fig. 8.8.1. In this diagram, K_T refers to an individual day and is defined as

$$K_T = H/H_0. \tag{8.8.2}$$

After verification with data from many locations, it can be concluded that these generalized curves are applicable everywhere. The following example illustrates the use of Fig. 8.8.1.

□ EXAMPLE 8.8.1. Determination of the fractional time f during which, for the month of June (i.e., all Junes of several years) in Montreal, the daily global radiation is less than 689 cal cm^{-2} day^{-1}, which was the daily global radiation on a horizontal surface in Montreal on 9 June 1976 (see Table 8.4.1).

Solution. We first determine $\overline{K}_T = \overline{H}/\overline{H}_0$ for the month of June.

From Table 8.4.3, $\overline{H} = 20.70$ MJ m^{-2} day^{-1} for June. From Table 4.2.1, $\delta_c = 23.04°$ for June. From Eq. (4.2.20) we obtain the following:

$$\overline{H}_0|_{\delta=\delta_c} = 41.78 \text{ MJ m}^{-2} \text{ day}^{-1}.$$

Therefore $\overline{K}_T = 20.70/41.78 = 0.51$.

We now determine $K_T = H/H_0$ for that individual day, 9 June 1976.

$$H = 689 \text{ cal cm}^{-2} \text{ day}^{-1} = 28.84 \text{ MJ m}^{-2} \text{ day}^{-1},$$

$$H_0 = 41.66 \text{ MJ m}^{-2} \text{ day}^{-1}.$$

From Fig. 8.8.1, against $\overline{K}_T = 0.51$ and $K_T = 0.69$, we read the following:

$$f = 85\%.$$

From Table 8.4.1 we count that in June 1976, on 22 days out of 30, the insolation was less than 689 cal cm^{-2} day^{-1}. That is, the observed $f = 22/30 = 73\%$. The discrepancy between 73 and 85% arises from the fact that we employed the June data for only one year, namely, 1976. Had we counted the days of several June months, the result would have been closer to that predicted by the frequency chart. □

The hourly time distribution curves developed by Whillier [25] and Hottel and Whillier [30] are identical to the daily distribution curves (Fig. 8.8.1). The hourly time distribution curves are used to evaluate better the performance of solar collectors whose response to incident radiation is nonlinear.

In the Chapter 11, it is shown that calculation of insolation on inclined planes requires separate values of the direct and diffuse radiation on horizontal surfaces. A quantitative value of diffuse radiation is also required in many practical applications such as agriculture, architecture, climatology, and illumination.

As mentioned earlier, diffuse horizontal radiation is measured at very few stations, whereas global radiation is measured at many locations. In the remaining sections, methods of estimating diffuse horizontal radiation are treated: all the methods are based on measured or estimated values of the global horizontal radiation. The particular cases that will be treated are the estimation of

(a) the daily diffuse radiation H_d,
(b) the monthly average daily diffuse radiation $\bar{H}_d$,
(c) the monthly average hourly diffuse radiation $\bar{I}_d$, and
(d) the hourly diffuse radiation I_d.

Case a will be treated first.

8.9 Estimation of the Daily Diffuse Radiation on a Horizontal Surface

On a cloudy day, the global radiation received is an indication of the extent of cloudiness and should be an indication of the extent of the diffuse radiation. At a location, the parameter K_T, where $K_T = H/H_0$, is an indicator of the daily clearness. Correspondingly, the parameter K, where $K = H_d/H$, gives an indication of the amount of diffuse radiation. We would like to predict the value of H_d from a given value of H on an individual day. Although this is very difficult, it is, however, possible to predict the value of H_d given a certain value of H. Here we are dealing with the averages of the randomly chosen days, and we should not be confused with the monthly averages, which are long-term monthly means.

A correlation between H/H_0 and H_d/H was first developed by Liu and Jordan [26]. They used diffuse radiation data that were not compensated for shadow-band effects. Consequently, the Liu and Jordan correlation has been reexamined by a number of researchers. Two studies worthy of particular note are by Ruth and Chant [31] and Collares-Pereira and Rabl [28]. Both studies concluded that the Liu and Jordan correlation underestimates diffuse radiation. This is of course directly attributed to the negligence of shadow-band effects by Liu and Jordan.

Ruth and Chant utilized radiation data for several years from four stations of the Canadian network. The diffuse-radiation data in Canada are measured by pyranometers fitted with shadow bands and are duly corrected. An analytical expression that fits this correlation [32] is given as follows:

$$\frac{H_d}{H} = \begin{cases} 0.98 & K_T \leq 0.1, \\ 0.910 + 1.154K_T - 4.936K_T^2 + 2.848K_T^3, & 0.1 \leq K_T \leq 0.7. \end{cases} \tag{8.9.1}$$

Collares-Pereira and Rabl utilized pyrheliometer data of a very short period from five stations located in the United States. Their correlation in analytical form is given by

$$\frac{H_d}{H} = \begin{cases} 0.99, & K_T \le 0.17, \\ 1.188 - 2.272K_T + 9.473K_T^2 - 21.856K_T^3 + 14.648K_T^4, & \\ \quad 0.17 \le K_T \le 0.8. & \end{cases} \tag{8.9.2}$$

The two preceding correlations are plotted in Fig. 8.9.1. Liu and Jordan's correlation is also plotted in the same diagram. It is apparent that the Liu and Jordan correlation substantially underestimates diffuse irradiation. It is evident that correspondence between Ruth and Chant's correlation and the Collares-Pereira and Rabl's correlation is very good. The small differences are only in the region described by very cloudy skies, $K_T < 0.3$. Since the two correlations are based on data of different durations and from different instruments, it is debatable which one is the most accurate.

The combined data of the two studies, [28] and [31], spanned stations from 31° to 53° N latitude. The closeness of the two correlations demonstrates that within this range, latitude effects on any one of the two correlations are minimal. At high latitudes, the average air mass is large, and air mass is a very important parameter for irradiance on cloudless days (see Chapters 6 and 7). Because correspondence between the two correlations is exact for partly cloudy and cloudless days, it further strengthens the argument that at least within 31° to 53° N (or south) latitudes, these correlations are independent of latitude effects. However, at very high latitudes, perennial

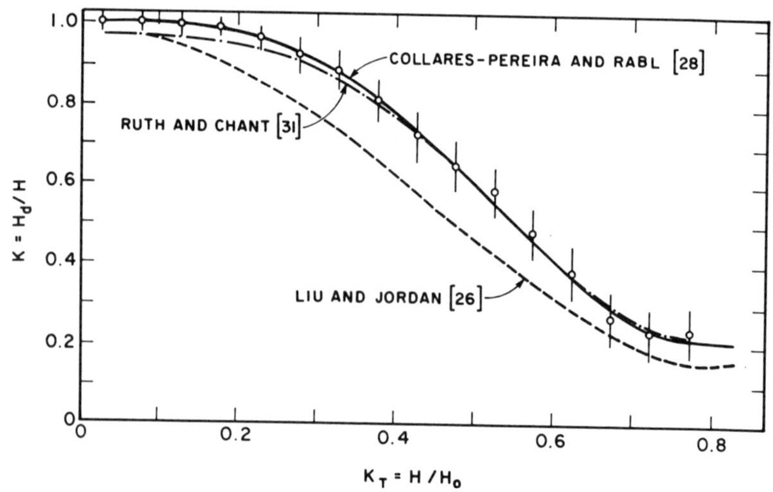

Figure 8.9.1 Variation of the daily diffuse radiation as a function of the clearness index. Adapted with permission from Collares-Pereira and Rabl [28], copyright 1979, Pergamon Press, Ltd.

snow cover could significantly increase diffuse radiation caused by multiple reflections between the earth and cloud cover.

Figure 8.9.1 shows that under very cloudy conditions ($K_T \to 0$), global radiation is composed of mainly diffuse radiation, but on very clear days ($K_T > 0.7$) about 20% of the daily radiation is diffuse. Another way of studying daily diffuse radiation is through the correlation of H/H_0 with H_d/H_0. Such a plot is shown in Fig. 8.9.2 and is more instructive than the previous one. It shows three sky conditions: very cloudy, partly cloudy, and clear skies. During very cloudy days, diffuse radiation is equal to global radiation. During partly cloudy days, diffuse radiation is between 20 and 35% of its extraterrestrial value. On clear days, it is about half of what it is on partly cloudy days.

We now study the limitations of correlations (8.9.1) and (8.9.2) through the following example.

□ EXAMPLE 8.9.1. Prediction of the daily diffuse radiation and comparison of these values with the measured data in Table 8.4.2 for 9 and 11 June 1976, when the daily global radiation on a horizontal surface in Montreal is 689 and 223 cal cm^{-2} day^{-1}, respectively (Table 8.4.1).

Solution. (a) On 9 June 1976

$$H = 689 \text{ cal cm}^{-2}\text{ day}^{-1} = 28.84 \text{ MJ m}^{-2}\text{ day}^{-1},$$

$$H_0 = 41.66 \text{ MJ m}^{-2}\text{ day}^{-1}.$$

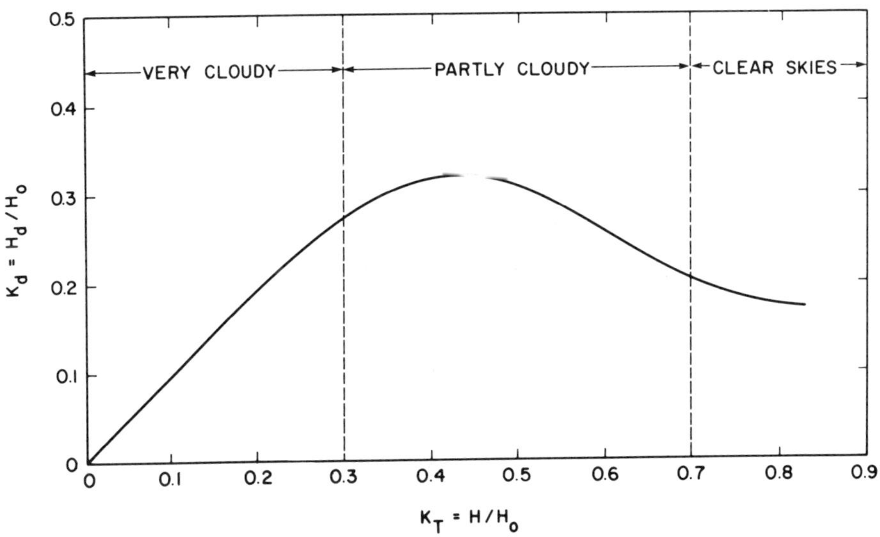

Figure 8.9.2 Variation of the daily diffuse radiation as a function of the clearness index.

Therefore $K_T = H/H_0 = 0.7$, which indicates a clear day.

From Fig. 8.9.1 it is evident that Eqs. (8.9.1) and (8.9.2) should yield identical results. From Eq. (8.9.2) we have

$$H_d/H = 1.188 - 2.272(0.7) + 9.473(0.7)^2 - 21.856(0.7)^3 + 14.648(0.7)^4$$
$$= 0.26.$$

Therefore, $H_d = 7.49$ MJ m^{-2} day^{-1}. From Table 8.4.2, the measured value is

$$H_d = 158 \text{ cal cm}^{-2} \text{ day}^{-1} = 6.61 \text{ MJ m}^{-2} \text{ day}^{-1}.$$

(b) On 11 June 1976

$$H = 223 \text{ cal cm}^{-2} \text{ day}^{-1} = 9.34 \text{ MJ m}^{-2} \text{ day}^{-1},$$
$$H_0 = 41.66 \text{ MJ m}^{-2} \text{ day}^{-1}.$$

Therefore, $K_T = H/H_0 = 0.22$, which indicates a cloudy day.

In this region, the two correlations, Eqs. (8.9.1) and (8.9.2), differ slightly. Let us use Eq. (8.9.2) again; we obtain

$$H_d/H = 0.94,$$

or

$$H_d = 8.79 \text{ MJ m}^{-2} \text{ day}^{-1}.$$

From Table 8.4.2 the measured value is

$$H_d = 177 \text{ cal cm}^{-2} \text{ day}^{-1}$$
$$= 7.41 \text{ MJ m}^{-2} \text{ day}^{-1}.$$

Referring back to Fig. 8.9.1 we note that the Ruth and Chant correlation will predict a slightly lower value of H_d compared to that obtained above and consequently will be closer to the measured value. □

8.10 Estimation of the Monthly Average Daily Diffuse Radiation on a Horizontal Surface

Solar energy processes are basically transient. However, a period of an hour is considered small enough to represent such transience. Therefore, the deduction of hourly radiation (monthly mean or for individual hours) is generally the ultimate aim. The topic treated in the previous section, estimation of the daily diffuse radiation, is of little value since it does not eventually lead to the deduction of its hourly component (treated in Section 8.12).

In this section, estimation of the monthly average daily diffuse radiation from known values of the monthly average daily global radiation is discussed. A number of approaches are followed:

(1) the method of Liu and Jordan,
(2) Page's linear regression correlation, and
(3) estimation through the number of bright sunshine hours.

In obtaining long-term averages, it is usual to take periods of calendar months. However, shorter periods such as 10-day periods have also been used.

A. *The Method of Liu and Jordan* [26]

In this method, use is made of the generalized distribution curves in Fig. 8.8.1 and the correlation between H_d/H and H/H_0 developed by Liu and Jordan, which is shown in Fig. 8.9.1. For illustration of the method, consider Fig. 8.10.1. This diagram has two sketches: H/H_0 vs. frequency and H_d/H vs. H/H_0.

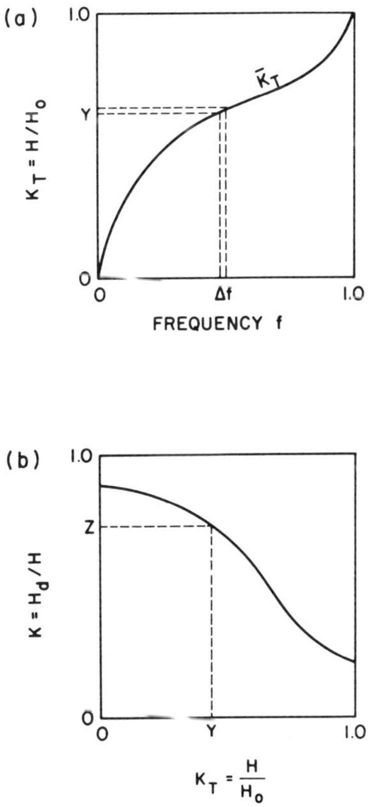

Figure 8.10.1 Development of the correlation for the monthly average daily diffuse radiation. (a) Frequency curve, (b) daily diffuse correlation curve.

In the previous section, $\bar{K}_T$ has been defined as follows

$$\bar{K}_T = \bar{H}/\bar{H}_0. \tag{8.8.1}$$

From Fig. 8.10.1a, $\bar{K}_T$ can also be written thus:

$$\bar{K}_T = \int_{f=0}^{f=1} \frac{H}{H_0}\, df = \int_{f=0}^{f=1} K_T\, df. \tag{8.10.1}$$

Now, let us define another parameter, $\bar{K}_d$:

$$\bar{K}_d = \bar{H}_d/\bar{H}_0. \tag{8.10.2}$$

As above, by definition

$$\bar{K}_d = \int_{f=0}^{f=1} \frac{H_d}{H_0}\, df = \int_{f=0}^{f=1} \frac{H_d}{H}\frac{H}{H_0}\, df \tag{8.10.3}$$

$$= \sum ZY\, \Delta f \tag{8.10.4}$$

where Z, Y, and Δf are marked on Fig. 8.10.1. From above, against a fixed value of $\bar{K}_T$, a corresponding value of $\bar{K}_d$ can be obtained by integrating (8.10.4). Once $\bar{K}_d$ is evaluated for a given value of $\bar{K}_T$, $\bar{H}_d/\bar{H}$ is obtained by definition. Thus

$$\bar{K}_d/\bar{K}_T = \bar{H}_d/\bar{H}. \tag{8.10.5}$$

The Liu and Jordan relationship between $\bar{K}_T$ and $\bar{H}_d/\bar{H}$ is plotted in Fig. 8.10.2 and may be expressed algebraically as follows:

$$\bar{H}_d/\bar{H} = 1.39 - 4.027\bar{K}_T + 5.531\bar{K}_T^2 - 3.108\bar{K}_T^3, \quad 0.3 < \bar{K}_T < 0.7. \tag{8.10.6}$$

One aspect of Liu and Jordan's approach is particularly important. The distribution curves (Fig. 8.8.1), with a substantial range of $\bar{K}_T$ values (0.3–0.7), were based on global radiation data from several stations. The diffuse-radiation data, however, were obtained from only one station, where $\bar{K}_T$ ranged from 0.4 to 0.5. The method, nevertheless, permitted development of a relationship between $\bar{H}_d/\bar{H}$ and $\bar{K}_T$ with a large range of $\bar{K}_T$ indicated in Eq. (8.10.6). Klein and Duffie [33] have studied the generality of Liu and Jordan's method. With the correlations for the daily diffuse radiation based on data from India, Israel, Australia, and Canada, they recalculated the relationship between $\bar{H}_d/\bar{H}$ and $\bar{K}_T$. Their recalculated results did not yield a single (universal) correlation. This leads to a belief that individual correlations are required for different climatic and geographic regions.

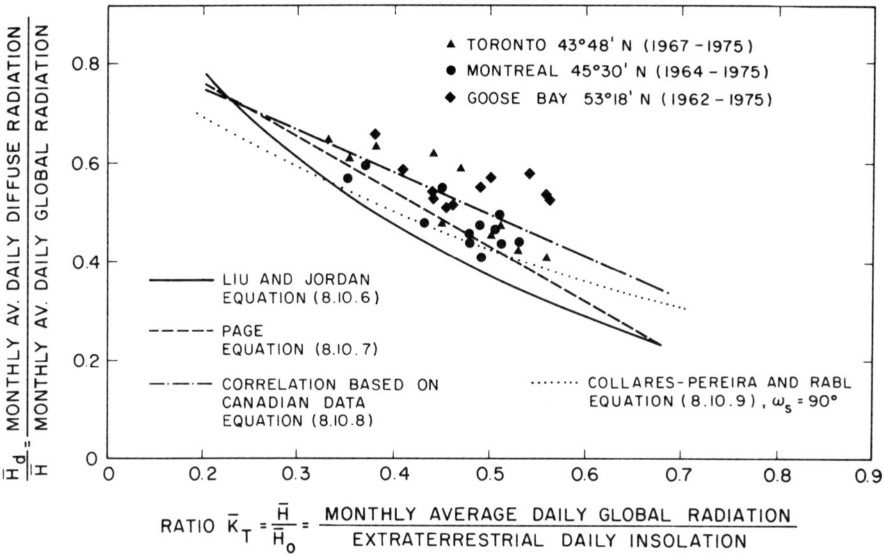

Figure 8.10.2 Ratio of the monthly average daily diffuse radiation to the monthly average daily global radiation. Adapted with permission from Iqbal [32], copyright 1979, Pergamon Press, Ltd.

B. *The Method of Page* [34]

Page proposed to develop a correlation based on direct regression analysis of data. Using data from ten locations situated between 40° N and 40° S latitudes, he recommended the following linear equation:

$$\overline{H}_d/\overline{H} = 1.00 - 1.13\overline{K}_T. \qquad (8.10.7)$$

Iqbal [32] proposed a slightly different correlation using Canadian data, which were corrected for shadow-band effects. This correlation

$$\overline{H}_d/\overline{H} = 0.958 - 0.982\overline{K}_T, \qquad 0.3 < \overline{K}_T < 0.6, \qquad (8.10.8)$$

is based on the NASA design value of the solar constant, 1353 W m^{-2}, whereas Page used an older value of this constant. Further, in Eq. (8.10.8), $\overline{H}_0$ in $\overline{K}_T$ was computed as a true average of the daily sums.

Collares-Pereira and Rabl [28] developed a correlation whose coefficients vary with season. Their correlation, based on data from the United States, is

$$\begin{aligned}\overline{H}_d/\overline{H} &= 0.775 + 0.347(\pi/180)(\omega_s - 90)\\ &\quad - [0.505 + 0.261(\pi/180)(\omega_s - 90)]\cos[2(\overline{K}_T - 0.9)], \qquad (8.10.9)\end{aligned}$$

where ω_s is the sunset-hour angle in degrees and is equal to 90° from February to April and from August to October, 100° from May to July, and 80° from November to January and where $2(\bar{K}_T - 0.9)$ is in radians.

Equations (8.10.6)–(8.10.9) are plotted in Fig. 8.10.2 and confirm Duffie and Klein's conclusion that further work is necessary before a universal correlation can be developed.

C. *Correlation of Average Diffuse Radiation with Number of Bright Sunshine Hours*

When the sky is completely covered by clouds, the number of bright sunshine hours recorded will be nil. Under such a condition, all radiation received on the earth will be of a diffuse nature. On the other hand, on a completely cloudless day, a certain number of bright sunshine hours will be recorded, and diffuse radiation will be quite minimal. Therefore, one suspects a correlation between the diffuse radiation and the number of bright sunshine hours.

A linear correlation of the type

$$\bar{H}_d/\bar{H} = C_3 - C_4(\bar{n}/\bar{N}_d) \tag{8.10.10}$$

was proposed by Iqbal [35]. In this equation, the coefficients C_3 and C_4 are obtained through regression analysis of the measured data. The symbol $\bar{n}$ is the average number of bright sunshine hours per day and $\bar{N}_d$ is the average day length [Eq. (8.5.3)]. Based on data from three Canadian stations, the following correlation was obtained:

$$\bar{H}_d/\bar{H} = 0.791 - 0.635(\bar{n}/\bar{N}_d). \tag{8.10.11}$$

All the foregoing methods of estimation require an *a priori* value of the global horizontal radiation $\bar{H}$. It would, therefore, be useful to develop a method in which this restriction is removed. This is made possible by multiplying Eq. (8.5.5) by Eq. (8.10.10), yielding

$$\frac{\bar{H}_d}{\bar{H}_0} = C_5 + C_6\frac{\bar{n}}{\bar{N}_d} + C_7\left(\frac{\bar{n}}{\bar{N}_d}\right)^2. \tag{8.10.12}$$

In the above equation, the coefficients C_5, C_6, and C_7 should be obtained through regression analysis using actual data and not by algebraic manipulation of the constants in Eqs. (8.5.5) and (8.10.10). Using the Canadian data, Iqbal [35] recommended the following correlation:

$$\frac{\bar{H}_d}{\bar{H}_0} = 0.163 + 0.478\frac{\bar{n}}{\bar{N}_d} - 0.655\left(\frac{\bar{n}}{\bar{N}_d}\right)^2, \tag{8.10.13}$$

where the diffuse-radiation data from three widely spread locations were utilized.

All the previous predictive correlations are essentially site specific. That is, they are valid only for the location or for the region where the clearness parameter or the sunshine records were obtained. It is known that the clearness parameter or the number of bright sunshine hours does not adequately represent all the climatological factors. Optical properties of the sky and the ground cover are also important. Higher latitudes are generally characterized by larger air mass and larger ground albedo; both of them increase the ratio of diffuse to global radiation. It is difficult to separate the effect of latitude from other parameters. Nevertheless, from Fig. 8.10.3 (which is based on the data in Table 8.4.4) we see that although latitude is not the most dominant parameter, its underlying effect is obvious. On the other hand, local climate can have a more dominant effect than latitude. This can be observed by comparing the plot of Montreal with that of Carpentras (in sunny and warm southern France), both of which are at about the same latitude: we find that the latter has a substantially smaller fraction of diffuse radiation.

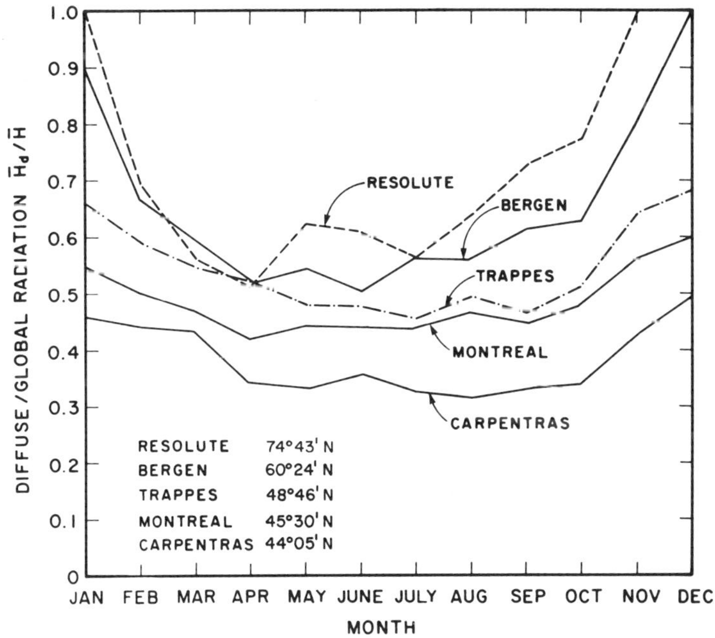

Figure 8.10.3 Annual variation of the ratio of the daily diffuse radiation to the daily global radiation for different stations.

Hay [18, 36] generalized the procedure by considering radiation before and after multiple reflections between the earth and cloud cover. Let us assume $\overline{H}'$ and $\overline{H}'_{\mathrm{d}}$ are the global and diffuse radiation emerging from the atmosphere before striking the ground. After multiple reflections, both these quantities will increase in magnitude. However, this increase in magnitude will depend on the albedo of the earth, of the clouds, and of the cloudless portion of the sky. From Hay, the relationship between the radiation before and after multiple reflections is given below:

$$\overline{H} - \overline{H}' = \overline{H}\rho[\rho_{\mathrm{a}}(\bar{n}/\overline{N}_j) + \rho_{\mathrm{c}}(1 - \bar{n}/\overline{N}_j)] \tag{8.10.14a}$$

and also

$$\overline{H}_{\mathrm{d}} - \overline{H}'_{\mathrm{d}} = \overline{H}\rho[\rho_{\mathrm{a}}(\bar{n}/\overline{N}_j) + \rho_{\mathrm{c}}(1 - \bar{n}/\overline{N}_j)], \tag{8.10.14b}$$

where ρ is ground albedo, ρ_{c} is cloud albedo, and ρ_{a} is clear-sky albedo. The quantities $\bar{n}$ and $\overline{N}_j$ have already been defined in Section 8.5. Assuming $\rho_{\mathrm{c}} = 0.6$, $\rho_{\mathrm{a}} = 0.25$, and using actual data from a number of Canadian stations, Hay proposed the following expression, which correlates diffuse and global radiation before multiple reflections:

$$\frac{\overline{H}'_{\mathrm{d}}}{\overline{H}'} = 0.970\,2 + 1.668\,8\frac{\overline{H}'}{\overline{H}_0} - 21.303\left(\frac{\overline{H}'}{\overline{H}_0}\right)^2 + 51.288\left(\frac{\overline{H}'}{\overline{H}_0}\right)^3 - 50.081\left(\frac{\overline{H}'}{\overline{H}_0}\right)^4 + 17.551\left(\frac{\overline{H}'}{\overline{H}_0}\right)^5. \tag{8.10.14c}$$

The calculation procedure is the following: from a given $\overline{H}$, $\overline{H}'$ is obtained from (8.10.14a), $\overline{H}'_{\mathrm{d}}$ from (8.10.14c), and finally $\overline{H}_{\mathrm{d}}$ from (8.10.14b).

The following detailed example presents a comparison between the various correlations treated in this section.

□ EXAMPLE 8.10.1. Prediction of the yearly variation of the monthly average daily diffuse radiation on a horizontal surface in Montreal and comparison of the results with the measured values, employing all correlations discussed in this section; results are presented in tabular form and illustrated graphically.

Solution. Montreal is 45°30′ N. $\overline{H}$ values are obtained from Table 8.4.3. $\overline{H}_0|_{\delta=\delta_c}$ values are calculated from Eq. (4.2.18). The sunshine hours for Montreal are listed in Table 8.5.1.

Table 8.10.1 lists the predicted values, which are plotted in Fig. 8.10.4. The percentage differences between the measured and the predicted results, the RMSEs, and the MBEs are given in Table 8.10.2. From an examination of this table it is obvious that the error from the Liu and Jordan correlation is consistently high. Among the remaining equations, the correlation (8.10.11)

Table 8.10.1

Predicted Values of the Monthly Average Daily Diffuse Radiation on a Horizontal Surface in Montreal[a]

Month	$\overline{H}$ (MJ m^{-2} d^{-1})	$\bar{n}/\overline{N}_d$ [b]	$\overline{K}_T$	Predicted $\overline{H}_d$ (MJ m^{-2} day^{-1}), from Eq. (8.10.6)	(8.10.7)	(8.10.8)	(8.10.9)	(8.10.11)	(8.10.13)	(8.10.14)	Measured
Jan	5.27	0.35	0.45	2.00	2.35	2.51	2.11	3.00	2.67	2.49	2.89
Feb	8.73	0.43	0.52	3.03	3.50	3.82	3.51	4.52	4.05	3.82	4.39
Mar	12.50	0.47	0.51	4.74	5.58	5.96	5.38	6.16	6.15	5.34	5.88
Apr	15.81	0.47	0.48	5.99	7.06	7.54	6.81	7.79	7.88	6.62	6.61
May	18.54	0.53	0.48	7.52	8.90	9.39	8.99	8.43	9.31	7.92	8.19
June	20.70	0.54	0.51	7.51	8.77	9.46	9.19	9.28	9.52	8.14	9.19
Jul	21.19	0.56	0.53	7.86	9.22	9.90	9.58	9.23	9.58	8.33	9.26
Aug	17.18	0.54	0.49	6.51	7.67	8.19	7.40	7.70	8.13	6.95	8.01
Sep	13.22	0.53	0.48	5.36	6.35	6.69	5.99	6.01	6.68	5.68	5.88
Oct	8.26	0.45	0.43	3.75	4.43	4.59	4.08	4.17	4.98	4.13	3.96
Nov	4.47	0.27	0.35	2.29	2.65	2.70	2.23	2.77	3.11	2.70	2.56
Dec	3.78	0.30	0.37	1.89	2.20	2.25	1.85	2.27	2.56	2.28	2.28

[a] See Example 8.10.1.
[b] For Eq. (8.10.14), $\bar{n}/\overline{N}_j$ was employed.

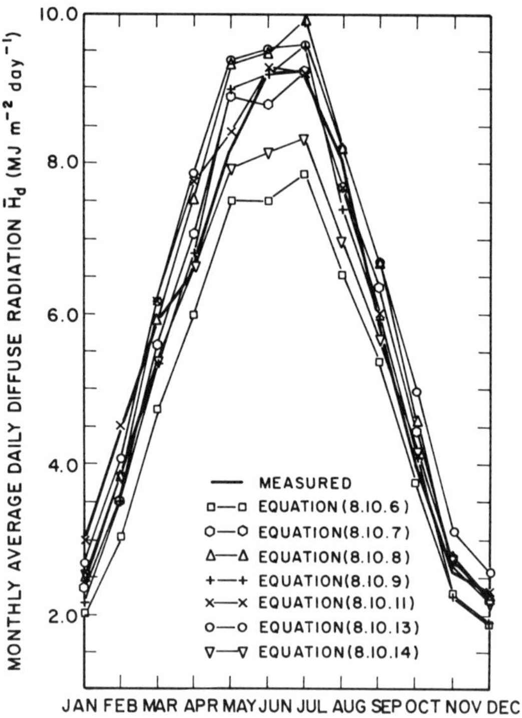

Figure 8.10.4 Comparison of the various correlations to predict monthly average daily diffuse radiation at Montreal (45°30′ N).

has the lowest RMSE. A slight disadvantage with (8.10.11) is that the global radiation $\bar{H}$ and the number of bright sunshine hours $\bar{n}$ are both required to compute $\bar{H}_d$. In some instances, when the bright sunshine records may not be readily available, recourse must be made to one of Eqs. (8.10.7)–(8.10.9). In such instances, this author recommends Page's correlation, Eq. (8.10.7). This recommendation is based on solutions of similar examples carried out with data from several stations located in Canada and Europe (see footnote 4). However, Page's correlation is not always superior to the other correlations. □

Although it is correct that the Liu and Jordan correlation generally underestimates diffuse radiation, the congruence with the measured data of the remaining correlations varies with the month. Some of the reasons for the differences are as follows.

(a) The clearness parameter $\bar{K}_T$ and the number of bright sunshine hours do not describe all the climatic variables, such as the optical properties of clouds and ground albedo.

Table 8.10.2

Percentage Difference between the Measured and Predicted Results and a Comparison of the Root-Mean-Square Error and Mean Bias Error[a]

Month	(8.10.6)	(8.10.7)	(8.10.8)	(8.10.9)	(8.10.11)	(8.10.13)	(8.10.14)
Jan	30.87	18.62	13.05	26.88	3.71	7.68	13.68
Feb	31.07	20.24	12.99	20.04	3.00	7.79	13.00
Mar	19.41	5.12	1.36	8.49	4.71	4.63	9.23
Apr	9.32	6.75	14.05	2.96	17.81	19.28	0.22
May	8.22	8.70	14.61	9.74	2.88	13.68	3.28
Jun	18.33	4.56	2.98	0.05	0.93	3.59	11.47
Jul	15.16	0.46	6.87	3.43	0.37	3.49	10.02
Aug	18.69	4.28	2.27	7.67	3.89	1.53	13.17
Sep	8.84	7.96	13.83	1.90	2.17	13.61	3.37
Oct	5.20	11.95	15.84	2.98	5.39	25.77	4.37
Nov	10.58	3.58	5.55	13.06	8.18	21.58	5.53
Dec	17.14	3.53	1.41	18.79	9.44	12.22	0.22
RMSE	1.006	0.417	0.554	0.426	0.386	0.666	0.620
MBE	−0.855	−0.003	0.355	−0.143	0.211	0.489	−0.371

[a] See Example 8.10.1.

(b) Not all authors utilized diffuse-radiation data corrected for shadow-band effects.

(c) Different values of the solar constant were used.

(d) The method of calculating $\overline{H}_0$ varied slightly from author to author.

It seems necessary, however, to point out that all correlations that are based on measured values of $\overline{H}_d$ and $\overline{H}$ naturally include multiply reflected diffuse radiation. A pyranometer, with or without a shadow band, cannot avoid the multiply reflected diffuse radiation in its reading. Any correlation that separates the multiply reflected diffuse radiation from the rest is not inherently superior to other types of correlations, and this has been demonstrated by the results in Example 8.10.1.

8.11 Estimation of the Monthly Average Hourly Diffuse Radiation on a Horizontal Surface

Many of the transient processes in solar energy applications can be approximated by using long-term averages of hourly radiation. Estimation of the average hourly global radiation from the average daily global radiation has been already treated in Section 8.7. Here two methods of estimating average hourly diffuse radiation are discussed: Liu and Jordan's method and Hay's method.

A. *Liu and Jordan's Method* [26]

This method has essentially the same approach as that followed by Whillier (Section 8.7) in estimating the average hourly global radiation. Consider Fig. 8.11.1, where the daily variation in $\bar{I}_d$ is plotted for Montreal. The area under the curve represents $\bar{H}_d$. Comparing Fig. 8.11.1 with Fig. 8.7.2 and assuming that the ratio of hourly to daily radiation in the two cases is identical, we have the following:

$$\bar{I}_d/\bar{H}_d = \bar{I}_0/\bar{H}_0. \tag{8.11.1}$$

The right-hand side of the above equation is identical to the right-hand side of Eq. (8.7.2). In the numerator of (8.7.2),

$$(24/\pi)\sin(\pi/24) = 0.997\,2,$$

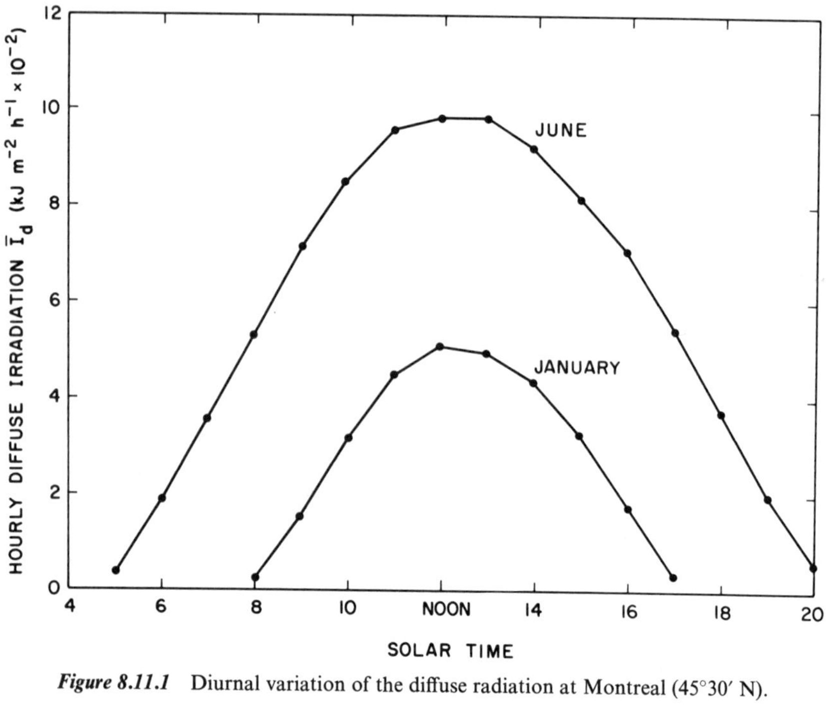

Figure 8.11.1 Diurnal variation of the diffuse radiation at Montreal (45°30′ N).

and by assuming it equal to 1, (8.11.1) can be rewritten as below:

$$\frac{\bar{I}_d}{\bar{H}_d} = \frac{\pi}{24} \frac{\cos \omega_i - \cos \omega_s}{\sin \omega_s - \omega_s(\pi/180) \cos \omega_s}. \tag{8.11.2}$$

The above equation holds very well under verification with measured data. Figure 8.11.2 demonstrates the verification of Eq. (8.11.2) where the solid lines are obtained from the right-hand side of this equation. The measured data in this diagram represent mean values of the hour pairs around solar noon. Under cloudy skies there are always some asymmetries; see Fig. 8.11.3.

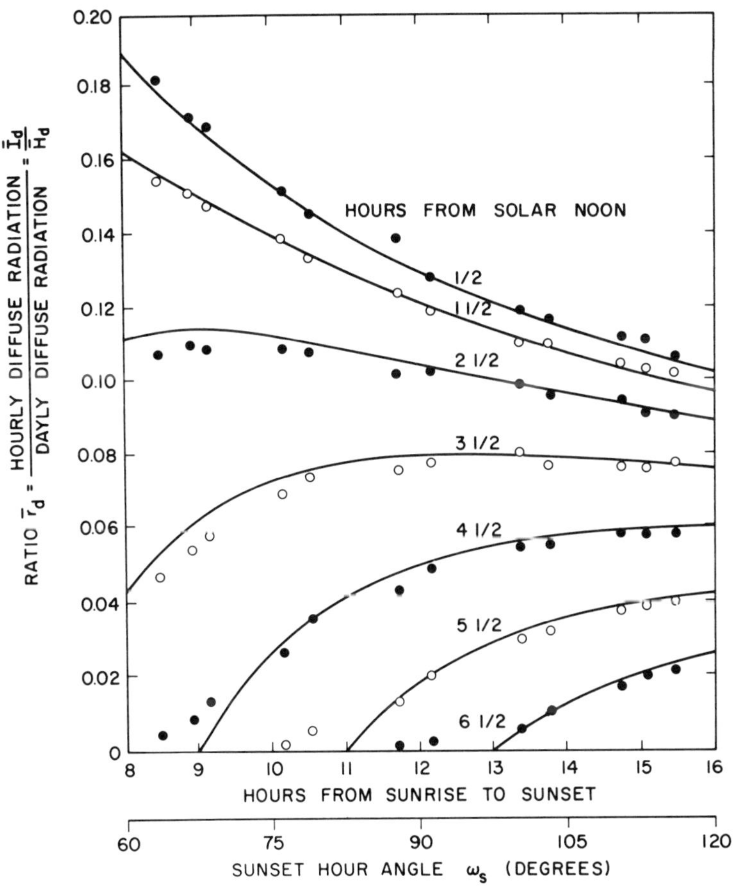

Figure 8.11.2 Distribution of the monthly average hourly diffuse radiation at Montreal (45°30′ N) for 1964–1975. Adapted with permission from Iqbal [27], copyright 1979, Pergamon Press, Ltd. (Note: these plots are valid all over the world.)

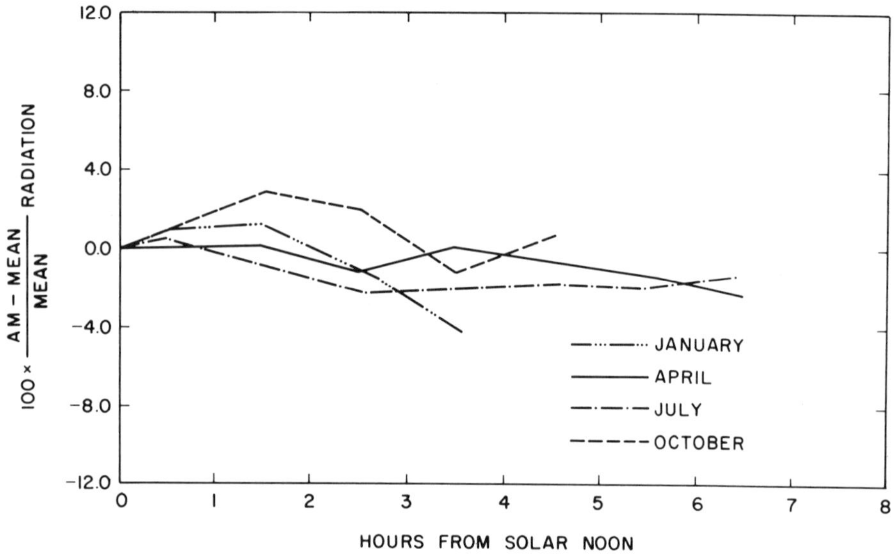

Figure 8.11.3 Hourly variation of the diffuse radiation from its mean value at Montreal (45°30′ N). Adapted with permission from Iqbal [27], copyright 1979, Pergamon Press, Ltd.

Recapitulating the preceding sections, it may be said that given a measured (or estimated) value of the average global radiation $\overline{H}$, its hourly components $\bar{I}$ and $\bar{I}_{\mathrm{d}}$ can be estimated. However, the resulting hourly values are symmetric around solar noon.

B. *Hay's Method* [18, 36]

Hay's method of estimating mean hourly diffuse radiation from the mean hourly global radiation is identical to his method of predicting mean daily diffuse radiation described in Eq. (8.10.14). For the sake of completeness, we rewrite this equation replacing the terms representing daily radiation with their hourly counterparts. Consequently, we have

$$\bar{I} - \bar{I}' = \bar{I}\rho[\rho_{\mathrm{a}}(\bar{n}/\overline{N}_j) + \rho_{\mathrm{c}}(1 - \bar{n}/\overline{N}_j)] \tag{8.11.3a}$$

and also

$$\bar{I}_{\mathrm{d}} - \bar{I}_{\mathrm{d}}' = \bar{I}\rho[\rho_{\mathrm{a}}(\bar{n}/\overline{N}_j) + \rho_{\mathrm{c}}(1 - \bar{n}/\overline{N}_j)], \tag{8.11.3b}$$

where $\rho_{\mathrm{a}} = 0.25$, $\rho_{\mathrm{c}} = 0.6$, and $\bar{I}'$ and $\bar{I}_{\mathrm{d}}'$ are the global and diffuse hourly irradiation before multiple reflections between the ground and the sky. The relationship between these two quantities is written

$$\frac{\bar{I}'_d}{\bar{I}'} = 0.970\,2 + 1.668\,8\frac{\bar{I}'}{\bar{I}_0} - 21.303\left(\frac{\bar{I}'}{\bar{I}_0}\right)^2 + 51.288\left(\frac{\bar{I}'}{\bar{I}_0}\right)^3 - 50.081\left(\frac{\bar{I}'}{\bar{I}_0}\right)^4 + 17.551\left(\frac{\bar{I}'}{\bar{I}_0}\right)^5. \quad (8.11.3c)$$

To predict $\bar{I}_d$, $\bar{I}'$ is estimated from (8.11.3a) with known values of $\bar{I}$. Then $\bar{I}'_d$ is obtained from (8.11.3c) and finally $\bar{I}_d$ from (8.11.3b).

Hay's method is particularly useful when measured values of $\bar{I}$ are available. Since the hourly radiation is rarely, if ever, symmetric around solar noon, this method brings out such asymmetries of the diffuse radiation. However, in most practical applications in engineering and architecture, the effects of the asymmetry of solar radiation around solar noon are of very minor importance. Consequently, the procedure of Whillier/Liu and Jordan appears adequate and quite accurate as far as the prediction of hourly global and diffuse radiation is concerned. This is particularly so considering that measured values of hourly global radiation are generally not readily available; the same applies to ground albedo. Also, the cloud albedo varies with the type, thickness, and number of cloud layers. Moreover, as far as the albedo of the cloudless sky is concerned, we have already seen in Chapter 7 that it too is a strong function of the aerosols in the atmosphere. Many of these parameters are not easily available to the users of such correlations. Nevertheless, Hay's study does lead to a general understanding that development of accurate correlations should include many environmental parameters.

☐ EXAMPLE 8.11.1. Calculation of the corresponding hourly diffuse radiation during April at hour ending 14:00 (LAT) by (a) Liu and Jordan's method and, (b) Hay's method (when the measured hourly global radiation is 1 827 kJ m^{-2} h^{-1}) when the monthly average daily diffuse radiation on a horizontal surface in Montreal is 6.61 MJ m^{-2} day^{-1}.

Solution. (a) Liu and Jordan's method.

For this method, we employ Eq. (8.11.2). For the hour ending at 14:00, $\omega_i = -22.5°$. In April, $\delta_c = 9.46°$. From Eq. (1.5.4), $\omega_s = 99.71°$. Therefore

$$\frac{\bar{I}_d}{6\,610} = \frac{\pi}{24}\frac{\cos(-22.5) - \cos(99.71)}{\sin(99.71) - [2\pi(99.71)/360]\cos(99.71)},$$

or $\bar{I}_d = 740$ kJ m^{-2} h^{-1}.

(b) Hay's method.

For this method, we employ Eq. (8.11.3). From Table (8.5.1), $\bar{n} = 188/30 = 6.27$. From Table (9.4.4), $\rho = 0.22$.

$$\overline{N}_j = \frac{1}{75}\cos^{-1}\frac{\cos 85 - \sin(45.5)\sin(9.46)}{\cos(45.5)\cos(9.46)}$$

$$= 12.33 \text{ h}.$$

From Eq. (8.11.3a),

$$\bar{I}' = 1\,827\{1 - 0.22[0.25(6.27/12.33) + 0.60(1 - 6.27/12.33)]\}$$
$$= 1\,657 \text{ kJ m}^{-2}\text{ h}^{-1}.$$

From Eq. (4.2.9), $\bar{I}_0 = 3\,693$ kJ m^{-2} h^{-1}. Therefore $\bar{I}'/\bar{I}_0 = 0.45$.
From Eq. (8.11.3c), we obtain

$$\bar{I}'_d/\bar{I}' = 0.9702 + 1.668\,8(0.45) - 21.303(0.45)^2 + 51.288(0.45)^3 - 50.081(0.45)^4 + 17.551(0.45)^5$$
$$= 0.35.$$

Therefore, $\bar{I}'_d = 0.35(1\,657) = 580$ kJ m^{-2} h^{-1}.
From Eq. (8.11.3b) we finally obtain $\bar{I}_d$:

$$\bar{I}_d = 580 + (1\,827 - 1\,657) = 750 \text{ kJ m}^{-2}.$$

From Table 8.4.3, the measured value is 729 kJ m^{-2} h^{-1}. It is apparent that Liu and Jordan's correlation gives very accurate results in this particular instance. In fact, their method is very accurate and in addition is very simple. □

8.12 Estimation of the Hourly Diffuse Radiation on a Horizontal Surface

The foregoing methods dealing with the mean values of the daily and the hourly radiation are very useful tools for engineers and architects. However, for research purposes and, for instance, mathematical simulations of solar energy processes, values of global and diffuse radiation for individual hours are needed. In Section 8.9, estimation of the daily diffuse radiation H_d was treated. Unfortunately, a knowledge of H_d does not lead further to its hourly distribution. In this section, estimation of I_d from measured values of I is studied.

The hourly global radiation on horizontal surfaces is now recorded at many stations in the industrialized world. These records are mostly available on magnetic tapes in machine-compatible form. Although the hourly global radiation is recorded at many places, stations measuring hourly diffuse radiation are very few. It is therefore necessary to develop methods of predicting the diffuse component of the hourly global radiation.

The amount of diffuse radiation depends on both the solar altitude and the cloud cover. Under clear-sky conditions and for a given atmosphere, solar altitude is the governing parameter. The effect of cloud cover increases as the amount of sky covered by clouds increases and also depends on the geometric distribution of the clouds with respect to the solar beam. The problem appears to be complicated. However, it is possible to develop a statistical correlation between I and I_d along the lines of that Liu and Jordan [26] developed between H and H_d (Section 8.9).

Along the lines of the daily ratio, let us define the new hourly ratio as

$$M_T = I/I_0. \tag{8.12.1}$$

It seems that, in general, the ratio M_T is indicative of the clearness of the sky. As such, the variation of I_d/I with M_T should be of a similar nature to that of H_d/H with K_T, except that in the present case the solar altitude should be an additional parameter. Because of this expected similarity, a number of researchers have developed correlations linking I_d/I with I/I_0. We first present three correlations which do not take into account the solar altitude. These studies are by Orgill and Hollands [37], by Erbs *et al.* [38], and by Spencer [39].

A. *Orgill and Hollands Correlation* [37]

This correlation, the first of its kind, is based on four years of data from Toronto, Canada (43°48′ N). The diffuse-radiation data were measured by shadow-band pyranometer. This correlation divides the sky cover into three parts and is given below:

$$\begin{aligned} I_d/I &= 1.0 - 0.249M_T, \qquad 0 \le M_T \le 0.35, \\ &= 1.577 - 1.84M_T, \qquad 0.35 \le M_T \le 0.75, \\ &= 0.177, \qquad M_T > 0.75. \end{aligned} \tag{8.12.2}$$

This correlation is shown in Fig. 8.12.1.

B. *Erbs et al. Correlation* [38]

Because Eq. (8.12.2) was based on data from a somewhat high latitude, Erbs *et al.* reworked Orgill and Hollands' procedure with data from five U.S.

stations, with latitudes ranging from 31° to 42° N. The data were of short duration, ranging from one to four years. The U.S. data were based on pyrheliometric measurements where the diffuse radiation is obtained by subtraction from the global measurement [see Eq. (8.2.4)]. The correlation by Erbs *et al.* is given below:

$$\begin{aligned} I_d/I &= 1.0 - 0.09M_T, \qquad 0 \le M_T \le 0.22, \\ &= 0.951\,1 - 0.160\,4M_T + 4.388M_T^2 \\ &\quad - 16.638M_T^3 + 12.336M_T^4, \qquad 0.22 \le M_T \le 0.80, \\ &= 0.165, \qquad M_T > 0.80. \end{aligned} \tag{8.12.3}$$

C. *Spencer Correlation* [39]

We have seen in Fig. 8.10.3 that as far as the mean daily diffuse radiation is concerned, the latitude does indeed have some bearing on it, because, as the latitude increases, the average air mass also increases. In order to explore this latitude dependence, Spencer proposed the following correlation:

$$I_d/I = a_4 - b_4M_T, \qquad 0.35 \le M_T \le 0.75. \tag{8.12.4}$$

It is presumed that I_d/I has constant values beyond the above range of M_T. The coefficients a_4 and b_4 are latitude dependent. These regression coefficients were obtained through data from five stations in Australia. The coefficients, valid for 20°–45° S latitudes are

$$a_4 = 0.940 + 0.011\,8|\phi| \tag{8.12.5}$$

and

$$b_4 = 1.185 + 0.013\,5|\phi| \tag{8.12.6}$$

where ϕ is the latitude in degrees, indicating an increasing proportion of diffuse radiation at higher latitudes.

The correlations of Orgill and Hollands, Erbs *et al.*, and Spencer are plotted in Fig. 8.12.1. The correlations of Orgill and Hollands and of Erbs *et al.* are almost identical. On the other hand, Spencer's equation plotted at $\phi = 40°$ yields consistently lower results.

We now present two studies in which the solar altitude directly comes into play. These studies are by Boes *et al.* [40] and Iqbal [41].

D. *Boes et al. Correlation* [40]

In the U.S. radiation network, pyrheliometer measurements of the direct normal radiation are recorded as a matter of routine. Therefore, it seems

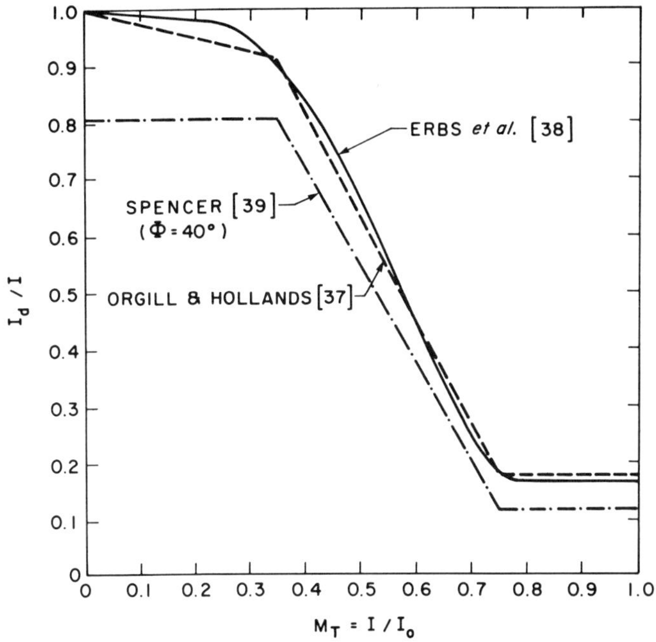

Figure 8.12.1 Comparison of three correlations.

logical that the direct normal irradiance be correlated with global irradiance, and the diffuse irradiance be obtained by subtraction. Boes and colleagues present the following correlation which is based on one year of data from three U.S. stations, at latitudes varying from 35° to 42° N:

$$\begin{aligned} \dot{I}_n &= 400, \quad \alpha \leq 10^\circ, \quad M_T > 0.5, \\ &= -520 + 1\,800 M_T, \quad 0.3 \leq M_T \leq 0.85, \\ &= 1\,000, \quad M_T > 0.85, \end{aligned} \tag{8.12.7}$$

where $\dot{I}_n$ is in watts per square meter, and $\alpha \leq 10^\circ$ combined with $M_T > 0.5$ represents low sun and bright clouds. In the range $M_T > 0.85$, representing almost cloudless skies, the original method provides for values of $\dot{I}_n$ ranging from 950 to 1050 W m^{-2} in different months. However, in the above equation we have used one constant value, 1 000 W m^{-2}.

Boes *et al.*'s work has been utilized to develop contour maps of the average daily direct normal radiation for the United States [6]. These maps are useful for a preliminary assessment of concentrating systems.

E. *Iqbal Correlation* [41]

We have seen in Fig. 8.9.2 that the sky cover can be categorized as very cloudy, partly cloudy, or very clear. It seems obvious that under the last two categories, the solar altitude should have an important bearing on the prediction of I_d from I. Keeping this in mind, Iqbal developed a correlation utilizing data from two French and three Canadian stations. The latitudes ranged from 43° to 53° N and the data from all stations were for a minimum period of nine years. The resulting correlation, in a graphical form, is shown in Fig. 8.12.2. In this figure, M_T is plotted against M_d, where

$$M_d = I_d/I_0. \tag{8.12.8}$$

The general variation of M_d with M_T is of a similar nature to that of K_d with K_T of the individual daily values.

The correlation in Fig. 8.12.2 indicates that under overcast conditions ($M_T < 0.35$), M_d increases linearly with M_T. Further, in this region, the solar altitude has no bearing on the fraction of diffuse radiation. Up to this point, the correspondence with a similar study by Bugler [42] is significant.

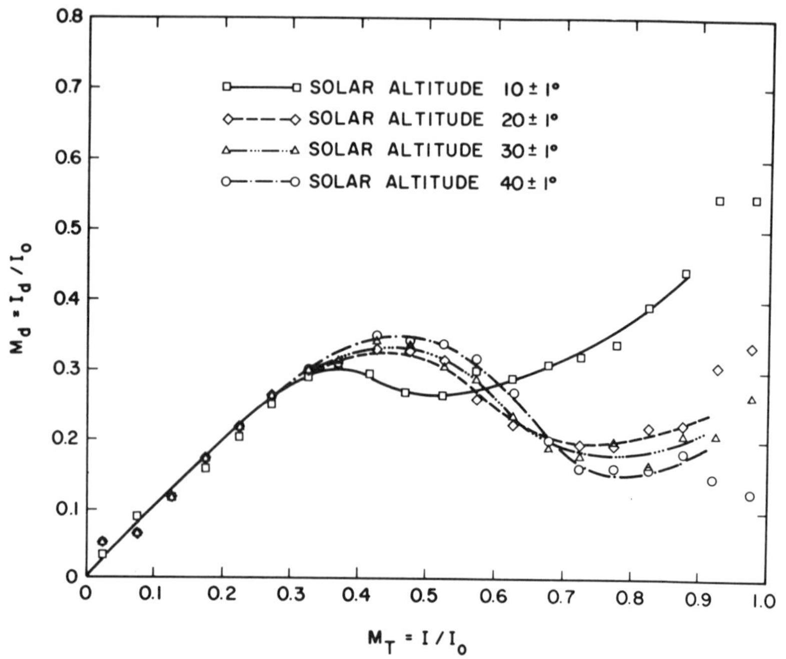

Figure 8.12.2 Variation of the hourly diffuse radiation with the hourly global radiation as a function of solar altitudes. Adapted with permission from Iqbal [41], copyright 1980, Pergamon Press, Ltd.

As M_T goes beyond 0.35, the effect of solar altitude begins to emerge, and the region immediately following this may be considered as having partly cloudy skies. At the edge of this region, the diffuse component increases briefly with increase in global radiation, and then begins to decrease as the partly cloudy skies become clearer. At particular solar altitudes, a minimum value of the diffuse radiation is reached. The value of M_T at which M_d reaches its minimum value varies with the solar altitude.

The region beyond which a minimum value of M_d is reached may be considered as having mainly clear-sky conditions. In this region, M_d increases again with M_T, lower solar altitudes giving a higher percentage of diffuse radiation. Under partly cloudy skies and under clear skies, solar altitudes lower than 30° have a marked effect on the fraction of diffuse radiation.

When skies are completely cloudless, it is only natural that we should compute I_d (and also I) through the material presented in Chapter 7, provided data on atmospheric parameters are available.

In the foregoing, we have summarized five studies treating the prediction of hourly diffuse radiation from hourly global radiation for individual hours. Let us now spot-check their accuracy through the data in Tables 8.4.1 and 8.4.2. We present this spot-check in the following example.

□ EXAMPLE 8.12.1. Hourly diffuse radiation predicted from hourly global radiation data for June 1976 in Montreal (Table 8.4.1) for the following hours:

Date	Hour ending
7 June	7
10 June	16
11 June	7

The results are compared with the measured values in Table 8.4.2.

Solution. We first present solution details of all correlations for one of the individual hours and then present a table of results for all three hours. We choose hour ending 16 on 10 June for the worked solution.

From Table 8.4.1, $I = 52$ cal cm^{-2} h^{-1}, or $I = 2\,176$ kJ m^{-2} h^{-1}. I_0 is obtained from Eq. (4.2.9),

$$I_0 = 3\,204 \text{ kJ m}^{-2}\,\text{h}^{-1}.$$

Therefore

$$M_T = I/I_0 = 0.679.$$

The solar altitude is obtained from Eq. (1.5.1), and $\alpha = 42.2°$. In order to predict I_d, we utilize the various correlations one by one.

(1) Orgill and Hollands' correlation, Eq. (8.12.2)

$$I_d/I = 1.557 - 1.84(0.679) = 0.308.$$

The predicted I_d is

$$I_d = 0.308(2\,176) = 670 \text{ kJ m}^{-2}\text{ h}^{-1}.$$

From Table (8.4.2), the measured value is

$$I_d = 16 \text{ cal cm}^{-2}\text{ h}^{-1} = 670 \text{ kJ m}^{-2}\text{ h}^{-1}.$$

The percentage difference is zero.

(2) Erbs *et al.* correlation, Eq. (8.12.3).

From this correlation we have

$$\begin{aligned} I_d/I &= 0.951\,1 - 0.160\,4(0.679) + 4.388(0.679)^2 \\ &\quad - 16.638(0.679)^3 + 12.333\,6(0.679)^4 \\ &= 0.279. \end{aligned}$$

The predicted I_d is

$$I_d = 0.279(2\,176) = 607 \text{ kJ m}^{-2}\text{ h}^{-1}.$$

The percentage difference from the measured value indicates 9.4% underestimation. We have noticed from Fig. 8.12.1 that the agreement between Eqs. (8.12.2) and (8.12.3) is very close. However, there are some differences at certain values of the clearness index M_T.

(3) Spencer correlation, Eq. (8.12.4)

This correlation is presumably valid for latitudes between 20° and 45° N. The latitude of Montreal lies at the edge of this limit. We calculate the coefficients a_4 and b_4 as

$$\begin{aligned} a_4 &= 0.94 + 0.011\,8|45.5| = 1.477, \\ b_4 &= 1.185 + 0.013\,5|45.5| = 1.799. \end{aligned}$$

Therefore

$$I_d/I = 1.477 - 1.799(0.679) = 0.256.$$

The predicted I_d is

$$I_d = 0.256(2\,176) = 557 \text{ kJ m}^{-2}\text{ h}^{-1}.$$

The percentage difference from the measured value indicates 16.9% underestimation.

(4) Boes *et al.* correlation, Eq. (8.12.7)

This equation yields the direct normal irradiance in watts per square meter:

$$\dot{I}_n = -520 + 1\,800(0.679) = 702 \text{ W m}^{-2}$$

or

$$I_n = 702(3.6) = 2\,528 \text{ kJ m}^{-2} \text{ h}^{-1}$$

$$I_b = 2\,528 \cos(90 - 42.2) = 1\,698 \text{ kJ m}^{-2} \text{ h}^{-1}.$$

The diffuse radiation is obtained from Eq. (8.2.4):

$$I_d = I - I_b = 2\,176 - 1\,698 = 478 \text{ kJ m}^{-2} \text{ h}^{-1}.$$

The percentage difference from the measured value indicates 28.7% underestimation.

(5) Iqbal correlation

The predicted value is read from Fig. 8.12.2. Against I/I_0 of 0.679 and $\alpha = 42{,}2°$, we read

$$I_d/I_0 = 0.21.$$

Therefore, the predicted I_d is

$$I_d = 0.21(3\,204) = 673 \text{ kJ m}^{-2} \text{ h}^{-1}.$$

The percentage difference from the measured value is $\simeq 0\%$.

Calculations similar to those shown above can be carried out for the other two hours. We present below the final results for all three hours.

Date (June)	Hour ending	Measured ($\text{kJ m}^{-2} \text{h}^{-1}$) I	Measured ($\text{kJ m}^{-2} \text{h}^{-1}$) I_d	Percentage difference from correlation of: Orgill and Hollands	Erbs *et al.*	Spencer	Boes *et al.*	Iqbal
7	7:00	711	401	−39.9	−43.6	−48.2	−59.9	−45.6
10	16:00	2 176	670	0	−9.4	−16.9	−28.7	0
11	7:00	335	329	−4.8	−1.8	−15.2	0	−1.5

This table shows the difficulty inherent in predicting the diffuse component from global radiation for individual hours. Spencer's correlation, which may not be applicable to Montreal data, consistently underpredicts. Among the remaining correlations, none is entirely satisfactory, except under completely overcast conditions, when they all yield good results. On a long-term basis (not on a monthly average basis) it is possible that a particular correlation may be most suitable for a specific application.

It is obvious that considerable work and a completely different approach is required to solve this problem. □

8.13 Further Reading

The previous sections have dealt with methods of predicting the global and diffuse radiation on horizontal surfaces. In this section, a further bibliographic review is presented as supplementary reading.

A global classification of solar radiation was carried out by Terjung [43], who prepared maps showing regional patterns on a global basis. Studies dealing with the frequency of daily global radiation have been made by Fritz and MacDonald [44], Bennett [45], and Kalma [46]. Bendt *et al.* [47] and Theilacker and Klein [48] have reconfirmed the essential validity of the fractional time distribution curves developed by Hottel and Whillier [30] and Liu and Jordan [26].

Schulze [49] critically reviewed insolation–sunshine correlations. Perrin de Brichambaut [50] used the insolation–sunshine correlations in mapping solar radiation in France. Kimura and Stephensen [51] developed a method of estimating mean hourly beam and diffuse radiation from the cloud-cover data. This is the ASHRAE [52] cloudy-sky model and should not be confused with the ASHRAE clear-sky model discussed in Chapter 7.

A number of researchers have tried to verify the Liu and Jordan correlations: Choudhury [53] with data from India, Stanhill [54] with data from Israel, and Kalma and Fleming [55] with data from several countries. Paltridge and Proctor [56] presented monthly mean solar radiation statistics for Australia. Mani and Chacko [57] discussed the diurnal and seasonal variations of diffuse solar radiation from measurements made at two stations in India.

Lacis and Hansen [58], Hoyt [59], Atwater and Ball [60], and Barbaro *et al.* [61] have presented parametric models to compute solar radiation under cloudy skies.

In Section 8.6, we considered correlations based only on the total cloud amount. The cloud types and their properties, number of cloud layers, and positions were not considered. Studies that take into consideration these parameters are called *layer models*. Davies *et al.* [62] and Suckling and Hay [63] have developed such layer models. Comparisons of the various cloudy-sky models are available in [64] and [65]. An overview of the U.S. solar radiation network and data development efforts is presented by Boes [66].

Nomenclature[6]

a, b, c	Coefficients in the various equations
C	Cloudiness index or cloud normal, or fraction of sky covered by clouds (dimensionless)
E_o	Eccentricity correction factor of the earth (dimensionless)
f	Fractional time during which the daily total radiation $\leq H$ (dimensionless)
H	Daily global radiation incident on a horizontal surface (MJ m^{-2} day^{-1})
H_b	Daily beam radiation incident on a horizontal surface (MJ m^{-2} day^{-1})
H_c	Clear-sky daily global radiation incident on a horizontal surface (MJ m^{-2} day^{-1})
H_d	Daily diffuse radiation incident on a horizontal surface (MJ m^{-2} day^{-1})
H_0	Extraterrestrial daily radiation incident on a horizontal surface (MJ m^{-2} day^{-1})
I	Hourly global radiation incident on a horizontal surface (kJ m^{-2} h^{-1})
I_b	Hourly beam radiation incident on a horizontal surface (kJ m^{-2} h^{-1})
I_d	Hourly diffuse radiation incident on a horizontal surface (kJ m^{-2} h^{-1})
I_0	Extraterrestrial hourly radiation incident on a horizontal surface (kJ m^{-2} h^{-1})
I_n	Hourly direct radiation incident on a surface normal to solar rays (kJ m^{-2} h^{-1})
j	Fraction of the total daylight period during which the zenith angle is greater than 85° (dimensionless)
K	H_d/H
K_d	H_d/H_0
K_T	H/H_0
M_d	I_d/I_0
M_T	I/I_0
N_d	Day length (h)
N_j	Modified day length which excludes the fraction during which the sun is less than 5° above the horizon (dimensionless)
n	Number of bright sunshine hours per day
n_d	Number of days in the month under consideration
n_y	Number of years of data utilized
S	n/N_d or $\bar{n}/\bar{N}_d$
δ	Declination (degrees), north positive, south negative
δ_c	Declination on characteristic days (degrees) (see Table 4.2.2)
θ_z	Zenith angle, the angle between the beam from the sun and the vertical (degrees)
ρ	Ground albedo
ρ_a	Cloudless sky albedo (dimensionless)
ρ_c	Cloud albedo (dimensionless)
$\tau_b(\omega)$	Atmospheric transmittance to beam radiation variable with the hour angle (dimensionless)
ϕ	Latitude (degrees), north positive
ω	Hour angle (degrees), solar noon zero and mornings positive
ω_i	Hour angle at the middle of an hour
ω_s	Sunset-hour angle for a horizontal surface (degrees)

[6] Overdots are used to indicate instantaneous values, overbars to indicate monthly average values, and primes to indicate values before multiple reflections.

References

1. A. J. Drummond, On the measurement of sky radiation. *Arch. Meteorol. Geophys. Bioklimatol.* **7,** 413–436 (1956).
2. Monthly Radiation Summary, Atmospheric Environment Service, Downsview, Ontario, Canada.
3. H. E. Landsburg (ed.), "World Survey of Climatology," Vols. 1–15. Elsevier, Amsterdam.
4. G. O. G. Löf, J. A. Duffie, and C. O. Smith, World Distribution of Solar Radiation. Rep. No. 21, Solar Energy Laboratory, Univ. of Wisconsin (1966). See also *Sol. Energy* **10** (1), 29–37 (1966).
5. D. de Jong, "Net Radiation Received by a Horizontal Surface at the Earth." Delft Univ. Press, Delft, 1973.
6. Solar Radiation Energy Resource Atlas of the United States. SERI/SP-642-1037. Superintendent of Documents, U.S. Government Printing Office, Washington, D.C. (1981). For the sunshine data, see Climatic Atlas of the United States. National Climatic Center, Asheville, North Carolina (1979).
7. D. W. Phillips and D. Aston, Canadian solar radiation data normals, abstracts, charts and other specialized information. Atmospheric Environment Service, Downsview, Ontario, Canada (1980).
8. V. Cinquemani, J. R. Owenby, and R. G. Baldwin, Input data for solar systems. Report prepared for the U.S. Department of Energy by the National Oceanic and Atmospheric Administration (1978).
9. H. H. Kimball, Variations in the total and luminous solar radiation with geographical position in the United States. *Mon. Weather Rev.* **47** (11), 769–793 (1919).
10. A. Ångström, Solar and Terrestrial radiation. *Q. J. R. Meteorol. Soc.* **50,** 121–126 (1924).
11. S. Fritz and T. H. MacDonald, Average solar radiation in the United States. *Heat. Vent.* **46** (7), 61–64 (1949).
12. C. L. Mateer, A preliminary estimate of the average insolation in Canada. *Can. J. Agr. Sci.* **35,** 579–594 (1955).
13. J. A. Prescott, Evaporation from a water surface in relation to solar radiation. *Trans. R. Soc. S. Aust.* **64,** 114–118 (1940).
14. B. J. Yorke and G. R. Kendall, Daily bright sunshine, Atmospheric Environment Service, Downsview, Ontario, Canada (1972).
15. M. R. Rietveld, A new method for estimating the regression coefficients in the formula relating solar radiation to sunshine. *Agr. Meteorol.* **19,** 243–252 (1978).
16. J. Glover and J. S. G. McCulloch, The empirical relation between solar radiation and hours of sunshine. *Q. J. R. Meteorol. Soc.* **84,** 172–175 (1958).
17. I. Bennett, Monthly maps of mean daily insolation for the United States. *Sol. Energy* **9** (3), 145–159 (1965).
18. J. E. Hay, Calculation of monthly mean solar radiation for horizontal and inclined surfaces. *Sol. Energy* **23** (4), 301–307 (1979).
19. J. N. Black, The distribution of solar radiation over the earth's surface. *Arch. Meteorol. Geophys. Bioklimatol.* **7,** 165–189 (1956).
20. D. J. Norris, Correlation of solar radiation with clouds. *Sol. Energy* **12** (1), 107–112 (1968).
21. I. Bennett, Correlation of daily insolation with daily total sky cover, opaque sky cover and percentage of possible sunshine. *Sol. Energy* **12** (3), 391–393 (1969).
22. Solmet Vol. 1, Hourly solar radiation–surface meteorological observations. Final Report TD-9724. National Climatic Center, Asheville, North Caroline (1978).
23. Solmet Vol. 2, Hourly solar radiation–surface meteorological observations. Final Report TD-9724, National Climatic Center, Asheville, North Carolina (1979).

24. A. Whillier, The determination of hourly values of total solar radiation from daily summation. *Arch. Meteorol. Geophys. Bioklimatol. Ser. B* **7** (2), 197–204 (1956).
25. A. Whillier, Solar energy collection and its utilization for house heating. Sc.D. Thesis, Dept. of Mech. Eng., Massachusetts Institute of Technology, (1953).
26. B. Y. H. Liu and R. C. Jordan, The interrelationship and characteristic distribution of direct, diffuse and total solar radiation. *Sol. Energy* **4** (3), 1–19 (1960).
27. M. Iqbal, A study of Canadian diffuse and total solar radiation data, II. Monthly average hourly horizontal radiation. *Sol. Energy* **22** (1), 87–90 (1979).
28. M. Collares-Pereira and A. Rabl, The average distribution of solar radiation correlations between diffuse and hemispherical and between daily and hourly insolation values. *Sol. Energy* **22** (2), 155–164 (1979).
29. A. Whillier, Solar radiation graphs. *Sol. Energy* **9** (3), 164–165 (1965).
30. H. C. Hottel and A. Whillier, Evaluations of flat plate solar collector performance. *Trans. Conf. Use Sol. Energy* Vol. 2, pp. 74–104. Univ. of Arizona Press, Phoenix, 1958.
31. D. W. Ruth and R. E. Chant, The relationship of diffuse radiation to total radiation in Canada. *Sol. Energy* **18** (2) 153–154 (1976).
32. M. Iqbal, A study of Canadian diffuse and total solar radiation data, I. Monthly average daily horizontal radiation. *Sol. Energy* **22** (1), 81–86 (1979).
33. S. A. Klein and J. A. Duffie, Estimation of monthly average diffuse radiation. *Proc. Ann. Meet. Am. Section, Int. Sol. Energy Soc., Denver, Colorado* (1978).
34. J. K. Page, The estimation of monthly mean values of daily total short-wave radiation on vertical and inclined surfaces from sunshine records for latitudes 40°N–40°S. *Proc. U.N. Conf. New Sources Energy* Paper No. S98, Vol. 4, pp. 378–390 (1961).
35. M. Iqbal, Correlation of average diffuse and beam radiation with hours of bright sunshine. *Sol. Energy* **23** (2), 169–173 (1979).
36. J. Hay, A revised method for determining the direct and diffuse components of the total short-wave radiation. *Atmosphere* **14** (4), 278–287 (1976).
37. J. F. Orgill and K. G. Hollands, Correlation equation for hourly diffuse radiation on a horizontal surface. *Sol. Energy* **19** (4), 357–359 (1977).
38. D. G. Erbs, S. A. Klein, and J. A. Duffie, Estimation of the diffuse radiation fraction for hourly, daily and monthly-average global radiation. *Sol. Energy* **28** (4), 293–304 (1982).
39. J. W. Spencer, A comparison of methods for estimating hourly diffuse solar radiation from global solar radiation. *Sol. Energy* **29** (1), 19–32 (1982).
40. E. C. Boes, H. E. Anderson, I. J. Hall, R. R. Prairie, and R. T. Stromberg, Availability of direct, total and diffuse solar radiation to fixed and tracking collectors in the U.S.A. Sandia Rep. SAND77-0885 (1977).
41. M. Iqbal, Prediction of hourly diffuse solar radiation from measured hourly global radiation on a horizontal surface. *Sol. Energy* **23** (5), 491–503 (1980).
42. J. W. Bugler, The determination of hourly insolation on an inclined plane using a diffuse irradiation model based on hourly measured global horizontal insolation. *Sol. Energy* **19** (5), 477–491 (1977).
43. W. H. Terjung, A global classification of solar radiation. *Sol. Energy* **13** (1), 67–81 (1970).
44. S. Fritz and T. H. MacDonald, The number of days with solar radiation above or below specified values, *Sol. Energy* **4** (1), 20–22 (1960).
45. I. Bennett, Frequency of Daily insolation in anglo North America during June and December. *Sol. Energy* **11** (1), 41–55 (1967).
46. J. D. Kalma, Estimating frequencies of daily amounts of global radiation over Australia. *Aust. Meteorol. Mag.* **18,** 134–145 (1970).
47. P. Bendt, M. Collares-Pereira, and A. Rabl, The frequency distribution of daily insolation values. *Sol. Energy* **27** (1), 1–5 (1981).

48. J. C. Theilacker and S. A. Klein, Improvements in the utilizability relationships. *Proc. Ann. Meeting ISES U.S. Section, Phoenix, Arizona (June 2–6 1980).*
49. R. E. Schulze, A physically based method of estimating solar radiation from sunshine records. *Agr. Meteorol.* **16,** 85–101, (1976).
50. Chr. Perrin de Brichambaut, Estimation des resources énergetiques solaires en France. *Supplément au cahiers E.F.E.D.E.S.* No. 1 (1975).
51. K. Kimura and D. G. Stephenson, Solar radiation on cloudy days. *Trans ASHRAE* **75** (2), 227–239 (1969).
52. ASHRAE, Procedure for determining heating and cooling loads for computerized energy calculations. Algorithm for building heat transfer subroutines. New York (1975).
53. N. K. O. Choudhury, Solar radiation at New Delhi. *Sol. Energy* **7** (2), 44–52 (1963).
54. G. Stanhill, Diffuse sky and cloud radiation in Israel. *Sol. Energy* **10** (2), 96–101 (1966).
55. J. D. Kalma and P. M. Fleming, A note on estimating the direct and diffuse components of global radiation. *Arch. Meteorol. Geophys. Bioklimatol. Ser. B* **2,** 191–205 (1972).
56. G. W. Paltridge and D. Proctor, Monthly mean solar radiation statistics for Australia. *Sol. Energy* **18** (3), 235–243 (1976).
57. A. Mani and O. Chacko, Measurements of diffuse solar radiation at Delhi and Poona. *Indian J. Meteorol. Geophys.* **14** (4), 416–432 (1963).
58. A. A. Lacis and J. E. Hansen, A parameterization for the absorption of solar radiation on the earth's atmosphere. *J. Atmos. Sci.* **31,** 118–132 (1974).
59. D. V. Hoyt, A model for the calculation of solar global insolation. *Sol. Energy* **21** (1), 27–35 (1978).
60. M. A. Atwater and J. T. Ball, A numerical solar radiation model based on standard meteorological observations. *Sol. Energy* **21** (3), 163–170 (1978).
61. S. Barbaro, S. Coppolino, C. Leone, and E. Sinagra, An atmospheric model for computing direct and diffuse solar radiation. *Sol. Energy* **22** (3), 225–228 (1979).
62. J. A. Davies, W. Schertzer, and M. Nunez, Estimating global solar radiation. *Boundary-Layer Meteorol.* **9,** 33–52 (1975).
63. P. W. Suckling and J. E. Hay, A cloud layer-sunshine model for estimating direct, diffuse and total solar radiation. *Atmosphere* **15** (4), 194–207 (1977).
64. J. A. Davies, Models for estimating incoming solar irradiance. Report to A.E.S. Department of Geography, McMaster Univ., Hamilton, Ontario (1980).
65. J. F. MacLaren Ltd., Hooper and Angus Associates Ltd., J. Hay, and H. Davies, Define, develop and establish a merged solar and meteorological computer data base. Atmospheric Environment Service, Downsview, Ontario (1980).
66. E. C. Boes, Insolation medeling overview. *Energy* **4,** 523–529 (1979).

Chapter 9

GROUND ALBEDO

9.1 Introduction

When radiant energy is incident on a surface, it may be partly absorbed, partly reflected, and partly transmitted. The properties of a surface or a material associated with these three functions are called its *absorptivity*, *reflectivity*, and *transmissivity*. The fraction of the total incident energy associated with these properties is termed *absorptance*, *reflectance*, and *transmittance*. When the source of incident radiation is the sun, the term "albedo,"[1] is commonly used instead of "reflectance." Here, the two terms are interchangeable. It is usual to express reflectance as a fraction of one, whereas albedo is sometimes expressed as a percentage and sometimes as a fraction of one. In this text we express it in the latter form.

Determining an accurate value of albedo is very important in evaluating the total insolation on a building or on a solar energy collecting device. It is also important in studies dealing with thermal balance in the atmosphere. In spite of the importance of albedo, this subject has not been well studied. Part of the problem lies in defining the characteristics of a surface covering a large land mass; a possible exception is a water basin.

[1] From the Latin *albus*, which means light, white, or pale color. Albedo means the illumination property of the earth and its atmosphere especially just before dawn and just after sunset. In meteorology, it means reflectance to solar radiation of the ground or of the earth and/or its atmosphere.

9.2 Elementary Definitions

In very general simple terms, the albedo ρ may be defined thus:

$$\rho = \frac{\text{Radiation reflected from a surface}}{\text{Radiation incident on the surface}}.$$

The incident energy may involve beam and/or diffuse radiation, whereas the nature of the reflected energy strongly depends upon the surface properties. Further, the direction of the incident beam and the hemispherical distribution of the incoming sky diffuse radiation have a strong bearing on the reflectance of a surface: a number of definitions can, therefore, be associated with the albedo of a surface. With respect to the reflective properties, two limiting characteristics of a surface are called *diffuse* and *specular.*

A perfectly diffuse surface[2] is one in which the intensity of radiation leaving the surface is uniform in all angular directions. Thus, after reflection from a diffuse surface, the history of the incident radiation is completely obliterated. Consider Fig. 9.2.1a: the hemispherically reflected radiation and the corresponding diffuse intensity (see Section 2.4) are related by the equality

$$\dot{I}\rho = \pi i, \tag{9.2.1}$$

where $\dot{I}$ is the irradiance, whether global, beam, or diffuse; ρ the albedo; and i the uniform intensity of reflected radiation. Such a surface may also be called an isotropic reflector. When reflection is not perfectly diffuse, as is shown in Fig. 9.2.1b, i_{d}, the reflected anisotropic intensity, can be related to the total reflected radiation as follows:

$$\dot{I}\rho = \int_0^{\frown} i_{\mathrm{d}}\, d\omega \cos\Phi, \tag{9.2.2}$$

where $d\omega$ is the solid angle related to the intensity of reflected radiation and $\frown$ indicates that the integration is to be carried out over a hemisphere.

Surfaces that are very smooth with respect to the wavelength of the incident radiation are called specular surfaces. A specularly reflecting surface is illustrated in Fig. 9.2.2a. The incoming intensity is contained within a solid angle $d\omega_{\mathrm{i}}$ and is incident at an angle Φ_{i} to the surface normal. All the reflected intensity is contained within a solid angle $d\omega_{\mathrm{r}}$ such that

$$d\omega_{\mathrm{i}} = d\omega_{\mathrm{r}} \qquad \text{and} \qquad \Phi_{\mathrm{i}} = \Phi_{\mathrm{r}}.$$

[2] A perfectly diffuse surface is also called "Lambertian" surface, named after Johann Heinrich Lambert (see Section 6.23).

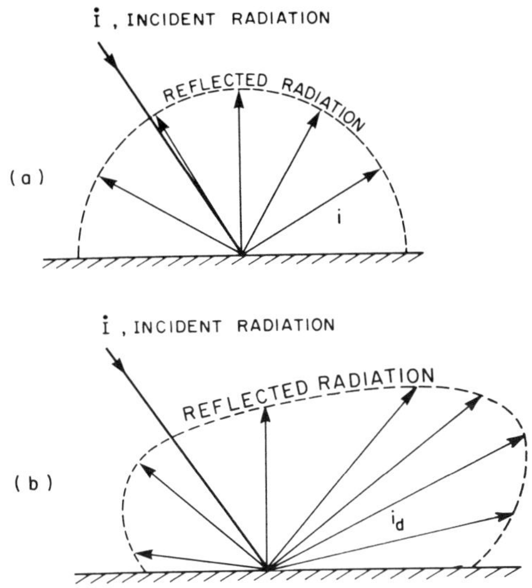

Figure 9.2.1 Ground reflection: (a) isotropic diffuse reflection and (b) anisotropic diffuse reflection.

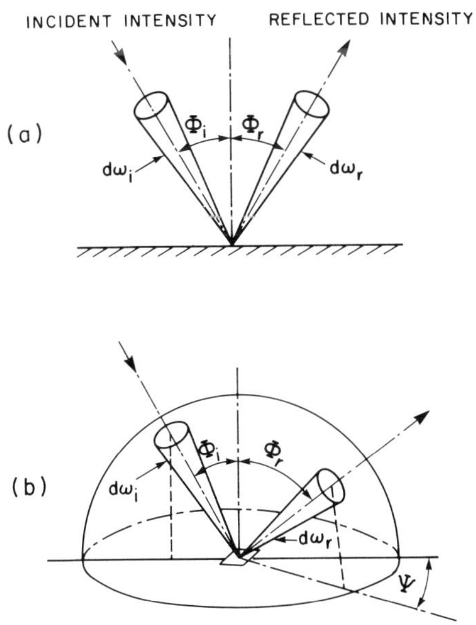

Figure 9.2.2 Specular reflection: (a) perfectly specular, and (b) bidirectionally specular.

Further, the incident and reflected intensities are in one plane containing the surface normal. The reflectance of such a surface can be computed theoretically from an equation by Fresnel (the s is mute in Fresnel), which is based on electromagnetic theory (for further details see Refs. 3 and 4 in Chapter 2).

A surface that is not perfectly specular is illustrated in Fig. 9.2.2b, where any one of the following conditions may hold:

$$d\omega_i \neq d\omega_r, \qquad \Phi_i \neq \Phi_r, \qquad \psi \neq 0,$$

where ψ is the azimuth of the reflected radiation.

Such a surface has some importance in solar energy applications, and the associated albedo is called bidirectional reflectance. For example, under clear skies, the albedo of the sea, a lake, ice, or a field of vegetation with glossy leaves generally would be bidirectional in nature. Apart from the two limiting surface characteristics described above, the reflectivity in general will be a function of the distribution of incoming radiation. Two practical cases are those of the anisotropic diffuse irradiance and the beam irradiance, each of which may be reflected back anisotropically into the hemisphere.

The foregoing description of albedo referred to its directional aspects; albedo is also strongly dependent on the spectral distribution of incoming radiation. For instance, the white paper of this page has a very high albedo for the visible band of the solar spectrum, although it absorbs almost all radiation at higher wavelengths, such as that emanating from the reader's body; it has, consequently, a low reflectance for long-wave radiation. Here we shall consider reflectance only with respect to solar radiation in the 0.2–4 μm wavelength range.

In the following sections, the albedo of a number of natural surfaces is discussed.

9.3 Spectral Albedo and the Effect of Solar Height

In this section, the spectral albedo of a number of ground covers is discusses. The surfaces considered are (a) bare soil, (b) vegetative cover, (c) snow cover; and (d) water surface. In the next section, the values of albedo integrated over all wavelengths are presented.

A. *Bare Soil*

The spectral albedo of bare soil increases with wavelength. For some soils, the albedo may begin to decrease after reaching a peak at a certain wavelength. The albedo increases as the solar height decreases to about 10°. A further decrease in solar height decreases the albedo [1].

B. *Vegetative Cover*

The spectral albedo of vegetative covers, especially those with rich green leaves, exhibits some peculiar selective features. The albedo is generally low in the visible spectrum and high in the infrared region. The selective feature is with respect to the chlorophyll absorption band interval near $\lambda = 0.65\ \mu m$, where the albedo is minimum. Spectral albedo has a strong dependence on the type of vegetation, its age, and the season. For vegetative covers, the albedo increases as solar height decreases and is always at its minimum at solar noon. At low solar height, the albedo is higher at higher wavelengths.

C. *Snow Cover*

The albedo of snow cover is strongly dependent on the surface conditions of the snow, as well as on the distribution of global irradiation from the sky hemisphere. The spectral albedo of clean snow under clear skies is almost uniform for the wavelength band 0.50–0.80 μm and decreases slowly on either side of this band. Surface roughness of the snow cover has an important bearing on the albedo: for all wavelengths, the albedo decreases with an increase in surface roughness. During a period of a day, albedo varies with solar height as well as with hourly ratio of diffuse to beam radiation. During a completely cloudy day, the albedo remains constant.

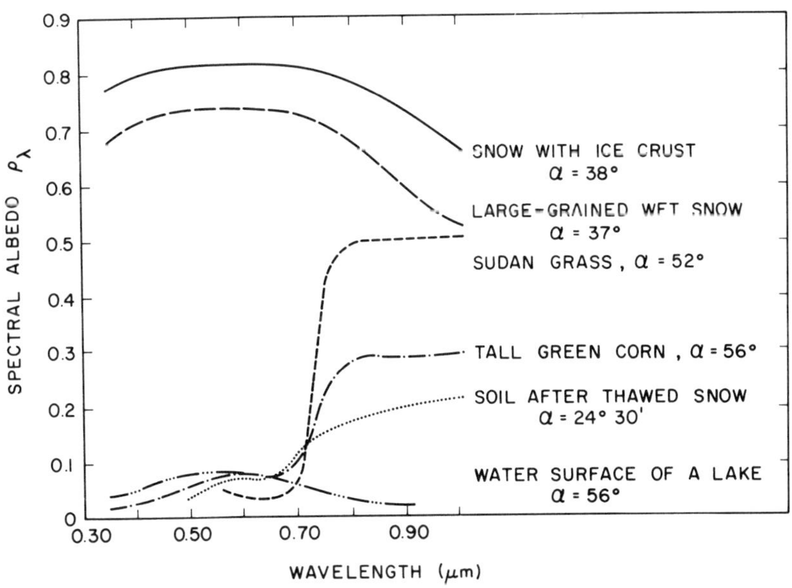

Figure 9.3.1 Spectral albedo of various ground covers. Adapted from Paltridge and Platt [2].

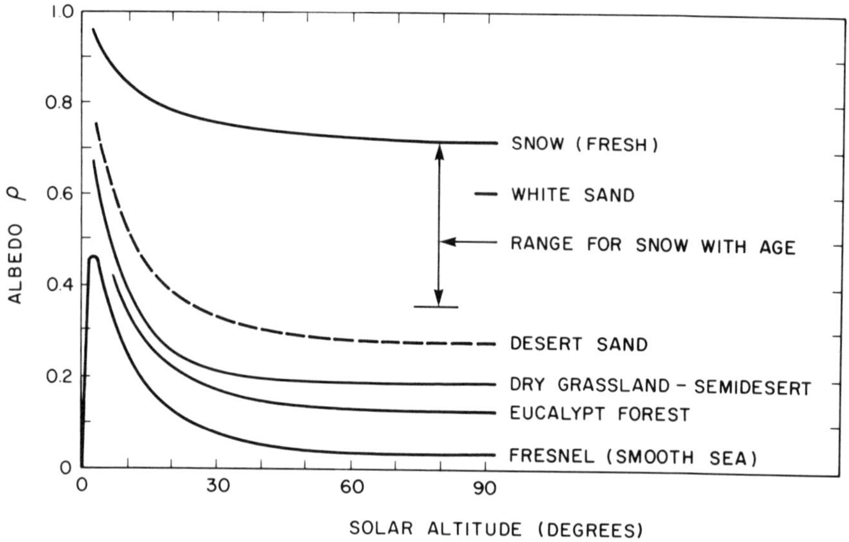

Figure 9.3.2 Angular variation of the albedo of various ground covers. Adapted from Paltridge and Platt [2].

D. *Water Surface*

The spectral albedo of pure water having a calm surface can be determined theoretically through Fresnel's equation. The albedo is dependent on the angle of incidence and the refractive index of water. This index decreases steadily with wavelength. As the solar altitude decreases, the albedo increases. However, as far as the wavelength dependence is concerned, within the visible band, the albedo is almost invariant with wavelength.

Measured values of the spectral albedo for a number of surfaces are shown in Fig. 9.3.1 [2]. Except for ice, snow, and water, all surfaces exhibit low values of albedo in the visible wavelengths and a rapid increase after $\lambda = 0.7\ \mu m$. The dependence of wavelength-integrated albedo on solar height has been theoretically examined by Paltridge and Platt [2] and is given in Fig. 9.3.2. An interesting feature of this diagram is that most surfaces exhibit Fresnel reflection characteristics at low solar altitudes.

9.4 The Overall Albedo

In this section, the basic features of albedo values for a number of ground covers integrated over the solar spectrum and solar height are briefly discussed.

The soil surface albedo has strong dependence on moisture content and tillage conditions: moisture decreases the albedo. The plant albedo can have large variations depending on the season and the phase of growth. The albedo of ice has not been well studied; however, it is known that the albedo of an ice layer over water can be substantially low and that of ice under snow can be very high. The albedo of snow and ice can also fluctuate a great deal depending on the freshness of the snow, the water under the ice, and the amount of dirt on the surface. Figure 9.3.2 may also be used to obtain the albedo of snow for clear-sky conditions; for completely cloudy skies and isotropic diffuse radiation, these values should be increased by 10–15%.

For solar height greater than 10°, the albedo of a smooth water surface under beam radiation may be obtained from Fig. 9.3.2. A water surface under diffuse radiation has an albedo of 0.08–0.11, the lower value being for clear-sky diffuse radiation and the higher value for diffuse radiation from cloudy skies. These albedo values need to be slightly modified to take into account the depth of the water basin and the fact that a large water surface is not horizontal because of the earth's curvature.

Albedo is generally measured by two back-to-back pyranometers mounted in a horizontal position a few meters above the underlying surface. The pyranometer facing the sky measures the global incident solar radiation and the one facing the earth measures the reflected energy (albedo is the ratio of the reflected to the incident energy). Average values of such measurements taken under various cloud conditions and solar heights are reported in the literature. Such measurements characterize the albedos of small underlying surfaces.

However, engineering and architectural applications require measurement of the weighted-average albedo of a large land mass that a receiving surface sees. This land mass may itself consist of patches of various surfaces, and for each patch the albedo for diffuse radiation and the variation of albedo with solar height must be determined. Although such detailed information for an individual situation may not be available, Hunn and Calafell [3], using a photographic method, have presented the albedo for some typical winter landscapes in the United States. However, it seems that recourse has still to be made to a subjective selection of monthly weighted-average albedo, representative of a large area composed of patches of different surfaces.

Table 9.4.1 lists the albedo of natural ground covers. Most of these data are based on Russian authors who have done extensive studies of this subject.

It is also useful to note an important element in average albedo data of a natural ground cover located at two different latitudes. Since the two locations will have different average solar altitudes, the albedo values will be different. Further, the albedo of a surface at two different locations of the same latitude but under different sunshine conditions may not be identical

Table 9.4.1

Albedo of Natural Ground Covers

Item	Albedo ρ	Reference
I. *Crops*		
Alfalfa	0.02–0.05	5
Beets (sugar)	0.18	1
	0.25	6
Cotton	0.20–0.22	1
Grass		
dry	0.15–0.25	8
dry, wizened in sun	0.19	8
dry, no sun	0.19–0.22	8
dry, high	0.31–0.33	8
green	0.26	1
high, fresh	0.26	8
wet, no sun	0.14–0.26	8
wet, sun	0.33–0.37	8
Heather	0.10	1
Lettuce	0.22	1
Lucerne	0.23–0.32	1
	0.22–0.24	6
Maize:		
15–20 cm, 40–50% cover	0.16	10
40–50 cm, 70–75% cover	0.18	10
140–200 cm, green cobs, 80% cover	0.20	10
200–250 cm, fully ripe	0.23	10
Rice	0.12	1
Rye		
winter	0.21	1
green	0.18	10
mass fluorescence	0.16	10
end of fluorescence	0.15	10
beginning of fading of leaves	0.13	10
faded leaves, $<50\%$ cover	0.11	10
Wheat		
summer	0.10–0.25	1
milky light green	0.13	10
yellow ripeness	0.17	10
full ripeness	0.21	10
II. *Other agricultural and waste (nonarable) lands*		
Soils		
chestnut soil, gray red:		
dry, leveled	0.20	10
moist, leveled	0.12	10
dry, ploughed	0.15	10
moist, ploughed	0.07	10
clay soils:		
blue, dry	0.23	1, 10
blue, moist	0.16	1, 10
gray sandy soils:		
level, dry	0.25	10
level, moist	0.18	10

Table 9.4.1 (*Continued*)

Item	Albedo ρ	Reference
ploughed, dry	0.20	10
ploughed, moist	0.11	10
black earth, dark gray:		
level, dry	0.13	10
level, moist	0.08	10
ploughed, dry	0.08	10
ploughed, moist	0.04	10
unspecified soil:		
dry, ploughed	0.20–0.25	8
Sand—Deserts		
fine, light sand	0.37	1
gray sand	0.21	1
death valley	0.25	8
Mojave Desert	0.24–0.28	8
quartz (white) sand	0.35–0.40	1
river sand	0.43	1
wet sand	0.09	8
yellow sand	0.35	1
white sand, New Mexico	0.60	11
valleys, plains and slopes	0.27	11
Forests (see also snow)		
exfoliating in dry season	0.24	1
exfoliating in wet season	0.18	1
green forest	0.03–0.06	8, 12
	0.04–0.10	8
bare ground—some trees	0.07	8, 12
coniferous forest	0.12	11
coniferous and deciduous:		
Jun $\theta_z = 39–55°$	0.14–0.17	8
Aug $\theta_z = 30–41°$	0.12–0.16	8
Sep $\theta_z = 26–38°$	0.19–0.10	8
steppes	0.40–0.52	8
tops of fir	0.10	1
tops of pine	0.14	1
tops of oak	0.18	1
tundra	0.11–0.23	8
marsh	0.10–0.18	8
III. *Snow and ice*		
Snow		
forests		
new, fallen	0.82	10
wet, fine grained	0.65	10
wet, medium gained	0.56	10
wet, large grained	0.47	10
against background of mixed landscape	0.34	10
separate spots	0.31	10
stable snow cover	0.65	1
unstable in spring	0.25	1
unstable in fall	0.30	1

(*Continued*)

Table 9.4.1 (*Continued*)

Item	Albedo ρ	Reference
fields		
new fallen	0.82	10
wet, fine grained	0.73	10
wet, medium grained	0.64	10
wet, large grained	0.55	10
new fallen		
dry, bright, white, clean	0.72–0.98	10
wet, bright, white	0.80–0.85	10
compacted		
dry, clean	0.66–0.80	8
wet, gray, white	0.61–0.75	8
melting (soaked with water)	0.35	8
stable cover:		
above lat = 60°	0.80	1
below lat = 60°	0.70	1
Ice (general ice forms)		
coastal ice, no snow	0.4–0.5	10
melting pack ice, no snow	0.49–0.67	10
frozen puddles	0.42–0.50	10

Table 9.4.2

Albedo of Building Materials[a]

Item	Albedo ρ	Item	Albedo ρ
Bricks		Limestone	
clay, cream, glazed	0.64	anston	0.40
lime clay, French	0.54	bath	0.47
red	0.32	portland	0.64
stafford blue	0.11	White marble	0.56
white glazed	0.74	Reddish granite	0.45
Tiles		Slate	
clay, purple (dark)	0.18	blue-gray	0.13
red	0.33	gray dark	0.10
concrete, uncolored	0.35	purple	0.14
concrete, black	0.09	Wood	0.22
concrete, brown	0.15	Aluminum	0.85
Asphalt		Iron	
new	0.09	new galvanized	0.35
pavement	0.15	galvanized, very dirty	0.08
pavement, weathered	0.18	Steel	0.80
Roofing		Copper	0.74
bituminous felt	0.12	Paint	
sheet, green	0.14	aluminum	0.46
sheet, black matte surface	0.13	oil paint, cream, light	0.70
Asbestos cement		oil paint, green, light	0.50
aged	0.25		
red	0.31		
white	0.39		

[a] From Gubareff *et al.* [4].

Table 9.4.3

Monthly Average Ground Albedo for Some U.S. Locations[a]

	Jan	Feb	Mar	Apr	May	Jun	Jul	Aug	Sep	Oct	Nov	Dec
Albuquerque, NM	0.478	0.432	0.394	0.303	0.280	0.280	0.280	0.280	0.280	0.280	0.409	0.478
Apalachicola, FL	0.140	0.140	0.140	0.140	0.140	0.140	0.140	0.140	0.140	0.140	0.140	0.140
Bismark, ND	0.660	0.660	0.660	0.468	0.289	0.180	0.180	0.180	0.212	0.404	0.628	0.660
Brownsville, TX	0.280	0.280	0.280	0.280	0.280	0.280	0.280	0.280	0.280	0.280	0.280	0.280
Boston, MA	0.365	0.331	0.287	0.160	0.140	0.140	0.140	0.140	0.140	0.140	0.192	0.322
Cape Hatteras, NC	0.180	0.180	0.180	0.180	0.180	0.180	0.180	0.180	0.180	0.180	0.180	0.180
Caribou, ME	0.660	0.660	0.660	0.539	0.244	0.140	0.140	0.140	0.192	0.400	0.573	0.660
Charleston, SC	0.140	0.140	0.140	0.140	0.140	0.140	0.140	0.140	0.140	0.140	0.140	0.140
Columbia, MO	0.608	0.539	0.435	0.192	0.140	0.140	0.140	0.140	0.140	0.157	0.400	0.556
Dodge City, KS	0.644	0.564	0.500	0.276	0.180	0.180	0.180	0.180	0.180	0.212	0.468	0.628
El Paso, TX	0.280	0.280	0.280	0.280	0.280	0.280	0.280	0.280	0.280	0.280	0.280	0.280
Ely, NV	0.660	0.660	0.660	0.556	0.365	0.209	0.140	0.157	0.296	0.539	0.628	0.660
Ft. Worth, TX	0.215	0.206	0.190	0.180	0.180	0.180	0.180	0.180	0.180	0.180	0.186	0.206
Fresno, CA	0.140	0.140	0.140	0.140	0.140	0.140	0.140	0.140	0.140	0.140	0.140	0.140
Great Falls, MT	0.591	0.573	0.573	0.365	0.192	0.140	0.140	0.140	0.175	0.279	0.487	0.556
Lake Charles, LA	0.140	0.140	0.140	0.140	0.140	0.140	0.140	0.140	0.140	0.140	0.140	0.140
Madison, WI	0.660	0.608	0.573	0.296	0.140	0.140	0.140	0.140	0.140	0.209	0.469	0.643
Medford, OR	0.469	0.400	0.313	0.192	0.140	0.140	0.140	0.140	0.140	0.192	0.365	0.418
Miami, FL	0.140	0.140	0.140	0.140	0.140	0.140	0.140	0.140	0.140	0.140	0.140	0.140
Nashville, TN	0.310	0.279	0.227	0.149	0.140	0.140	0.140	0.140	0.140	0.149	0.227	0.310
New York, NY	0.339	0.309	0.244	0.149	0.140	0.140	0.140	0.140	0.140	0.140	0.175	0.296
N. Omaha, NE	0.660	0.591	0.521	0.244	0.140	0.140	0.140	0.140	0.140	0.192	0.487	0.643
Phoenix, AZ	0.280	0.280	0.280	0.280	0.280	0.280	0.280	0.280	0.280	0.280	0.280	0.280
Raleigh, NC	0.296	0.270	0.227	0.149	0.140	0.140	0.140	0.140	0.140	0.149	0.218	0.309
Santa Maria, CA	0.140	0.140	0.140	0.140	0.140	0.140	0.140	0.140	0.140	0.140	0.140	0.140
Seattle, WA	0.279	0.209	0.175	0.140	0.140	0.140	0.140	0.140	0.140	0.140	0.157	0.227
Washington, DC	0.313	0.287	0.227	0.149	0.140	0.140	0.140	0.140	0.140	0.140	0.192	0.300

[a] From SOLMET Vol. 2 [13].

Table 9.4.4

Monthly Average Ground Albedo for Some Canadian Locations[a]

Month	Vancouver	Edmonton	Saskatoon	Winnipeg	Toronto	Ottawa	Montreal
Jan	0.18	0.58	0.49	0.54	0.50	0.62	0.32
Feb	0.17	0.57	0.50	0.55	0.50	0.60	0.33
Mar	0.17	0.46	0.42	0.47	0.38	0.43	0.25
Apr	0.14	0.32	0.24	0.31	0.28	0.12	0.22
May	0.14	0.26	0.18	0.20	0.25	0.13	0.20
Jun	0.14	0.25	0.20	0.21	0.25	0.19	0.20
Jul	0.14	0.25	0.21	0.21	0.25	0.19	0.20
Aug	0.14	0.25	0.22	0.23	0.25	0.20	0.20
Sep	0.14	0.26	0.22	0.23	0.25	0.20	0.20
Oct	0.14	0.28	0.22	0.24	0.25	0.25	0.21
Nov	0.15	0.39	0.27	0.32	0.29	0.21	0.23
Dec	0.18	0.53	0.27	0.45	0.39	0.56	0.28

[a] Courtesy of Dr. John Hay.

owing to unequal ratios of beam and diffuse radiation. It is useful to keep in mind these factors while applying the tabulated data.

Table 9.4.2 contains albedos of building materials. These data are obtained from Gubareff *et al.* [4], who carried out an extensive survey of the radiative properties of different kinds of engineering and building materials.

Regional albedos of some U.S. and Canadian sites are listed in Tables 9.4.3 and 9.4.4, respectively.

Nomenclature

- $\dot{I}$ Irradiance ($\mathrm{W\,m^{-2}}$)
- i Intensity of isotropically reflected radiation ($\mathrm{W\,m^{-2}\,sr^{-1}}$), as a subscript i indicates incidence
- i_{d} Intensity of anisotropically reflected radiation ($\mathrm{W\,m^{-2}\,sr^{-1}}$)
- r A subscript r indicates reflection
- α Solar altitude or solar height (degrees)
- λ Wavelength (μm), as a subscript λ indicates monochromaticity
- ψ Azimuth of the specularly reflected radiation (deg)
- ρ Albedo, reflectance (dimensionless)
- Φ Incident or reflected angle of radiation with respect to the surface normal (degrees)
- $\int^{\circ}$ Integration over a hemisphere

References

1. K. Ya. Kondratyev, "Radiation in the Atmosphere." Academic Press, New York, 1969.
2. G. W. Paltridge and C. M. R. Platt, "Radiative Processes in Meteorology and Climatology." American Elsevier, New York, 1976.

3. B. D. Hunn and D. D. Calafell, Determination of average ground reflectivity for solar collectors. *Solar Energy* **19** (1), 87–89 (1977).
4. G. G. Gubareff, J. E. Janssen and R. H. Torborg, Thermal Radiation Properties Survey. Honeywell Research Center, Minneapolis, Minnesota (1960) (available from University Microfilms, Inc., Ann Arbor, Michigan).
5. E. V. Ashburn and R. G. Weldon, Spectral diffuse reflectance of desert surfaces. *Opt. Soc. Am.* **46,** 583–586 (1956).
6. J. L. Monteith, Reflection of short wave radiation by vegetation. *Q. J. R. Meteorol. Soc.* **85,** 386–392 (1959).
7. R. Tousey and E. O. Hulburt, Brightness and polarization of the daylight sky at various altitudes above sea level. *Opt. Soc. Am.* **37,** 78–92 (1947).
8. P. J. List, Smithsonian Meteorological Tables. *Smith. Misc. Coll.* pp. 442–443 (1966).
9. N. N. Kalitin, The measurements of the albedo of a snow cover. *Mon. Weather Rev.* **58,** 59–61 (1930).
10. K. Ya. Kondratyev, Radiation Processes in the Atmosphere. World Meteorological Organization, W.M.O. No. 309 (1972).
11. J. H. Conover, Cloud and terrestrial albedo determinations from Tiros satellite pictures. *Appl. Meteorol.* **4,** 378–386 (1965).
12. H. H. Kimball and I. F. Hand, Reflectivity of different kinds of surfaces. *Mon. Weather Rev.* **58,** 280–282 (1930).
13. SOLMET Vol. 2, Hourly solar radiation–surface meteorological observations. Final Rep., TD-9724, National Climatic Center, Ashville, North Carolina (1979).

Chapter 10

CONFIGURATION FACTORS

10.1 Introduction

The next chapter treats calculation of radiation incident on inclined surfaces. This will require determination of the values of configuration factors between the surface and the sky and between the surface and the ground. To provide this background, this chapter presents the basic formulas related to these configuration factors (also called shape factors, geometric factors, etc.).

10.2 Mathematical Formulation

Consider two surfaces A_1 and A_2 (Fig. 10.2.1). Let A_1 be the radiating surface and A_2 be the irradiated surface. Assume A_1 is perfectly diffuse and A_2 receives radiation from A_1 only. By definition, the configuration factor between A_1 and A_2 is as follows:

$$F_{A_1 \to A_2} = \frac{\text{Radiation incident on } A_2}{\text{Total radiation leaving } A_1}. \tag{10.2.1}$$

In the above definition it is useful to emphasize two points: radiation incident on A_2 is not necessarily the radiation absorbed by it, and the radiation

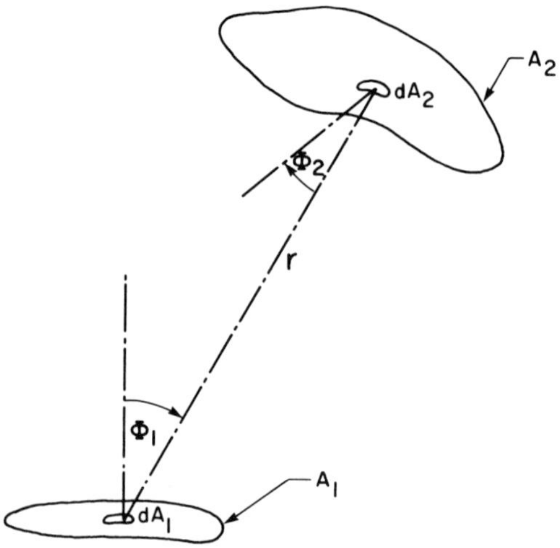

Figure 10.2.1 Geometrical relationship between two surfaces.

leaving A_1 is not necessarily the radiation emitted by this surface. Similar to the above, configuration factor definitions can be written between the surfaces $dA_1 \rightarrow dA_2$, $dA_1 \rightarrow A_2$, and $A_1 \rightarrow dA_2$.

We start by developing a configuration factor relationship between the elementary surfaces dA_1 and dA_2 at a distance r from each other. Let i_1 be the radiation intensity leaving dA_1 such that the total energy leaving this surface is given by the expression

$$i_1 \pi \, dA_1.$$

The energy leaving dA_1 that is incident on dA_2 is written

$$i_1 \cos \Phi_1 \, dA_1 \, d\omega_{12}.$$

The solid angle $d\omega_{12}$ is given thus:

$$d\omega_{12} = (dA_2 \cos \Phi_2)/r^2. \tag{10.2.2}$$

Φ_1 and Φ_2 are the angles between the normals to the elemental surfaces and the line joining dA_1 and dA_2, respectively. By definition, the configuration factor between dA_1 and dA_2 can be written

$$F_{dA_1 \rightarrow dA_2} = \frac{\text{Energy incident on } dA_2}{\text{Total energy leaving } dA_1} = \frac{i_1 \, dA_1 \cos \Phi_1 \, d\omega_{12}}{i_1 \pi \, dA_1}$$

$$= \frac{dA_2 \cos \Phi_1 \cos \Phi_2}{\pi r^2} \tag{10.2.3}$$

Multiplying both sides of (10.2.3) by dA_1, we obtain

$$dA_1\, F_{dA_1 \to dA_2} = dA_2 (dA_1 \cos \Phi_2 \cos \Phi_1)/\pi r^2$$

or

$$dA_1\, F_{dA_1 \to dA_2} = dA_2\, F_{dA_2 \to dA_1}. \tag{10.2.4}$$

Equation (10.2.4) shows the reciprocal nature of the configuration factors and is valid regardless of the sizes of the surfaces involved. In a general form, this relationship between any two surfaces A_j and A_k can be written

$$A_j F_{A_j \to Ak} = A_k F_{A_k \to A_j} \tag{10.2.5}$$

and is called the reciprocity theorem.

Continuing the above process, the configuration factor between the elementary surface dA_1 and the surface A_2 is as follows:

$$\begin{aligned} F_{dA_1 \to A_2} &= \left(\int_{A_2} i_1 \cos \Phi_1\, dA_1\, d\omega_{12} \right) \Big/ i_1 \pi\, dA_1 \\ &= \int_{A_2} \frac{\cos \Phi_1\, d\omega_{12}}{\pi} \\ &= \int_{A_2} \frac{\cos \Phi_1 \cos \Phi_2\, dA_2}{\pi r^2} \qquad (10.2.6) \\ &= \int_{A_2} F_{dA_1 \to dA_2}. \qquad (10.2.7) \end{aligned}$$

Equation (10.2.7) shows the additive nature of the configuration factors: since configuration factors are fractions, these fractions can be added. Consider a surface A_j completely enclosed by n other surfaces. For such an enclosure, the following equality can be written:

$$\sum_{k=1}^{n} F_{A_j \to A_k} = 1. \tag{10.2.8}$$

Analytical solution of equations such as (10.2.6) can be quite complicated. Here one method of solution is represented, which is valid when at least one surface is very small and the other very large. This method is called the unit-sphere method and is especially suitable for evaluating configuration factors between an inclined surface and the sky and between the inclined surface and the ground.

10.3 Unit-Sphere Method of Evaluating Configuration Factors

Consider an elemental area dA_1 at a distance r from the elemental area dA_2 in Fig. 10.3.1. The surface dA_1 is enclosed by a hemisphere of unit radius with its center at point 0. The term dA_2' represents a projection of the surface area dA_2 on the surface of the hemisphere such that the solid angle subtended by the surface dA_2 at point 0 is equal to the solid angle subtended by dA_2' at point 0. Therefore

$$d\omega_{12} = (dA_2 \cos \Phi_2)/r^2 = dA_2'/1^2$$

or

$$d\omega_{12} = dA_2'. \tag{10.3.1}$$

The area dA_2'' represents a projection of dA_2' on the base of the hemisphere. Therefore

$$dA_2'' = dA_2' \cos \Phi_1. \tag{10.3.2}$$

From Eq. (10.2.3), the configuration factor between the two elemental areas is as follows:

$$F_{dA_1 \to dA_2} = (\cos \Phi_1 \, d\omega_{12})/\pi.$$

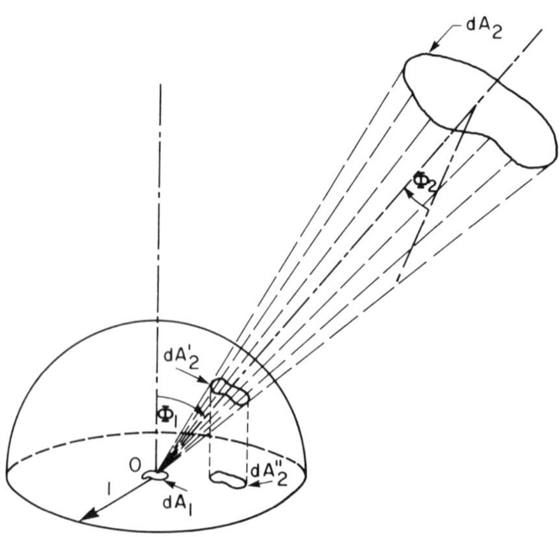

Figure 10.3.1 Projection of a surface onto a unit hemisphere.

Substituting the value of $d\omega_{12}$ from (10.3.1) into the above, we obtain

$$F_{dA_1 \to dA_2} = (\cos \Phi_1 \, dA_2')/\pi$$

or

$$F_{dA_1 \to dA_2} = dA_2''/\pi \tag{10.3.3}$$

$$= \frac{\text{Projected area on the base of a unit sphere}}{\text{Area of the base of a unit sphere}}. \tag{10.3.4}$$

The above equation also holds when determination of the configuration factor between an elemental area dA_1 and a large area A_2 is required. In the next section we apply this procedure to flat-plate collectors or any inclined surface.

10.4 Configuration Factors between an Inclined Flat-Plate Collector and the Sky and between the Collector and the Ground

Consider Fig. 10.4.1, showing a flat-plate collector inclined at an angle β from the ground. Let the collector be at the center of a hemisphere of unit radius. The surface of this hemisphere above the ground is the portion of the sky dome seen by the collector. The shaded area in the lower portion of this diagram is a projection of the sky dome seen by the collector on the base of a hemisphere of unit radius.

Let $F_{c \to s}$ be designated as the configuration factor between the collector and the sky; then

$$F_{c \to s} = \frac{\text{Shaded area in Fig. 10.4.1}}{\text{Area of a circle of unit radius}}$$

$$= [\tfrac{1}{2}\,(\text{area of a circle of unit radius}) + \tfrac{1}{2}\,(\text{area of an ellipse with axes of unity and } \cos \beta)]/\pi.$$

The above reduces to

$$F_{c \to s} = \tfrac{1}{2}(1 + \cos \beta). \tag{10.4.1}$$

Let A_c be the surface area of the collector and A_s that of the sky dome seen by it. We utilize these areas in the reciprocity relationship [Eq. (10.2.5)], and write

$$A_c F_{c \to s} = A_s F_{s \to c}. \tag{10.4.2}$$

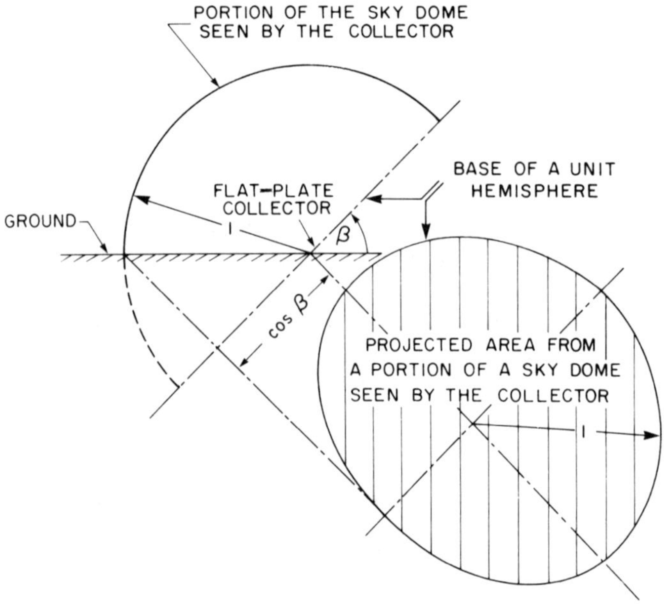

Figure 10.4.1 Projection of an inclined surface onto a unit hemisphere.

In the above, $F_{s \to c}$ is a configuration factor between the sky and the collector which may now be written as follows:

$$F_{s \to c} = (A_c/A_s)\tfrac{1}{2}(1 + \cos\beta). \tag{10.4.3}$$

The configuration factor between the collector surface and the ground, $F_{c \to g}$, can also be obtained from the unit-sphere method or from the enclosure equation [Eq. (10.2.8)]. Using the enclosure equation, we obtain

$$F_{c \to s} + F_{c \to g} = 1. \tag{10.4.4}$$

Therefore

$$F_{c \to g} = \tfrac{1}{2}(1 - \cos\beta). \tag{10.4.5}$$

Using the reciprocity theorem again, we find

$$F_{g \to c} = (A_c/A_g)\tfrac{1}{2}(1 - \cos\beta), \tag{10.4.6}$$

where $F_{g \to c}$ is the configuration factor between the ground and the collector and A_g is the area of the ground the collector sees.

In the above formulas, it is assumed that the radiation coming from the sky and that coming from the ground are perfectly diffuse. In reality, the sky diffuse radiation is usually anisotropic. If the distribution of its anisotropy were known, the sky dome could be divided into several parts in such a way that the radiation from each part might be assumed isotropic. The configuration factors from such parts of the sky and the collector, and hence the irradiation from such parts of the sky on the collector, could be computed.

For ground-reflected radiation, a process similar to the foregoing can be carried out to account for the anisotropy in reflected diffuse radiation. However, the matter is complicated if the reflections are specular. To handle such a situation, recourse has to be made to specialized literature such as Siegel and Howell [1] and Sparrow and Cess [2].

The above completes all the necessary configuration factor formulas required for inclined surfaces. The procedure laid down here can also be applied to determine shading effects of adjoining buildings on collectors or other surfaces.

Nomenclature

A	Area (m^2)
F	Configuration factor (dimensionless)
i	Intensity (W m^{-2} sr^{-1})
r	Distance from one elemental area to another (m)
β	Surface inclination from the horizontal (degrees)
Φ	Angle between the normal to a surface and the line joining another surface (degrees)
ω	solid angle (sr)

Subscripts

$c \rightarrow s$	From the inclined surface to the sky
$c \rightarrow g$	From the inclined surface to the ground
$s \rightarrow c$	From the sky to the inclined surface
$g \rightarrow c$	From the ground to the inclined surface
j, k	jth or kth surface
1, 2	Area 1 or 2 or at area 1 or 2

References

1. R. Siegel and J. R. Howell, "Thermal Radiation Heat Transfer." McGraw-Hill, New York, 1981.
2. E. M. Sparrow and R. D. Cess, "Radiation Heat Transfer." McGraw-Hill, New York, 1978.

Chapter 11

SOLAR RADIATION INCIDENT ON TILTED PLANES ON THE EARTH'S SURFACE

11.1 Introduction

Quantitative assessment of radiation arriving on an inclined surface is very important to engineers designing solar energy collecting devices, to architects designing buildings, and to agronomists studying insolation on foliage and vegetation on mountain slopes. In some solar energy devices, such as concentrators, determination of the instantaneous value of direct radiation incident on a reflecting surface is required. Moreover, in analyzing flat-plate collectors, for instance, direct and diffuse radiation quantities must be determined. In the latter case, values during a minimum period of 1 h need to be established.

In this chapter, we begin with measured or predicted global and diffuse radiation received on a horizontal surface and then transpose these quantities onto an inclined plane. Basically, the same procedure applies to theoretically calculated clear-sky irradiation on horizontal planes. The main difference lies in the treatment of sky diffuse radiation.

Consider a plane inclined at an angle β from the horizontal position (Fig. 11.1.1). For the time being, we assume that the surface is inclined in such a way

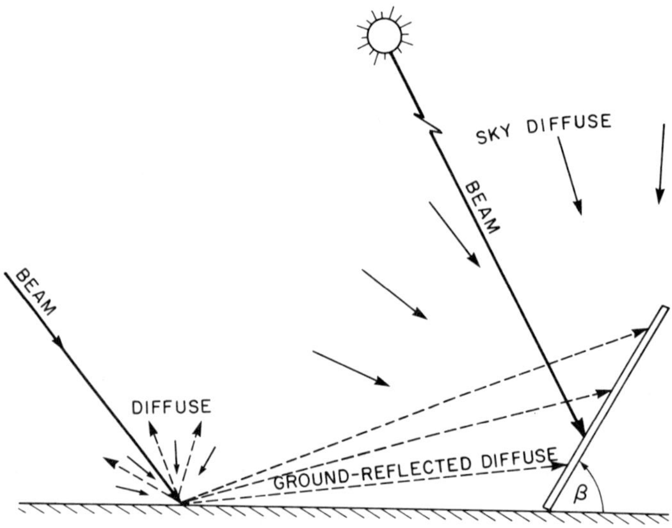

Figure 11.1.1 Incidence of beam, sky diffuse, and ground-reflected radiation on an inclined surface.

that it faces the sun; that is, direct radiation is not striking the back of this plane. As illustrated in this diagram, total radiation incident on the plane is composed of three components:

(1) direct radiation,
(2) radiation reflected from the ground, and
(3) sky diffuse radiation.

In the following sections, we develop procedures to compute each component separately. These procedures apply equally to monochromatic and spectrally integrated values. We begin with direct radiation.

11.2 Beam Radiation Incident on an Inclined Surface

In this section, instantaneous, hourly, and daily values of beam radiation incident on inclined planes are considered.

A. *Beam Irradiance on an Inclined Plane*

Suppose the beam irradiance $\dot{I}_b$ on a horizontal surface is known. This quantity can also be expressed in terms of the extraterrestrial irradiance on a horizontal surface and the (spectrally integrated) total instantaneous transmittance of the earth's atmosphere to beam radiation. Let this transmission factor be $\tau_b(\omega)$, where ω within parentheses indicates that τ_b varies with the hour angle. This $\tau_b(\omega)$ applies to cloudy as well as cloudless skies. The beam irradiance on a horizontal surface can be written as follows:

$$\dot{I}_b = \tau_b(\omega)\dot{I}_0. \tag{11.2.1}$$

In a manner similar to the above, beam irradiance on an arbitrarily inclined plane can be written thus:

$$\dot{I}_{b\beta\gamma} = \tau_b(\omega)\dot{I}_{0\beta\gamma}. \tag{11.2.2}$$

We would like to express $\dot{I}_{b\beta\gamma}$ in terms of $\dot{I}_b$. Multiplying and dividing the right-hand side of Eq. (11.2.2) by each side of (11.2.1), we have

$$\dot{I}_{b\beta\gamma} = \dot{I}_b\left[\tau_b(\omega)\dot{I}_{0\beta\gamma}/\tau_b(\omega)\dot{I}_0\right]$$

or

$$\dot{I}_{b\beta\gamma} = \dot{I}_b(\dot{I}_{0\beta\gamma}/\dot{I}_0). \tag{11.2.3}$$

In other words, the above can be written as follows:

$$\begin{array}{l}\text{Beam irradiance}\\ \text{on an inclined}\\ \text{surface on earth}\end{array} = \begin{array}{l}\text{Beam irradiance}\\ \text{on a horizontal}\\ \text{surface on earth}\end{array} \times \frac{\begin{array}{l}\text{Extraterrestrial}\\ \text{irradiance on an}\\ \text{inclined surface}\end{array}}{\begin{array}{l}\text{Extraterrestrial}\\ \text{irradiance on a}\\ \text{horizontal surface}\end{array}}. \tag{11.2.4}$$

Equation (11.2.3) can be written

$$\dot{I}_{b\beta\gamma} = \dot{I}_b(\cos\theta/\cos\theta_z) = \dot{I}_b\dot{r}_b, \tag{11.2.5}$$

where $\dot{r}_b$ is a conversion factor for beam irradiance discussed in Chapter 4. Equation (11.2.5) is valid for instantaneous values of radiative fluxes.

B. *Hourly Beam Irradiation on an Inclined Plane*

Integrating Eqs. (11.2.1) and (11.2.2) over a period of an hour and following the same procedure as before, we have

$$I_{b\beta\gamma} = I_b \int^{1\,\text{h}} \tau_b(\omega)\dot{I}_{0\beta\gamma}\,d\omega \bigg/ \int^{1\,\text{h}} \tau_b(\omega)\dot{I}_0\,d\omega. \tag{11.2.6}$$

The variation of $\tau_b(\omega)$ is usually not known; assuming it to be constant over a period of an hour, the above reduces to

$$I_{b\beta\gamma} = I_b \int^{1\,h} \dot{I}_{0\beta\gamma}\, d\omega \Big/ \int^{1\,h} \dot{I}_0\, d\omega, \tag{11.2.7}$$

$$I_{b\beta\gamma} = I_b(I_{0\beta\gamma}/I_0). \tag{11.2.8}$$

With the approximation of Eq. (4.4.5) and taking the time at midhour, we can write the above as follows:

$$I_{b\beta\gamma} = I_b(\cos\theta/\cos\theta_z) = I_b r_b. \tag{11.2.9}$$

Naturally, Eq. (11.2.9) is approximate, whereas Eq. (11.2.5) is exact.

C. *Daily Beam Irradiation on an Inclined Plane*

If we follow the above procedure, the daily beam irradiation on an inclined plane can be written as follows:

$$H_{b\beta\gamma} = H_b \int_{\omega_{sr}}^{-\omega_{ss}} \tau_b(\omega)\dot{I}_{0\beta\gamma}\, d\omega \Big/ \int_{\omega_s}^{-\omega_s} \tau_b(\omega)\dot{I}_0\, d\omega. \tag{11.2.10}$$

Assuming that $\tau_b(\omega)$ is constant throughout an individual day the above reduces to

$$H_{b\beta\gamma} = H_b \int_{\omega_{sr}}^{-\omega_{ss}} \dot{I}_{0\beta\gamma}\, d\omega \Big/ \int_{\omega_s}^{-\omega_s} \dot{I}_0\, d\omega \tag{11.2.11}$$

or

$$H_{b\beta\gamma} = H_b H_{0\beta\gamma}/H_0. \tag{11.2.12}$$

In other words,

$$H_{b\beta\gamma} = H_b R_b, \tag{11.2.13}$$

where R_b is a conversion factor for the daily beam radiation. For arbitrarily oriented surfaces, an expression for R_b is given by Eq. (4.4.18), and for a surface facing the equator, R_b may be obtained from Eq. (4.4.12) or Fig. 4.4.2. The monthly average values of beam radiation arriving on inclined planes may be computed by putting $\delta = \delta_c$ in r_b and R_b; the corresponding factors are called $\bar{r}_b$ and $\bar{R}_b$.

The daily value of beam radiation incident on an inclined plane can also be written in terms of the summation of its hourly values. That is,

$$H_{b\beta\gamma} = \sum^{\text{day}} I_{b\beta\gamma}. \tag{11.2.14}$$

D. Beam Radiation on an Inclined Plane at the Equinox

A special case arises for surfaces facing the equator: $\gamma = 0$. Let us reconsider, for example, Eq. (11.2.6) for hourly radiation. At the equinox, $\delta = 0$, and this equation reduces to

$$I_{b\beta} = I_b \int^{1\text{ h}} \tau_b(\omega)\cos(\phi - \beta)\, d\omega \Big/ \int^{1\text{ h}} \tau_b(\omega)\cos\phi\, d\omega \qquad (11.2.15)$$

or

$$I_{b\beta} = I_b \left(\cos(\phi - \beta) \int^{1\text{ h}} \tau_b(\omega)\, d\omega\right) \Big/ \left(\cos\phi \int^{1\text{ h}} \tau_b(\omega)\, d\omega\right). \qquad (11.2.16)$$

Hence

$$I_{b\beta} = I_b \cos(\phi - \beta)/\cos\phi. \qquad (11.2.17)$$

The above shows that at the equinox, for surfaces facing the equator, it is not necessary to assume that $\tau_b(\omega)$ is constant over a period of an hour. It can easily be shown that the above applies also to the daily values; thus

$$H_{b\beta} = H_b \cos(\phi - \beta)/\cos\phi. \qquad (11.2.18)$$

In the foregoing material, the final expressions for r_b and R_b have been obtained by assuming that the atmospheric transmittance of beam radiation remains constant throughout an hour or a day. The procedure followed was that laid down by Liu and Jordan [1] for surfaces facing the equator. The values of $\tau_b(\omega)$, in fact, may vary considerably during a particular day. Even on a long-term average basis, the variation of $\bar{\tau}_b$ may be substantial (Fig. 8.7.4). In spite of the hourly variation of τ_b or $\bar{\tau}_b$, it has been observed that the final computations of insolation on inclined planes through the foregoing procedure give very good results.

To determine mean hourly or daily beam irradiance on an inclined plane, all the foregoing equations can be repeated by placing bars over the radiation symbols.

11.3 Ground-Reflected Radiation Incident on an Inclined Plane

We treat separately (a) the hourly and (b) the daily amounts of ground-reflected radiation incident on inclined surfaces.

A. *Hourly Ground-Reflected Irradiation*

Consider again Fig. 11.1.1. The radiation arriving on the ground is composed of both beam and diffuse components. (The word "ground" here means a composite of all the earth's surface that the inclined plane "sees.") Depending on the earth's cover, the albedos for beam and diffuse radiation, ρ_b and ρ_d, respectively, may not be identical. Let A_g be the total ground area seen by the inclined plane. The total radiation reflected by the ground cover into an entire hemisphere can be written as follows:

$$(I_b\rho_b + I_d\rho_d)A_g.$$

We consider two particular cases: (1) perfectly diffuse reflection, called isotropic reflection; and (2) imperfectly diffuse reflection, called anisotropic reflection.

(I) Isotropic Reflection

Isotropic reflection usually occurs when the global radiation is composed of primarily diffuse radiation and/or when the ground cover is a perfectly diffuse reflector, such as a concrete floor. Under the isotropic condition, the fraction of the above energy incident on the inclined plane is given by the above quantity multiplied by the configuration factor from the ground to the inclined plane. Thus

$$I_r A_c = (I_b\rho_b + I_d\rho_d)A_g F_{g\to c}, \tag{11.3.1}$$

where I_r is the ground-reflected radiation per unit area arriving at the inclined plane, A_c is the surface area of the inclined plane, and $F_{g\to c}$ is the configuration factor from the ground to the inclined plane. Using reciprocity, Eq. (10.2.5), we have

$$I_r = (I_b\rho_b + I_d\rho_d)F_{c\to g}, \tag{11.3.2}$$

or

$$I_r = (I_b\rho_b + I_d\rho_d)\tfrac{1}{2}(1 - \cos\beta). \tag{11.3.3}$$

When the reflectances to beam and diffuse radiation are identical, we can use a common albedo. Under such a condition (11.3.3) reduces to the following:

$$I_r = \tfrac{1}{2}I\rho(1 - \cos\beta). \tag{11.3.4}$$

(II) Anisotropic Reflection

Under clean and cloudless skies, global radiation is composed primarily of direct radiation. When the ground is covered with a layer of water or with plants having glossy leaves, the reflection of such radiation is usually anisotropic (Fig. 9.2.1b). Temps and Coulson [2] recommended that under these conditions Eq. (11.3.4) should be multiplied by the following factor:

$$[1 + \sin^2(\theta_z/2)](|\cos \Delta|),$$

where Δ is the azimuth of the tilted surface with respect to that of the sun; this angle reduces to ω for surfaces tilted toward the equator. Consequently, with anisotropic reflection produced under clear skies, the ground-reflected radiation incident on an inclined surface can be written as follows:

$$I_r = \tfrac{1}{2} I\rho(1 - \cos \beta)[1 + \sin^2(\theta_z/2)](|\cos \Delta|). \qquad (11.3.5)$$

B. *Daily Ground-Reflected Irradiance*

Again, we treat separately the isotropic reflection and anisotropic reflection.

(I) Isotropic Reflection

The daily amount of ground-reflected radiation incident on an inclined plane follows directly from Eq. (11.3.4) and can be written as

$$H_r = \tfrac{1}{2} H\rho(1 - \cos \beta). \qquad (11.3.6)$$

(II) Anisotropic Reflection

The daily amount of anisotropically reflected radiation incident on an inclined plane cannot be expressed in terms of the daily horizontal global radiation. However, we can express it in terms of hourly summations as follows:

$$H_r = \sum^{\text{day}} \left[I\rho \left(\frac{1 - \cos \beta}{2} \right) \left(1 + \sin^2 \frac{\theta_z}{2} \right) (|\cos \Delta|) \right]. \qquad (11.3.7)$$

Before proceeding further it is necessary to emphasize that the Eqs. (11.3.5) and (11.3.7) apply only when the skies are clean and cloudless, and reflection is not perfectly diffuse.

Before we discuss the sky diffuse radiation incident on an inclined plane, it seems pertinent to study the angular distribution of diffuse radiation arriving on a horizontal surface.

11.4 Angular Distribution of Sky Diffuse Radiation Intensity

Consider an elemental horizontal surface dA on the base of an imaginary hemisphere respresenting the sky dome (Fig. 11.4.1). We assume that the sky is composed of elemental surfaces such as surface dS propagating diffuse solar radiant energy toward the horizontal surface. Let i_d be the intensity[1] of diffuse radiation associated with the surface dS. The rate of radiant energy arriving on the surface dA from dS is given by

$$d\dot{I}_d = i_d \, d\omega \cos \Phi, \tag{11.4.1}$$

where i_d is the intensity of radiation (W m^{-2} sr^{-1}), $d\omega$ the solid angle represented by the area dS (sr), and Φ the zenith angle of the surface dS (degrees). Because intensity is always expressed in terms of an area normal to the axis of the solid angle, multiplication by $\cos \Phi$ in Eq. (11.4.1) takes this aspect into account.

Equation (11.4.1) is similar to Eqs. (2.4.3) and (9.2.2). The latter two equations represent the rate of radiant energy emitted and reflected, respectively, from the surface dA. On the other hand, in Eq. (11.4.1) we are dealing with the rate of radiation incident on the surface dA.

When the sun is occulted, the integration of Eq. (11.4.1) yields us the total rate of sky diffuse radiation received at the surface dA; this can be written as

$$\dot{I}_d = \int^{\Omega} i_d \, d\omega \cos \Phi. \tag{11.4.2}$$

If the intensity were isotropic—that is, independent of its position on the sky—similar to Eq. (2.4.8), the preceding equation would have reduced to

$$\dot{I}_d = \pi i_d, \tag{11.4.3}$$

and because $\dot{I}_d$ is available from routine measurements, i_d could be easily derived and the problem of computing sky diffuse radiation incident on an inclined surface solved.

Under realistic skies, the diffuse intensity is not uniform and varies with the sky conditions. For an accurate assessment of the sky diffuse radiation incident on an inclined plane, it is imperative that the distribution of i_d over the dome of the sky be measured. The intensity of sky diffuse radiation can be mapped by traversing over the sky dome a radiometer with a small

[1] In meteorological literature this term is called *radiance* and has the same units as that of *intensity*. However, we use intensity because in engineering it is applied to radiation either leaving or incident on a surface.

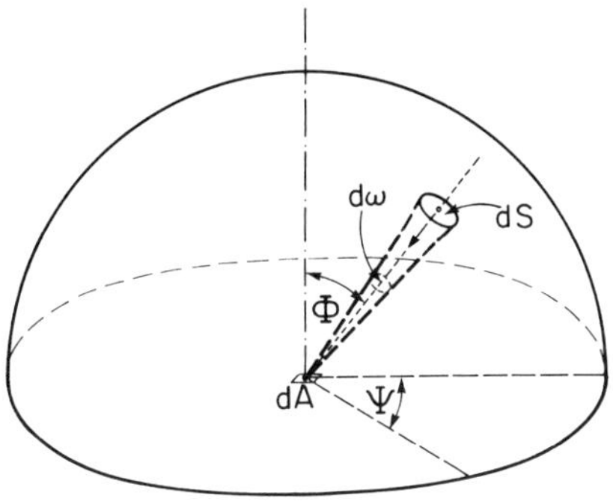

Figure 11.4.1 Coordinates of a small area on the surface of a hemisphere.

opening angle (a shading disk is used to occult the sun). Most investigations of this subject are based on radiometers, also called scanning telescopes. However, because of the time constant of a radiometer, scanning of the sky dome can take several minutes, during which time the sky conditions may change substantially. McArthur and Hay [3] have employed a photographic technique using a fish-eye lens. In this technique, densitometric evaluations of all-sky photographs are obtained.

All measurements, radiometric or photographic, have shown that under cloudless and clear skies the intensity from an element in the sky dome depends on the position of the element with respect to the sun and with respect to the zenith. Maximum intensities occur near the sun and the horizon and minimum intensities at an angle of 90° to the solar zenith. The radiation coming from a region around the solar disk is called circumsolar radiation or solar aureole and is caused by strong forward scattering of the aerosol particles. The extent of solar aureole depends on the atmospheric turbidity and the zenith angle of the sun. On the other hand, increased intensity near the horizon, called *horizon brightening*, results from the scattering by a larger air mass viewed by an observer in that direction. Figure 11.4.2a shows a densitometric contour map of the intensity distribution of sky diffuse radiation under a clear sky. In Figs. 11.4.2–11.4.4 the position of the sun is indicated by $*$. High circumsolar radiation and horizon brightening are apparent. Further, the intensity is symmetric about a plane containing the sun and the observer's zenith. Figure 11.4.2b shows a corresponding variation of i_d in a diagonal traverse through the sun and the observer's zenith. The position of a minimum at about 90° from the solar zenith is evident. With increased turbidity the curve will flatten out. The effect on intensity of atmospheric turbidity and of thin clouds is almost identical.

In Fig. 11.4.3a we show a contour map of intensity distribution for an overcast sky condition. It is apparent from this diagram that even in the case

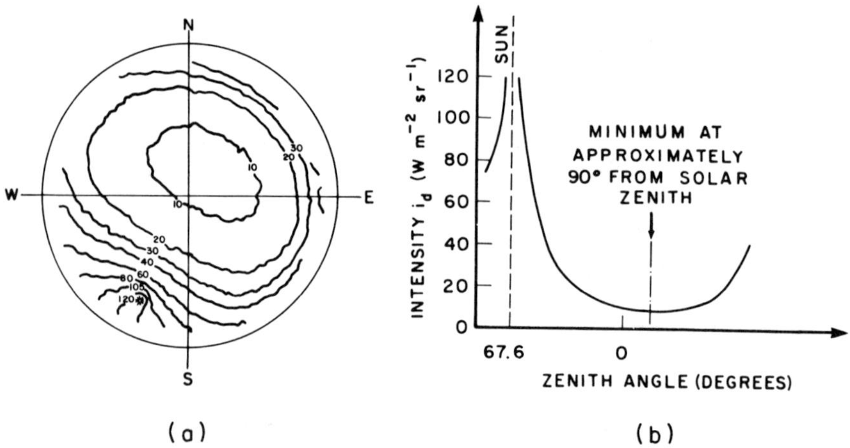

Figure 11.4.2 Angular variation of the sky diffuse intensity under clear skies. Adapted from McArthur and Hay [3].

of an overcast sky, the intensity is not completely isotropic over the sky dome. A diagonal traverse of intensity distribution is plotted in Fig. 11.4.3b and shows that in the case of an overcast sky, the intensity is maximum at about 90° from the solar zenith.

In Fig. 11.4.4a, partly cloudy sky conditions are presented. Because of strong reflections from the edges of scattered clouds, and from cloud patches, the distribution of intensity is highly complex. A diagonal traverse is presented in Fig. 11.4.4b and shows large variations in intensity from the solar zenith out to the horizon.

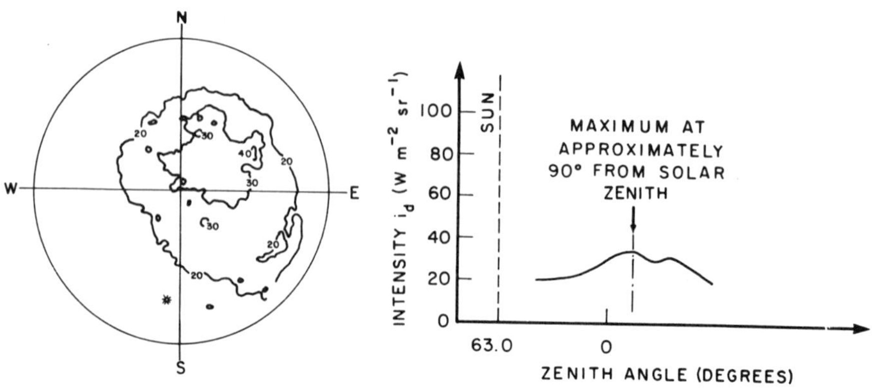

Figure 11.4.3 Angular variation of the sky diffuse intensity under overcast skies. Adapted from McArthur and Hay [3].

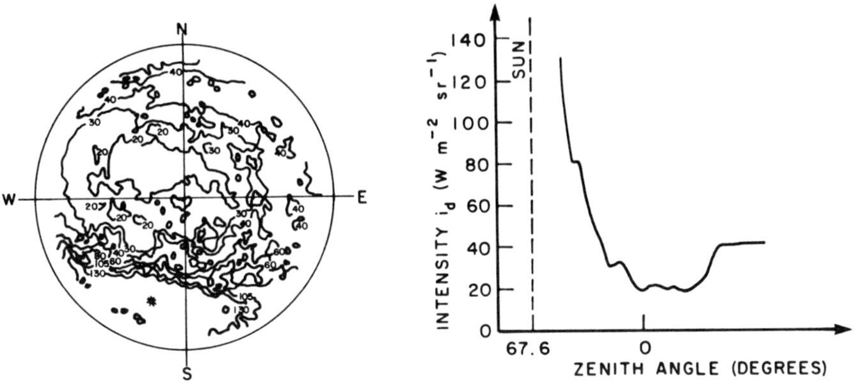

(a) (b)

Figure 11.4.4 Angular variation of the sky diffuse intensity under partly cloudy skies. Adapted from McArthur and Hay [3].

It is obvious that for an accurate assessment of radiation incident on an inclined plane, detailed information on the intensity distribution of sky diffuse radiation is required. However, these measurements are sparse. Moreover, the parameters influencing them, such as cloud characteristics, turbidity, and ground albedo, are often not measured simultaneously and hence do not allow one to draw generalizations and develop methods of predicting celestial distribution of intensity. Therefore, a number of empirical approximations have been developed and are presented in the next section.

11.5 Sky Diffuse Radiation Incident on an Inclined Plane

In the preceding section, we have seen that the sky conditions may be characterized under three broad categories—clear, overcast, and partly overcast. Empirical formulations have been developed for each category. We present below separate treatment for hourly and daily values.

A. *Hourly Sky Diffuse Radiation Incident on an Inclined Plane*

Following are three models for the three sky conditions.

(I) Circumsolar Model

This model applies to clean and cloudless skies. It assumes that all radiation arriving on a horizontal surface appears to come from the direction of the sun. Therefore, the diffuse radiation may be treated in the same manner as was beam radiation in Eq. (11.2.9) and is shown here in Fig. 11.5.1a. Consequently, the hourly sky diffuse radiation incident on an inclined plane can be written as follows:

$$I_s = I_d r_b. \tag{11.5.1}$$

This model in general overestimates sky diffuse irradiation.

(II) Isotropic Model

In this model the intensity of sky diffuse radiation is assumed uniform over the sky dome (Fig. 11.5.1b). It approximates the overcast sky condition shown in Fig. 11.4.3. Following the procedure laid down for ground-reflected radiation [Eqs. (11.3.1)–(11.3.4)], it can be shown that under the present assumption, the sky diffuse radiation incident on an inclined plane is given by the following expression:

$$I_s = \tfrac{1}{2} I_d (1 + \cos\beta). \tag{11.5.2}$$

This model underestimates sky diffuse irradiation on south-facing slopes.

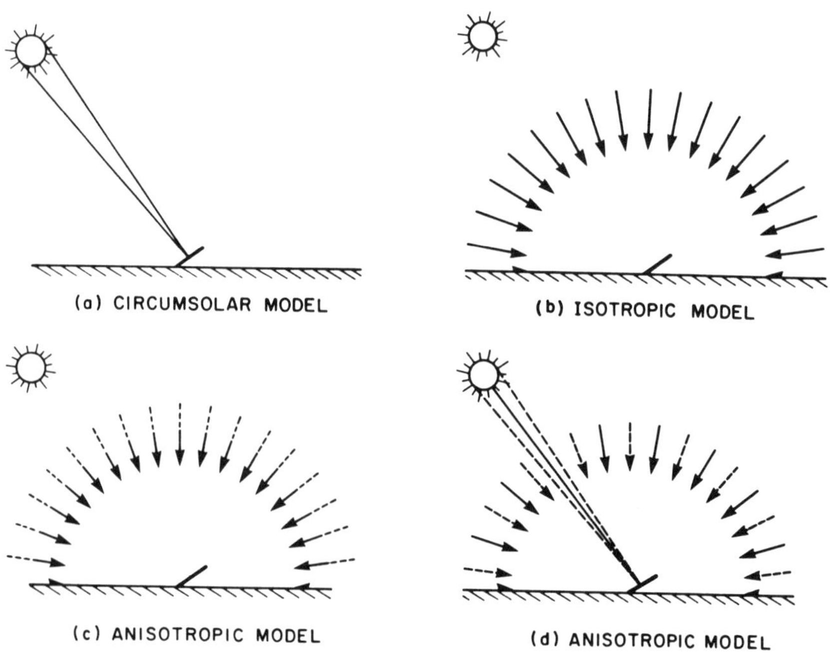

Figure 11.5.1 The various sky diffuse models: (a) circumsolar, (b) isotropic, (c) anisotropic (Klucher), and (d) anisotropic (Hay). Adapted from Hooper and Brunger [29], with permission of The American Society of Mechanical Engineers.

(III) Anisotropic Models

Here we try to model the conditions of partly cloudy skies shown in Fig. 11.4.4. This condition may vary from clear skies on one extreme to cloudy skies on the other extreme. We present below two empirical correlations, both of which require further validation.

(i) Anisotropic Model (Klucher). This model is based on a study of clear sky conditions by Temps and Coulson [2]. Their model was modified by Klucher [4], who incorporated conditions of cloudy skies. Temps and Coulson observed that the clear sky condition can be depicted by modifying the basic isotropic formulation, Eq. (11.5.2), by two factors: the factor

$$1 + \cos^2\theta \sin^3\theta_z$$

to take into account the increased intensity of sky diffuse radiation in the vicinity of the sun, and the factor

$$1 + \sin^3(\beta/2)$$

to take into account the brightening of the sky near the horizon. Consequently, for the clear sky conditions, the sky diffuse irradiation may be written thus:

$$I_s = \tfrac{1}{2} I_d(1 + \cos\beta)[1 + \sin^3(\beta/2)](1 + \cos^2\theta \sin^3\theta_z). \qquad (11.5.3)$$

Klucher modified the Temps and Coulson formulation as given below:

$$I_s = \tfrac{1}{2} I_d(1 + \cos\beta)[1 + F\sin^3(\beta/2)](1 + F\cos^2\theta \sin^3\theta_z), \qquad (11.5.4a)$$

where F is a modulating function given by

$$F = 1 - (I_d/I)^2 \qquad (11.5.4b)$$

which incorporates the cloudy sky conditions. When the skies are completely overcast $F = 0$ and Eq. (11.5.4) reverts to the isotropic model, Eq. (11.5.2). When skies are clear, $F \to 1$, and Eq. (11.5.4) approaches Eq. (11.5.3). The Temps–Coulson–Klucher model may be represented by Fig. 11.5.1c.

(ii) Anisotropic Model (Hay). This model, proposed by Hay [5], can be described as follows.

Let the diffuse radiation on a horizontal surface be composed of a circumsolar component coming directly from the direction of the sun, and an isotropically distributed diffuse component from the rest of the sky dome. Let these two components be weighted according to an isotropy index. This index is essentially a ratio of the beam radiation of a horizontal surface on the earth, I_b, to the extraterrestrial radiation on a horizontal surface I_0. The total diffuse radiation on a horizontal surface is divided into the circumsolar and the isotropically distributed diffuse radiation in the ratios

$$I_b/I_0 \qquad \text{and} \qquad (1 - I_b/I_0).$$

Consequently, the diffuse circumsolar radiation on a horizontal surface is

$$I_{dcs} = I_d(I_b/I_0), \tag{11.5.5}$$

and the isotropically distributed diffuse radiation on the horizontal surface is

$$I_{dis} = I_d(1 - I_b/I_0). \tag{11.5.6}$$

The sky diffuse radiation on an inclined plane, I_s, can now be obtained by appropriately transposing the above two quantities. Thus

$$I_s = I_d\left[(I_b/I_0)\, r_b + \tfrac{1}{2}(1 + \cos\beta)(1 - I_b/I_0)\right]. \tag{11.5.7}$$

It is evident from the above that when the beam radiation approaches zero, Eq. (11.5.7) reduces to Eq. (11.5.2), and when it approaches its extraterrestrial value, Eq. (11.5.1) is obtained.

Since pyrheliometric measurements are usually not available, we should write Eq. (11.5.7) in terms of the horizontal diffuse and global radiation. Thus

$$I_s = I_d\Biggl\{\underbrace{\frac{I - I_d}{I_0} r_b}_{\text{circumsolar}} + \underbrace{\tfrac{1}{2}(1 + \cos\beta)\left[1 - \frac{I - I_d}{I_0}\right]}_{\text{isotropic}}\Biggr\}. \tag{11.5.8}$$

This model may be represented by Fig. 11.5.1d.

B. *Daily Sky Diffuse Radiation Incident on an Inclined Plane*

We present below the corresponding expressions for the daily amounts.

(I) Circumsolar Model

In a manner similar to the development of Eq. (11.2.13), the daily amount of the circumsolar sky diffuse radiation incident on an inclined surface is given by

$$H_s = H_d R_b. \tag{11.5.9}$$

(II) Isotropic Model

Similar to Eq. (11.3.6), the daily amount of sky diffuse irradiation on an inclined surface can be written

$$H_s = \tfrac{1}{2}H_d(1 + \cos\beta). \tag{11.5.10}$$

(III) Anisotropic Models

(i) Anisotropic Model (Klucher). Corresponding to Eq. (11.5.4), an expression for the daily quantity is difficult to formulate. Consequently, we write

$$H_s = \sum^{\text{day}} \left\{ I_d \left(\frac{1 + \cos\beta}{2} \right) \left[1 + F \sin^3 \left(\frac{\beta}{2} \right) \right] (1 + F \cos^2\theta \sin^3\theta_z) \right\}. \quad (11.5.11)$$

(ii) Anisotropic Model (Hay). For this model it can be shown that

$$H_s = H_d\{[(H - H_d)/H_0]R_b + \tfrac{1}{2}(1 + \cos\beta)[1 - (H - H_d)/H_0]\}. \quad (11.5.12)$$

Equations (11.5.1)–(11.5.12) can be repeated to obtain the monthly average radiation by placing bars over the corresponding quantities.

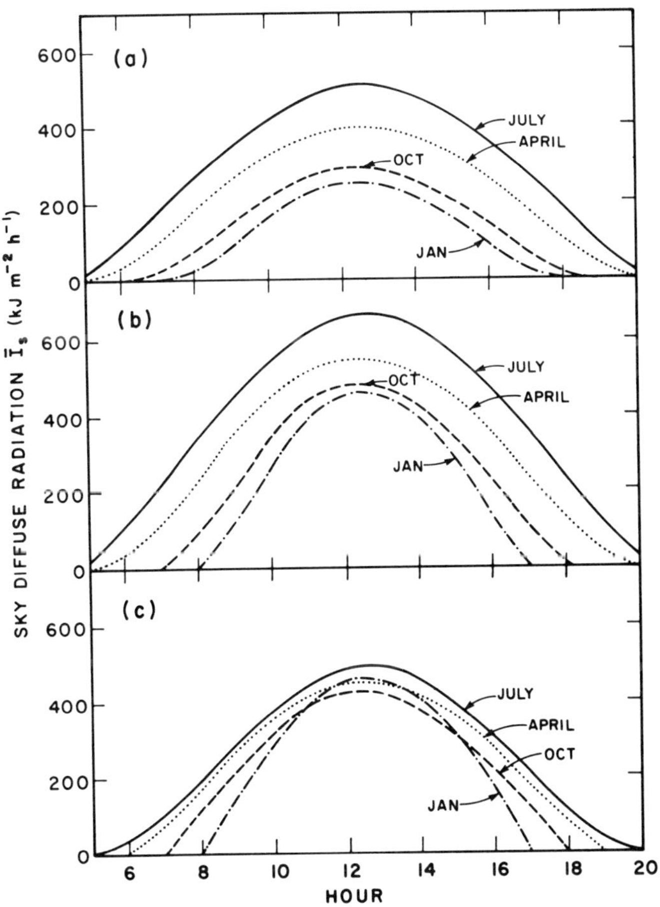

Figure 11.5.2 Diurnal variation of the sky diffuse radiation on south-facing vertical surfaces at Montreal. (a) Isotropic, (b) Klucher's model, and (c) Hay's model.

Let us examine the difference between the sky diffuse irradiation resulting from the isotropic model and the two anisotropic models. Using the measured mean hourly horizontal data for Montreal (Table 8.4.3), calculations of the hourly sky diffuse radiation were carried out for vertical surfaces. For the south-facing surfaces, results are shown in Fig. 11.5.2. Figure 11.5.2a represents the isotropic model for which the wall orientation has no bearing. The results from Klucher's model are presented in Fig. 11.5.2b, and those from Hay's model in Fig. 11.5.2c. The general shape of the three plots is similar. However, it can be observed from these diagrams that the isotropic approximation yields values somewhat lower than those obtained from the anisotropic models. For summer months, Klucher's model gives higher values than does Hay's model. For winter months, the two models yield almost identical results.

In Fig. 11.5.3, we consider sky diffuse irradiance on east-facing surfaces. Naturally, the isotropic model would yield the same results as in Fig. 11.5.2a, and consequently it is not shown here. However, the two anisotropic models yield very different results. Because Klucher's model is a modification of the basic isotropic model, the plots in Fig. 11.5.3a are almost symmetric around

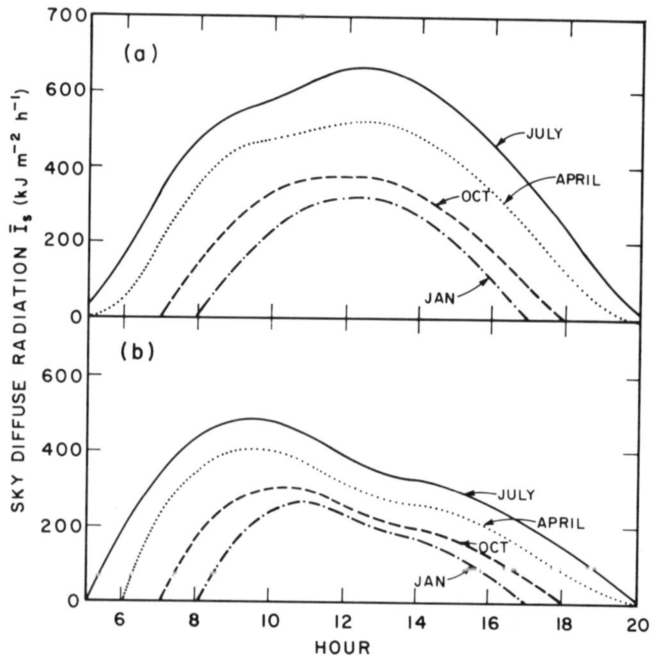

Figure 11.5.3 Diurnal variation of the sky diffuse radiation on east-facing vertical surfaces ($\beta = 90°$) at Montreal. (a) Klucher's model, (b) Hay's model.

noon. However, for an east-facing surface, the afternoon values should be lower than those for the forenoon—at least for the summer months—and they are not. The reason for the apparent failure of this method in this instance can be traced directly to Eq. (11.5.4a). The last term in parentheses disappears because the incidence angle does not exist in the afternoon. However, the term $1 + F\sin^3(\beta/2)$ increases the sky diffuse irradiation, and gives rise to the obvious anomaly. On the other hand, Hay's model appears to result in values being too low in the afternoon. This stems from the multiplier $1 - I_b/I_0$ in Eq. (11.5.7). In the afternoon, $r_b = 0$. However, this multiplier, which is always less than one, reduces the total amount of sky

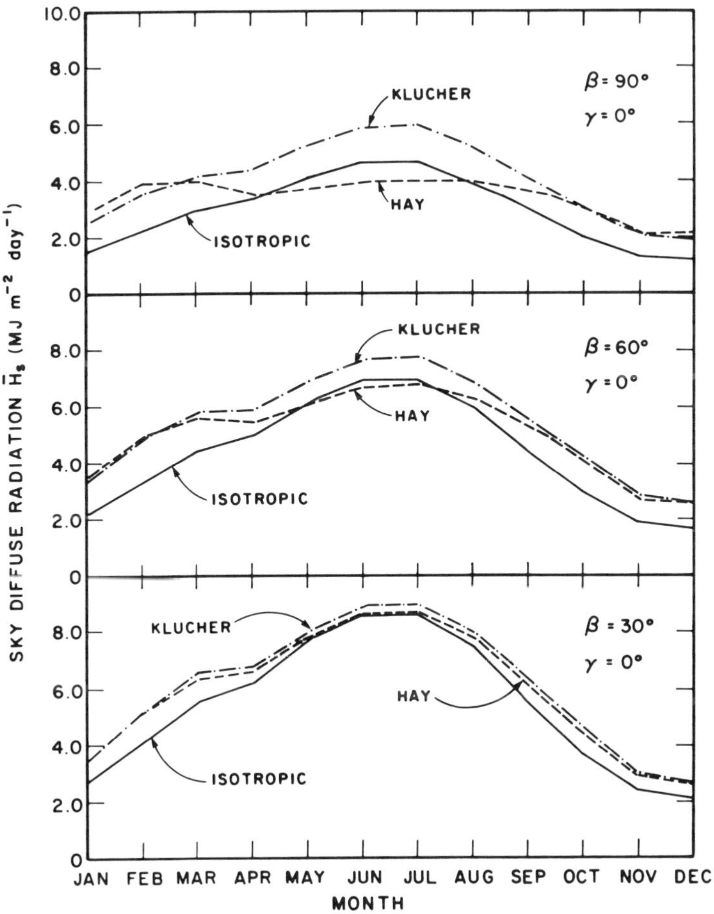

Figure 11.5.4 Annual variation of the sky diffuse radiation at various surface inclinations at Montreal.

diffuse radiation incident on the inclined plane. The comments made here on the two models for an east-facing surface apply equally to a west-facing surface. The conditions, however, will be reversed.

In Fig. 11.5.4 we present the results of calculations for the daily sky diffuse radiation incident on south-facing tilted surfaces. The monthly variations are shown for surfaces inclined at 30°, 60°, and 90° from the horizontal position. The two anisotropic models are compared to the isotropic model. At low inclinations, the two anisotropic models correspond very well. As tilt angle increases, differences emerge, but are confined to the months March to October.

From the foregoing, it is obvious that further studies are required to establish the superiority of one model over the other. For surfaces tilted toward the equator, and $\beta < 60°$, either of the two models may be used.[2] However, this author recommends Hay's model because it can be used in a straightforward manner for computing daily sky diffuse insolation on inclined planes when only the daily horizontal values are available.

11.6 Global Irradiation on an Inclined Plane

The total amount of radiation incident on an inclined plane is composed of beam, ground-reflected, and sky-diffuse components. In this section, these three components are put together for different time scales.

A. *Hourly Global Radiation Incident on an Inclined Plane*

At locations where the hourly global and diffuse radiation on horizontal surfaces are known or can be estimated, the global radiation on an inclined surface can be written

$$I_{\beta\gamma} = \underbrace{(I - I_{\mathrm{d}})r_{\mathrm{b}}}_{\text{beam}} + \underbrace{I_{\mathrm{r}} + I_{\mathrm{s}}}_{\text{diffuse}} \tag{11.6.1a}$$

or

$$I_{\beta\gamma} = I_{\mathrm{b}\beta\gamma} + I_{\mathrm{d}\beta}. \tag{11.6.1b}$$

[2] This statement has now been substantiated by measurements. See C. C. Y. Ma and M. Iqbal, Statistical comparison of models for estimating solar radiation on inclined surfaces, *Proc. ASES Ann. Meeting, Minneapolis/St. Paul, Minnesota, 1–3 June, 1983*, to be published in *Sol. Energy* **31** (3) (1983).

This equation is used extensively in research and in the development of simulation methods where separate values of the beam and diffuse components on inclined planes are required. With measured or estimated values of I and I_d, solar system analysis may be carried out hour by hour for several years. Such analysis can then lead to suitable design procedures that engineers and architects can directly employ.

Equation (11.6.1) applies to individual hours. The corresponding monthly mean quantities for inclined planes can be written as follows:

$$\bar{I}_{\beta\gamma} = \underbrace{(\bar{I} - \bar{I}_d)\bar{r}_b}_{\text{beam}} + \underbrace{\bar{I}_r + \bar{I}_s}_{\text{diffuse}} \tag{11.6.2a}$$

or

$$\bar{I}_{\beta\gamma} = \bar{I}_{b\beta\gamma} + \bar{I}_{d\beta}. \tag{11.6.2b}$$

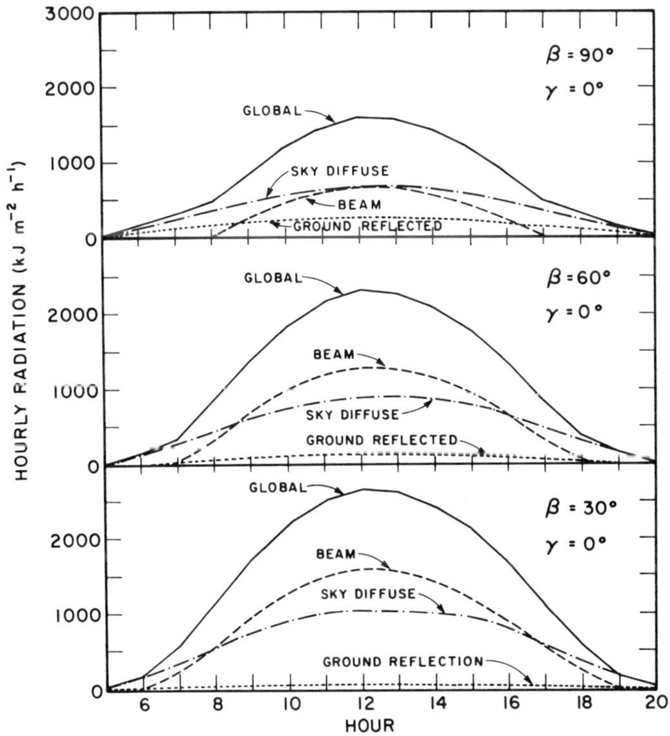

Figure 11.6.1 Diurnal variation of the different components of global radiation on inclined surfaces at Montreal in July. Ground-reflection is assumed isotropic.

This equation is very useful in analyzing such solar energy devices as flat-plate collectors. By utilizing the readily available quantities such as the monthly average radiation, temperature, and wind velocities, average collector efficiencies can be computed.

In Eqs. (11.6.1) and (11.6.2), the magnitudes of ground-reflected and sky diffuse radiation will depend on any particular model chosen, for example, isotropic or anisotropic. In Fig. 11.6.1 we show the magnitudes of global, beam, ground-reflected, and sky diffuse radiation incident on inclined planes. We utilize only one anisotropic model, and the horizontal data are for Montreal, again from Table 8.4.3. The results are shown for only the month of July, and the surface is tilted toward the equator at 30°, 60°, and 90° from the horizontal position. The ground-reflected component is minimum. The sky diffuse component is substantial; however, it is always less than the beam component except when the surface is almost vertical. The ratio of beam to sky diffuse components will, of course, vary with latitude and climate. In general, at high tilt angles it will increase as we move toward poles, and vice versa. At a given latitude, this ratio will be higher during winter than during summer.

B. *Daily Global Radiation Incident on an Inclined Plane*

Following the previous procedure, an expression for the daily value based on individual days can be written

$$H = (H - H_{\mathrm{d}})R_{\mathrm{b}} + H_{\mathrm{r}} + H_{\mathrm{s}} \tag{11.6.3a}$$

or

$$H_{\beta\gamma} = H_{\mathrm{b}\beta\gamma} + H_{\mathrm{d}\beta}. \tag{11.6.3b}$$

It may be pointed out that the daily values are seldom used in solar engineering either for research or for design purposes.

The methods used to obtain the horizontal average radiation $\overline{H}$ and $\overline{H}_{\mathrm{d}}$ were discussed in Chapter 8. In terms of these quantities, the slope radiation is given below:

$$\overline{H}_{\beta\gamma} = \underbrace{(\overline{H} - \overline{H}_{\mathrm{d}})\overline{R}_{\mathrm{b}}}_{\text{beam}} + \underbrace{\overline{H}_{\mathrm{r}} + \overline{H}_{\mathrm{s}}}_{\text{diffuse}} \tag{11.6.4a}$$

or

$$\overline{H}_{\beta\gamma} = \overline{H}_{\mathrm{b}\beta\gamma} + \overline{H}_{\mathrm{d}\beta}. \tag{11.6.4b}$$

The quantity $\overline{H}_{\beta\gamma}$ is most useful in estimating the long-term availability of solar energy on surfaces, whether these be mountain slopes or walls of buildings. In the design of buildings, the diffuse component in the above equation provides us with an estimate of the indirect illumination entering a building.

In the preceding equation the magnitude of the terms representing ground-reflected and sky diffuse radiation will vary according to the model under which they are computed. Under the assumption of isotropic ground reflection, isotropic sky diffuse radiation, and Hay's anisotropic model, the global radiation on inclined surfaces can be computed with the use of the daily horizontal values for global and diffuse radiation. Naturally, the daily global radiation on sloped surfaces can also be computed as a summation of the hourly slope radiation. For example,

$$\bar{H}_{\beta\gamma} = \sum^{\text{day}} \bar{I}_{\beta\gamma}. \tag{11.6.5}$$

The summation of the hourly values yields slightly, but only slightly, more accurate results than those obtained from Eq. (11.6.4), for instance. There is one main reason for this. The hourly procedure takes into account any asymmetry of radiation that usually exists around solar noon. At the equinox, however, $r_b = R_b$, etc., and consequently the hourly summations and the daily procedure yield identical values. As Fig. 11.6.2 shows, the results are identical for the equinox months, March and September. For other months, the two procedures give slightly different results. This particular diagram applies to mean values. For individual days, the differences between the daily and hourly summations may be substantial.

Before we proceed to the next topic, it will be useful to examine the effect of the two sky diffuse anisotropic models on the daily global radiation incident on inclined planes. In Fig. 11.6.3 we show these plots for south-facing surfaces inclined at 30°, 60°, and 90° from the horizontal position. The ground-reflected radiation has been assumed isotropic. It is apparent that at low tilt angles, the correspondence between the two anisotropic models is good.

□ EXAMPLE 11.6.1. The hourly global radiation incident on a collector on 4 June 1976 at the hour ending 11:00 (LAT) calculated for a flat-plate collector in Montreal (45°30′ N) tilted at 50° toward the equator. (Note: this appears to be a very clear day. It has high values of global and low values of diffuse radiation.)

Solution. On 4 June, $\delta = 22.41°$, $E_o = 0.970\,9$. For the hour ending at 11:00 (LAT), we obtain

$$I = 76 \text{ cal cm}^{-2}\text{ h}^{-1} = 3\,179 \text{ kJ m}^{-2}\text{ h}^{-1} \qquad \text{(Table 8.4.1)}$$

$$I_d = 9 \text{ cal cm}^{-2}\text{ h}^{-1} = 377 \text{ kJ m}^{-2}\text{ h}^{-1} \qquad \text{(Table 8.4.2)}.$$

We shall calculate the global radiation incident on the inclined plane under

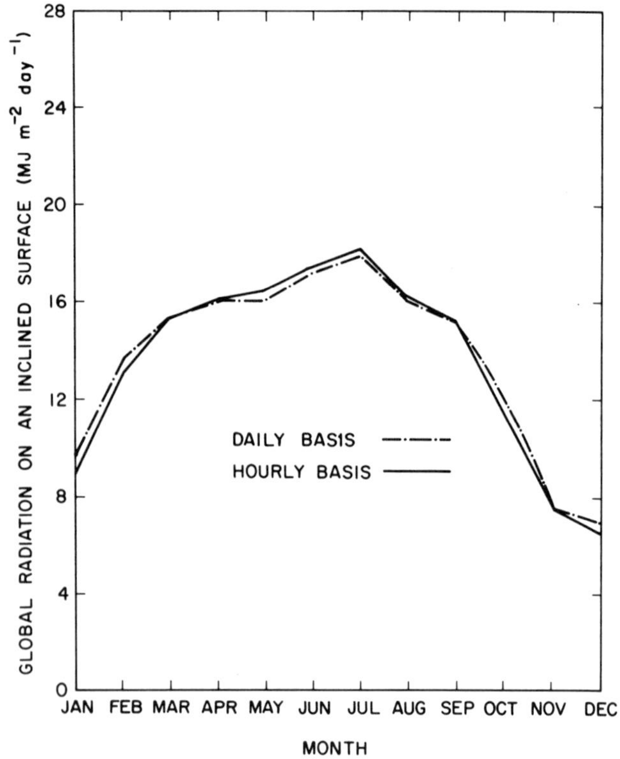

Figure 11.6.2 Comparison of daily versus hourly methods of calculating insolation on inclined planes at Montreal; slope = 50.0°. Adapted with permission from Iqbal [6], copyright 1978, Pergamon Press, Ltd.

the different sky diffuse models discussed above. Since the beam and the ground-reflected radiation are not influenced by the sky diffuse radiation models, we calculate first the former.

Hourly beam radiation incident on the inclined plane

$$\begin{aligned} I_{b\beta} &= I_b r_b \\ &= (I - I_d)(\cos\theta_0/\cos\theta_z) \qquad (\omega \text{ taken at midhour}) \\ &= (3\,179 - 377)(0.82/0.87) = 2\,641 \text{ kJ m}^{-2}\,\text{h}^{-1}. \end{aligned}$$

Hourly ground-reflected diffuse radiation incident on the inclined plane

$$I_r = \tfrac{1}{2} I \rho (1 - \cos\beta).$$

The albedos ρ given in Table 9.4.4 are for long-term average conditions. That is, all albedos with and without snow and rain. Under the circumstances, we

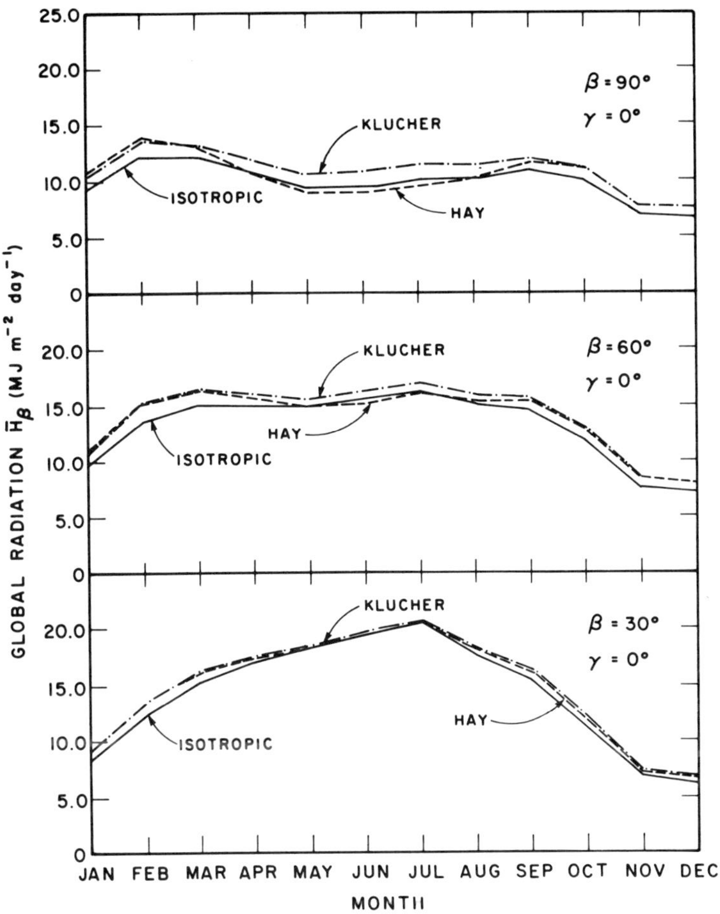

Figure 11.6.3 Comparison of the isotropic model with the two anisotropic models for computing global radiation on inclined planes at Montreal.

take $\rho = 0.2$. Therefore

$$I_r = (3\,179)\tfrac{1}{2}0.2\,(1 - \cos 50)$$
$$= 114\,\text{kJ m}^{-2}\,\text{h}^{-1}.$$

Because this appears to be a very clear day, let us recalculate the ground-reflected radiation assuming the reflection is anisotropic. The ground-reflected radiation incident on the inclined plane is given by Eq. (11.3.5),

$$I_r = (3\,179)\tfrac{1}{2}(0.2)(1 - \cos 50)\left[1 + \sin^2(29.5/2)\right](|\cos 22.5|)$$
$$= 112\,\text{kJ m}^{-2}\,\text{h}^{-1}.$$

There is very little difference between the isotropic and the anisotropic ground-reflected irradiation. The difference would have been slightly larger had the zenith angle been larger. Anisotropy of the ground-reflected radiation is large when the sun is near the horizon.

Hourly sky diffuse radiation incident on the inclined plane

1. Circumsolar model:

$$\begin{aligned} I_s &= I_d r_b \\ &= I_d(\cos\theta_0/\cos\theta_z) \\ &= 377(0.82/0.87) = 355\ \mathrm{kJ\,m^{-2}\,h^{-1}}. \end{aligned}$$

2. Isotropic model

$$\begin{aligned} I_s &= \tfrac{1}{2} I_d(1+\cos\beta) \\ &= \tfrac{1}{2} 377(1+\cos 50)\ \mathrm{kJ\,m^{-2}\,h^{-1}} \\ &= 310\ \mathrm{kJ\,m^{-2}\,h^{-1}}. \end{aligned}$$

3(a). Anisotropic model (Klucher)
from Eq. (11.5.4) we have

$$\begin{aligned} I_s &= \tfrac{1}{2} 377(1+\cos 50)\{1+[1-(377/3\,179)^2]\sin^3(50/2)\} \\ &\quad \times \{1+[1-(377/3\,179)^2](0.82)^2\sin^3 29.5\} \\ &= 359\ \mathrm{kJ\,m^{-2}\,h^{-1}}. \end{aligned}$$

3(b). Anisotropic model (Hay)
from Eq. (11.5.8) we can write

$$\begin{aligned} I_s &= 377\left[\frac{3\,179-377}{4\,921(0.84)}\frac{0.82}{0.87}+\frac{1}{2}(1+\cos 50)\left(1-\frac{3\,179-377}{4\,921(0.84)}\right)\right] \\ &= 340\ \mathrm{kJ\,m^{-2}\,h^{-1}}. \end{aligned}$$

The difference between the two sky diffuse anisotropic models is about 6%.
Global hourly radiation incident on the inclined plane is given by

$$I_\beta = I_{b\beta} + I_r + I_s.$$

From the four models, we have the following four values:

1. $I_\beta = 2\,641 + 114 + 355 = 3\,110\ \mathrm{kJ\,m^{-2}\,h^{-1}}$,

2. $I_\beta = 2\,641 + 114 + 310 = 3\,065\ \mathrm{kJ\,m^{-2}\,h^{-1}}$,

3(a). $I_\beta = 2\,641 + 114 + 359 = 3\,114\ \mathrm{kJ\,m^{-2}\,h^{-1}}$,

3(b). $I_\beta = 2\,641 + 114 + 340 = 3\,095\ \mathrm{kJ\,m^{-2}\,h^{-1}}$.

It is apparent from the above that the global radiation is dominated by its beam component. This is so because, for this individual hour, the horizontal diffuse radiation is only about 12% of the global radiation. Furthermore, the small differences in the sky diffuse radiation resulting from the four different models are insignificant as far as the global radiation incident on the inclined plane is concerned. □

□ EXAMPLE 11.6.2. Daily global radiation incident on the inclined plane calculated for the case of Example 11.6.1.

Solution. From Tables 8.4.1 and 8.4.2, on 4 June 1976

$$H = 710 \text{ cal cm}^{-2} \text{ day}^{-1} = 29.71 \text{ MJ m}^{-2} \text{ day}^{-1},$$

$$H_d = 113 \text{ cal cm}^{-2} \text{ day}^{-1} = 4.73 \text{ MJ m}^{-2} \text{ day}^{-1}.$$

The daily global radiation H_β incident on the inclined plane can be calculated from summation of the hourly values:

$$H_\beta = \sum^{\text{day}} I_\beta.$$

However, this process is tedious, and we shall employ the daily horizontal values directly wherever possible.

Daily beam radiation incident on the inclined plane

$$H_{b\beta} = H_b R_b = (H - H_d) R_b,$$

where R_b is obtained from Eq. (4.4.12) or from Figs. 4.4.3–4.4.7:

$$R_b = 0.77,$$

$$H_{b\beta} = (29.71 - 4.73)0.77 = 19.23 \text{ MJ m}^{-2} \text{ day}^{-1}.$$

Daily ground-reflected diffuse radiation incident on the inclined plane is obtained from

$$\begin{aligned} H_r &= \tfrac{1}{2} H\rho(1 - \cos\beta) \\ &= \tfrac{1}{2} 29.71(0.2)[1 - \cos(50)] \\ &= 1.06 \text{ MJ m}^{-2} \text{ day}^{-1}. \end{aligned}$$

In the event ground reflection is assumed anisotropic, we have to resort to the hourly summations given by

$$\begin{aligned} H_r &= \sum^{\text{day}} I\rho \frac{1 - \cos\beta}{2} \left[1 + \sin^2\left(\frac{\theta_z}{2}\right)\right](|\cos\Delta|) \\ &= 0.816 \text{ MJ m}^{-2} \text{ day}^{-1}. \end{aligned}$$

Daily sky diffuse radiation incident on the inclined plane is calculated from the following four models.

1. Circumsolar model:

$$H_s = H_d R_b = 4.73(0.77) = 3.64 \text{ MJ m}^{-2} \text{ day}^{-1}.$$

2. Isotropic model:

$$H_s = \tfrac{1}{2} H_d (1 + \cos\beta) = 4.73[1 + \cos(50)] = 3.89 \text{ MJ m}^{-2} \text{ day}^{-1}.$$

3(a). Anisotropic model (Klucher)

For this model, the daily value of the sky diffuse radiation incident on the inclined surface is given by

$$\begin{aligned} H_s &= \sum^{\text{day}} I_d \frac{1 + \cos\beta}{2}\left[1 + F\sin^3\left(\frac{\beta}{2}\right)\right](1 + F\cos^2\theta \sin^3\theta_z) \\ &= 4.39 \text{ MJ m}^{-2} \text{ day}^{-1}. \end{aligned}$$

3(b). Anisotropic model (Hay)

$$\begin{aligned} H_s &= H_d\left[\frac{H - H_d}{H_0} R_b + \frac{1}{2}(1 + \cos\beta)\left(1 - \frac{H - H_d}{H_0}\right)\right] \\ &= 4.73\left[\frac{24.98}{41.34} 0.77 + \frac{1}{2}[1 + \cos(50)]\left(1 - \frac{24.98}{41.34}\right)\right] \\ &= 3.74 \text{ MJ m}^{-2} \text{ day}^{-1}. \end{aligned}$$

It can be shown that summation of hourly values from Hay's model yields

$$H_s = 3.51 \text{ MJ m}^{-2} \text{ day}^{-1}.$$

This represents a difference of 6%, the hourly summation being more accurate.

The daily global radiation on the inclined plane is

$$H = H_{b\beta} + H_r + H_s.$$

From the four models we have the following four values:

1. $H_\beta = 19.23 + 1.06 + 3.89 = 24.18 \text{ MJ m}^{-2} \text{ day}^{-1}$,

2. $H_\beta = 19.23 + 1.06 + 3.64 = 23.93 \text{ MJ m}^{-2} \text{ day}^{-1}$,

3(a). $H_\beta = 19.23 + 1.06 + 4.39 = 24.68 \text{ MJ m}^{-2} \text{ day}^{-1}$,

3(b). $H_\beta = 19.23 + 1.06 + 3.74 = 24.03 \text{ MJ m}^{-2} \text{ day}^{-1}$.

The above shows that all the models yield results very close to each other. □

11.7 Ratio of Global Radiation on a Tilted Surface to That on a Horizontal Surface

After computing the total insolation on an inclined plane, it is useful to determine the effect of slope, albedo, and orientation by calculating the parameter R or $\bar{R}$, where

$$R = H_{\beta\gamma}/H \tag{11.7.1}$$

and

$$\bar{R} = \bar{H}_{\beta\gamma}/\bar{H}. \tag{11.7.2}$$

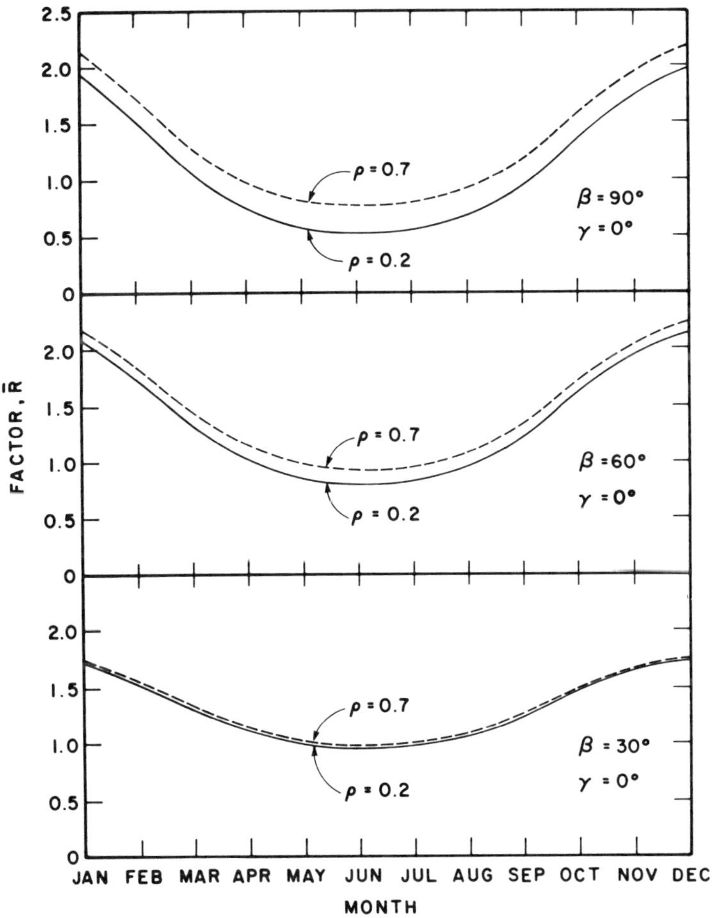

Figure 11.7.1 Effect of albedo on insolation on inclined planes at Montreal.

Further, as the horizontal diffuse radiation is in general an estimated quantity, and is determined in terms of a ratio H_d/H or $\overline{H}_d/\overline{H}$, it is sometimes convenient to use this ratio in computing R or $\overline{R}$, and eventually the slope radiation. For example, reconsider Eq. (11.6.4) and divide both sides by $\overline{H}$, which yields

$$\overline{R} = \frac{\overline{H}_{\beta\gamma}}{\overline{H}} = \left(1 - \frac{\overline{H}_d}{\overline{H}}\right)\overline{R}_b + \frac{\overline{H}_r + \overline{H}_s}{\overline{H}}. \tag{11.7.3}$$

An inclined surface will not necessarily always receive more radiation than a horizontal surface. In Fig. 4.4.2 we have noted that depending on latitude, month, and slope, $\overline{R}_b$ may be less than one, and $\overline{R}_b$ is an important parameter in Eq. (11.7.3). In Fig. 11.7.1 we have plotted $\overline{R}$ for south-facing surfaces inclined at 30°, 60°, and 90° from the horizontal position. All calculations are based on one anisotropic sky diffuse radiation model, and ground reflection is assumed perfectly diffuse. From this diagram we observe that at high tilt angles and during summer months, $\overline{R}$ may be less than one. The factor $\overline{R}$ will decrease further at low latitudes. In the same diagram the effect of an increase in ground albedo is shown by assuming a hypothetical constant value of $\rho_g = 0.7$. An increase in albedo substantially increases slope radiation when the tilt angle is high. At low tilt angles, ground albedo has a minimal effect.

11.8 Further Reading

In this chapter, elementary procedures to compute total irradiation on inclined planes have been laid down. In the literature, there are a large number of publications that present measured values, computational methods, and computer programs regarding insolation on tilted planes. A number of studies deal with clear-sky conditions only. The formulations dealing with the direct radiation on inclined planes are well understood. However, the diffuse irradiation on inclined planes and especially the calculation of its anisotropic distribution remain very intricate. In this section, a bibliography dealing with the above aspects of slope radiation is presented.

Dealing with the measured data on vertical and sloping surfaces are the studies conducted by Kimball and Hand [7], Hand [8, 9], and Cunniff [10]. Parmelee [11] presented a procedure to estimate the clear-sky diffuse radiation, and Thams [12] and Valko [13] presented data for oriented surfaces from Locarno-Monti. Based on the Russian data, Kondratyev and Manolova [14] and Kondratyev and Fedorova [15] analyzed the anisotropy of the sky diffuse radiation as well as that of the reflected radiation.

General calculation procedures for slope radiation have been presented by Becker and Boyd [16], Garnier and Ohmura [17, 18], Hay [19], Klein [20], and Klein and Theilacker [21]. Heywood [22, 23] has presented an empirical method based on actual measurements of radiation on inclined surfaces. Computer programs for clear-sky irradiation have been used by Ballantyne [24] and Dave *et al.* [25]. Methods to account for the anisotropy of sky diffuse radiation have been developed by Moon and Spencer [26], Weiss [27], and Stevens [28]. For clear skies, Hooper and Brunger [29] have presented a model in which a continuous variation of the sky diffuse radiation is assumed. Recently, Valko [30] conducted an excellent survey of literature dealing with the angular distribution of sky diffuse radiation, and the different models that treat the various aspects of insolation on tilted surfaces. Carter and Patel [31] presented an evaluation of methods of calculating solar radiation on inclined surfaces. In this evaluation they recommended that to compute sky diffuse radiation, the anisotropic model may be used when diffuse radiation is more than half the global radiation.

Nomenclature[3]

A_c	Area of the receiving surface (m^2)
A_g	Ground area seen by the receiving surface (m^2)
$F_{c\to g}$	Configuration factor from the receiving surface to the ground (dimensionless) [Eq. (10.4.5)]
$F_{g\to c}$	Configuration factor from the ground to the receiving surface (dimensionless)
H	Daily global radiation incident on a horizontal surface (MJ m^{-2} day^{-1})
H_b	Daily beam radiation incident on a horizontal surface (MJ m^{-2} day^{-1})
H_d	Daily diffuse radiation incident on a horizontal surface (MJ m^{-2} day^{-1})
H_r	Daily ground-reflected diffuse radiation incident on an inclined surface (MJ m^{-2} day^{-1})
H_s	Daily sky diffuse radiation incident on an inclined surface (MJ m^{-2} day^{-1})
$H_{b\beta}$	Daily beam radiation incident on a surface inclined toward the equator (MJ m^{-2} day^{-1})
$H_{d\beta}$	$H_r + H_s$, daily diffuse radiation incident on an inclined surface (MJ m^{-2} day^{-1}) (Note: the subscript γ is not added)
$H_{b\beta\gamma}$	Daily beam radiation incident on an inclined surface oriented in any direction (MJ m^{-2} day^{-1})
H_β	$H_{b\beta} + H_{d\beta}$, daily global radiation incident on a surface inclined toward the equator, (MJ m^{-2} day^{-1})
$H_{\beta\gamma}$	$H_{b\beta\gamma} + H_{d\beta}$, daily global radiation incident on an inclined surface oriented in any direction (MJ m^{-2} day^{-1})
H_0	Extraterrestrial daily radiation incident on a horizontal surface (MJ m^{-2} day^{-1})
$H_{0\beta}$	Extraterrestrial daily radiation incident on a surface inclined toward the equator (MJ m^{-2} day^{-1})
$H_{0\beta\gamma}$	Extraterrestrial daily radiation incident on an inclined surface oriented in any direction (MJ m^{-2} day^{-1})

[3] Overdots indicate instantaneous values; overbars indicate monthly average values.

I	Hourly global radiation incident on a horizontal surface (kJ m^{-2} h^{-1})
I_b	Hourly beam radiation incident on a horizontal surface (kJ m^{-2} h^{-1})
I_d	$I_{dcs} + I_{dis}$, hourly diffuse radiation incident on a horizontal surface (kJ m^{-2} h^{-1})
I_{dcs}	Hourly circumsolar diffuse radiation incident on a horizontal surface (kJ m^{-2} h^{-1})
I_{dis}	Hourly isotropic diffuse radiation incident on a horizontal surface (kJ m^{-2} h^{-1})
I_r	Hourly ground-reflected diffuse radiation incident on an inclined surface (kJ m^{-2} h^{-1})
I_s	Hourly sky diffuse radiation incident on an inclined surface (kJ m^{-2} h^{-1})
I_{sc}	Solar constant 4 871 kJ m^{-2} h^{-1} from NASA or 4 921 kJ m^{-2} h^{-1} from WRC
$I_{b\beta}$	Hourly beam radiation incident on a surface inclined toward the equator (kJ m^{-2} h^{-1})
$I_{d\beta}$	$I_r + I_s$, hourly diffuse radiation incident on an inclined surface oriented in any direction (kJ m^{-2} h^{-1})
$I_{b\beta\gamma}$	Hourly beam radiation incident on an inclined surface oriented in any direction (kJ m^{-2} h^{-1})
$I_{\beta\gamma}$	$I_{b\beta\gamma} + I_{d\beta}$, hourly global radiation incident on an inclined surface oriented in any direction (kJ m^{-2} h^{-1})
I_0	Extraterrestrial hourly radiation incident on a horizontal surface (kJ m^{-2} h^{-1})
$I_{0\beta}$	Extraterrestrial hourly radiation incident on a surface inclined toward the equator (kJ m^{-2} h^{-1})
$I_{0\beta\gamma}$	Extraterrestrial hourly radiation incident on an inclined surface oriented in any direction (kJ m^{-2} h^{-1})
R	$H_{\beta\gamma}/H$ or H_β/H
R_b	$H_{0\beta\gamma}/H_0$ or $H_{0\beta}/H_0$
r_b	$I_{0\beta\gamma}/I_0$ or $I_{0\beta}/I_0$
β	Inclination of a surface from the horizontal position (degrees)
γ	Surface azimuth angle (degrees), east positive, west negative
Δ	Azimuth of the tilted surface with respect to that of the sun (degrees)
δ	Declination (degrees), north positive, south negative
δ_c	Declination on characteristic days (degrees) (see Table 4.2.2)
θ	Angle between beam radiation and the surface normal (degrees)
θ_z	Zenith angle, the angle between the beam from the sun and the vertical (degrees)
ρ	Ground albedo
ρ_b	Ground albedo for beam irradiation
ρ_d	Ground albedo for diffuse irradiation
$\sum^{day}$	Summation over one day
$\tau_b(\omega)$	Atmospheric transmittance to beam radiation variable with the hour angle (dimensionless)
ϕ	Latitude (degrees), north positive
ω	Hour angle (degrees), solar noon zero and mornings positive
ω_s	Sunset-hour angle for a horizontal surface (degrees)
ω_{sr}	Sunrise-hour angle for an inclined surface oriented in any direction (degrees)
ω_{ss}	Sunset-hour angle for an inclined surface oriented in any direction (degrees)

References

1. B. Y. H. Liu and R. C. Jordan, Daily insolation on surfaces tilted toward the equator. *ASHRAE J.* **3** (10), 53–59 (1961).
2. R. C. Temps and K. L. Coulson, Solar radiation incident upon slopes of different orientations. *Sol. Energy* **19** (2), 179–184 (1977).

3. L. J. B. McArthur and J. E. Hay, A technique for mapping the distribution of diffuse solar radiation over the sky hemisphere. *J. Appl. Meteorol.* **20** (4), 421–429 (1981).
4. T. M. Klucher, Evaluation of models to predict insolation on tilted surfaces. *Sol. Energy* **23** (2), 111–114 (1979).
5. J. E. Hay, Study of Shortwave Radiation on Non-horizontal Surfaces. Rep. No. 79-12, Atmospheric Environment Service, Downsview, Ontario (1979).
6. M. Iqbal, Hourly vs daily method of computing insolation on inclined surfaces. *Sol. Energy* **21** (6), 485–489 (1978).
7. H. H. Kimball and I. F. Hand, Daylight illumination on horizontal, vertical and sloping surfaces. *Mon. Weather Rev.* **50** (12), 615–628 (1922).
8. I. F. Hand, Preliminary measurements of solar energy received on vertical surfaces. *Trans. Am. Geophys. Un.* **28** (5), 705–712 (1947).
9. I. F. Hand, Insolation on clear days at the time of solstices and equinoxes for latitude 42 N. *Heat. Vent.* **47** (1), 92–94 (1950).
10. C. V. Cunniff, Solar radiation on walls facing east and west, *Air Cond., Heat. Vent.* **55** (8), 82–88 (1958).
11. G. V. Parmelee, Irradiation of vertical and horizontal surfaces by diffuse solar radiation from cloudless skies. *Heat., Pip. Air Cond.* **26,** 129–137 (1954).
12. J. C. Thams, Die Globalstrahlung eines Sudhanges von 25 Neigung. *Arch. Meteorol. Geophys. Bioklimatol. Ser. B* **7** (2), 190–196 (1956).
13. P. Valko, Radiation load on buildings of different shape and orientation under various climatic conditions. WMO/SHO Tech. Note No. 109, pp. 87–109 (1970).
14. K. Ya. Kondratyev and M. P. Manolova, The radiation balance of slopes. *Sol. Energy* **4** (1), 14–19 (1960).
15. K. Ya. Kondratyev and M. P. Fedorova, Radiation regime of inclined surfaces. Proceedings of the UNESCO/WMO symposium. WMO/No. 477, pp. 36–61 (1976).
16. C. F. Becker and J. S. Boyd, Solar radiation availability on surfaces in the United States as affected by season, orientation, latitude, altitude and cloudiness. *J. Sol. Energy Sci. Technol.* **1** (1), 13–21 (1957).
17. B. J. Garnier and A. Ohmura, A method of calculating direct shortwave radiation income of slopes. *J. Appl. Meteorol.* **7,** 796–800 (1968).
18. B. J. Garnier and A. Ohmura, The evaluation of surface variations in solar radiation income. *Sol. Energy* **13** (1), 21–34 (1970).
19. J. E. Hay, Calculation of monthly mean solar radiation for horizontal and inclined surfaces. *Sol. Energy* **23** (4), 301–307 (1979).
20. S. A. Klein, Calculation of monthly average insolation on tilted surfaces, *Sol. Energy* **19** (4), 325–329 (1977).
21. S. A. Klein and J. C. Theilacker, An algorithm for calculating monthly-average radiation on inclined surfaces. *Trans. ASME J. Sol. Energy Eng.* **103,** 29–33 (1981).
22. H. Heywood, The computation of solar radiation intensities, Part 2—Solar radiation on inclined surfaces. *Sol. Energy* **10** (1), 46–52 (1966).
23. H. Heywood, A general equation for calculating total radiation on inclined surfaces. Paper No. 3/21, ISES Congress, Melbourne (March 2–6, 1970).
24. E. R. Ballantyne, The effect of orientation and latitude on the solar radiation received by test panels and fences during weathering studies. Paper E 124 *ISES Conf., Paris* (July 1973).
25. J. V. Dave, P. Halpern, and H. J. Myers, Computation of incident solar energy. *IBM J. Res. Dev.* **19** (6) (1975).
26. P. Moon and D. E. Spencer, Illumination from a non-uniform sky. *Trans. Illum. Eng. Soc.* **37,** 707–726 (1942).
27. T. A. Weiss, The estimation of a daily, clear-sky solar radiation intercepted by a tilted surface. M.Sc. Thesis, Colorado State Univ., Fort Collins, Colorado (1977).

28. M. D. Stevens, Standard distributions of clear sky radiance. *Q. J. R. Meteorol. Soc.* **103,** 457–465 (1977).
29. F. C. Hooper and A. P. Brunger, A model for the angular distribution of sky radiance. *Trans. ASME J. Sol. Energy Eng.* **102,** 196–202 (1980).
30. P. Valko, Some empirical properties of solar radiation and related parameters. International Energy Agency, DOE/ER-0084 (1980).
31. E. A. Carter and A. M. Patel, Evaluation of methods of calculating solar radiation on inclined surfaces. Final Rep. COO/4494-1, DOE (1978).

Chapter 12

SOLAR RADIATION MEASURING INSTRUMENTS

12.1 Introduction

In order to assess the availability of solar energy arriving on the earth, measurement of solar radiation at some locations is essential. From the measurements, empirical models are developed to predict the availability of solar energy at other locations. This chapter will discuss the instruments used to measure the various solar radiation fluxes arriving on the earth. Sources of errors and methods of calibration are discussed. Instruments for the measurement of broadband radiation only are presented.

In Chapter 8 we mentioned briefly the instruments associated with measuring the three principal solar radiation fluxes—direct, global, and diffuse. This chapter is devoted mainly to these three types of instruments. It seems useful to repeat these classifications here.

A *pyrheliometer* is an instrument for measurement of the direct solar radiation flux at normal incidence. The instrument is usually attached to an electrically driven equatorial mount for the sun.

A *pyranometer* is an instrument for measurement of the direct and diffuse irradiance arriving from the whole hemisphere. This hemisphere is usually the complete sky dome. A pyranometer can be used in a tilted position as well, in which case it will also receive the ground-reflected radiation.

A *pyranometer with a shading device* is an instrument that measures diffuse solar irradiance within a solid angle of 2π, with the exception of the solid angle subtended by the sun's disk. Again, the solid angle of 2π is usually the sky dome.

The heart of a radiometer (an instrument that measures radiant energy whether from the sun or from any other source) is its sensor, also called detector. We begin with radiation sensors.

12.2 Radiation Sensors

Detectors of the various instruments can be classified as *calorimetric*, *thermomechanical*, *thermoelectric*, or *photoelectric*.

A. *Calorimetric Sensors*

In the calorimetric instruments, the radiant energy is incident on a high-conductivity metal coated with a nonselective black paint of high absorptance. The radiant energy is converted into heat that can be measured by a variety of means, described below.

(1) The heat can be carried away by a flowing fluid whose change of enthalpy is measured. The change of enthalpy is an indication of the incident radiant flux.
(2) The heat gives rise to a change in the enthalpy of the absorbing metal (sensor). Again, this increase in enthalpy (or increase in temperature) can be measured easily.
(3) In the modern cavity-type instruments, the temperature difference across a transducer is maintained constant by additional electrical heating required between shielded and exposed phases. The irradiance is then proportional to the difference in cavity electrical heating in the two phases.

B. *Thermomechanical Sensors*

In the instruments based on the thermomechanical principle, the radiant flux is measured through bending of a bimetallic strip. In this system, two metal strips with different thermal expansion properties are rigidly held together. One end is fastened and the other is free to move. One strip is coated with a highly absorbent black paint and the other given a highly reflective coat. The blackened strip is exposed to solar radiation and the other is shielded from it. The two strips are insulated from each other to prevent heat flow from one to the other. The unequal temperatures and unequal coefficients of thermal expansion cause bending of the plates into a curve. The distortion is transmitted optically or mechanically to an indicator.

C. *Thermoelectric Sensors*

A thermoelectric device consists of two dissimilar metallic wires with their ends connected. An electromotive force (emf) is developed when the two junctions are at different temperatures (Fig. 12.2.1a). The emf developed

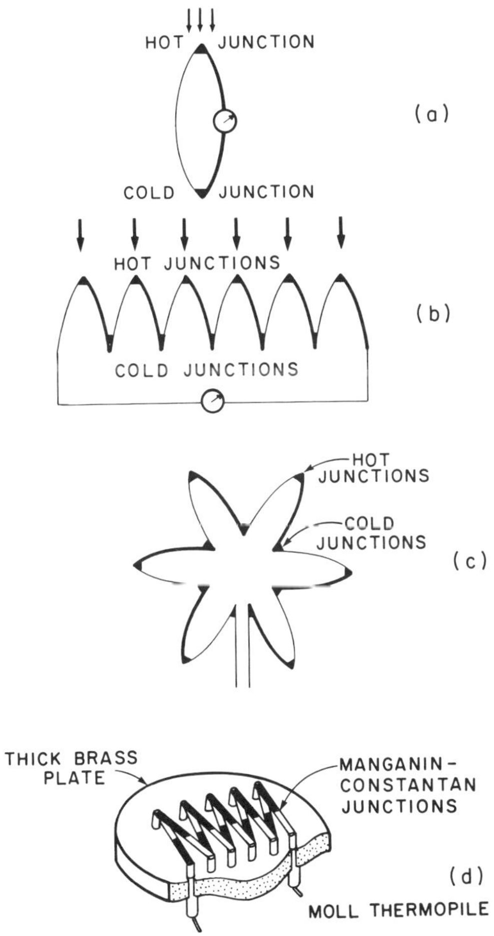

Figure 12.2.1 Various arrangements of thermoelectric sensors.

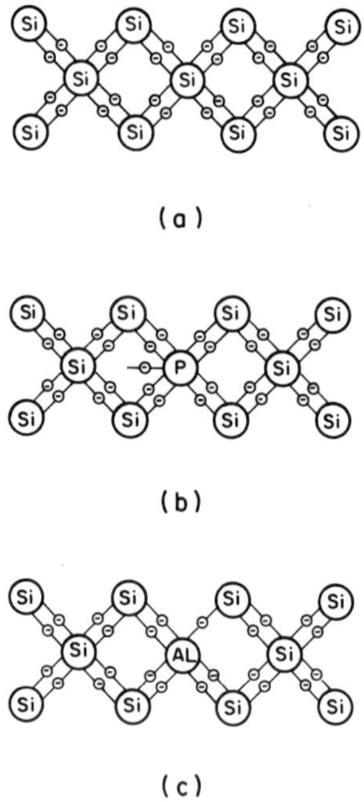

Figure 12.2.2 Silicon cell bondings.

is proportional to the temperature difference, and depends on the material of the two metals. A copper–Constantan pair is a popular combination for low-temperature applications. This device is applied in radiometry by exposing one junction to the incident radiation while the other is shielded from it.

The emf developed by a single thermocouple is very low. It can be increased by connecting a number of thermocouple junctions in series (Fig. 12.2.1b). An arrangement in which several thermocouple junctions are utilized is called a *thermopile*. In a number of solar radiation measuring instruments, the thermopile is arranged in a star-shaped flat grid (Fig. 12.2.1c). The hot junctions are coated with black paint, and the cold junctions are painted white to "shield" them from solar radiation.

In order to obtain stable conditions (usually called zero drift) it is necessary to maintain cold junctions at constant temperature. To achieve this objective, Moll [1] devised a thermopile in which the cold junctions are thermally attached to, but electrically insulated from, a massive brass plate (Fig. 12.2.1d). The thermal inertia of the plate absorbs short-period temperature variations forced by air currents. In the *Moll thermopile*, the thermocouples are made of very thin Manganin–Constantan strips connected to copper pins. The pins are in thermal contact with, but electrically insulated from, the brass plate.

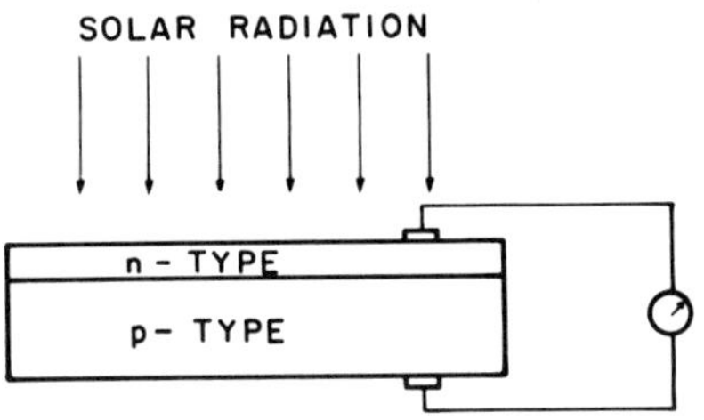

Figure 12.2.3 Semiconductor *p–n* junction.

D. *Photoelectric Sensors*

Among the photoelectric devices, *photovoltaic* instruments are most numerous in the field of solar radiation measurement. A photovoltaic device is made of a semiconducting material such as silicon. A silicon atom has four valence electrons forming a crystalline lattice with four neighboring atoms. When an impurity such as phosphorus, arsenic, or antimony having five valence electrons is added, the excess electron can be easily liberated to become a conducting electron. A semiconductor with an excess electron is called an *n*-type semiconductor. When an impurity such as aluminum, boron, or indium having three valence electrons is added to a silicon crystal, a "hole" is created for one of the normal bonding electrons. A semiconductor with an electron vacancy is called a *p*-type semiconductor. Figure 12.2.2 shows the three types of bondings described above.

A semiconductor *p–n* junction is formed by joining a material with a deficient electron to a material with an excess electron. When radiation at an energy level capable of ionizing the atoms is incident on the *p–n* junction, an electrical current arises from the continuous movement of excess electrons and holes. Figure 12.2.3 shows schematically a photovoltaic cell.

A major disadvantage of the silicon cell devices is their spectral response, which is strong only in the red and near-infrared portions of the solar spectrum. However, their advantages are lower cost and faster response times for instantaneous measurements.

We now study the instruments used to measure direct irradiance.

12.3 Measurement of Direct Irradiance: Pyrheliometers

Direct radiation is measured by a pyrheliometer, a telescopic type of instrument with a narrow opening called an *aperture*. This instrument faces the sun and follows its motion. The interest in establishing the value of the solar constant has been the main force behind development of this instrument. The Smithsonian Institution has played a leading role in this area.

Beginning at the turn of this century, Dr. Charles Greeley Abbot of the Smithsonian Institution developed three principal instruments called *water-flow*, *water-stir*, and *silver-disk* pyrheliometers. All three are based on the calorimetric principle.

A. *Abbot Water-Flow and Water-Stir Pyrheliometers*

A schematic diagram of one of the single-chamber versions [2, 3] of the water-flow pyrheliometer is shown in Fig. 12.3.1. The collimating tube is composed of two sections, AA and BB. The walls of the lower part, section AA, are coated with a highly absorbent black paint. Direct solar radiation from the sun is absorbed by the cone-shaped receiver E. Distilled water flows over the exterior walls of the thin metallic collimating tube, first over section AA and then over section BB. Inlet and outlet temperatures are measured at points D_1 and D_2. The exterior of the water channel is encased in a Dewar vacuum wall. Under stable conditions, accurate measurement of the water flow rate and of the temperature differences between inlet and outlet yield a measure of the incident radiative flux.

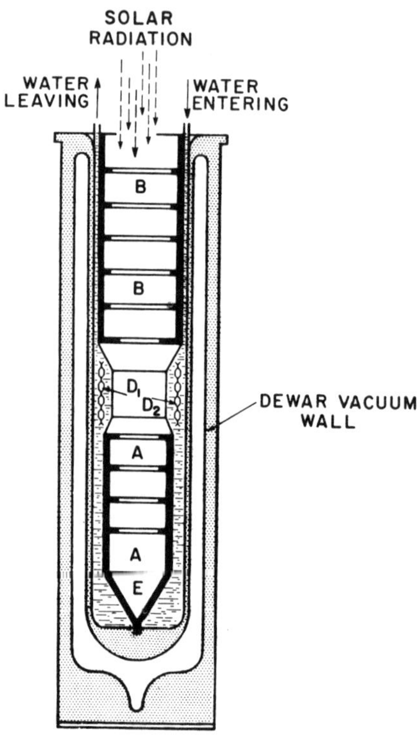

Figure 12.3.1 Abbott water-flow pyrheliometer. Adapted from Abbot [2, 3].

Before the double-chamber water-flow pyrheliometer described in the next section was built, a water-stir instrument was designed in which the principle of calorimetry by the method of mixtures was utilized [3]. In this single-chamber instrument, the sun's radiant heat is absorbed in a blackened conical cavity at the base of a collimating tube. An electric heater is used to simulate the solar radiant heat. A water tank insulated from the outside protects the calorimeter from ambient temperature variations. The water surrounding the absorption chamber is vigorously stirred by a stirrer mechanism. Solar radiant heat is equated to the electrical heat by balancing the temperatures.

In 1932 Abbot and Aldrich [4] improved the single-chamber water-flow pyrheliometer by utilizing two identical collimating chambers insulated from each other. Both chambers are equipped with an electrical heating system so that the role of the chambers can be easily interchanged. Each chamber has its own cooling water stream. When the instrument faces the sun, one chamber is exposed to sunlight and the other is shielded and heated by electric current. Under identical water-flow and temperature difference conditions the electrical power consumed represents the amount of solar irradiance. In practice the thermocouples of the two chambers are differentially connected to each other's absorbers, thereby eliminating the necessity of measuring temperature or flow rate. This technique significantly increased the accuracy of measurement compared to the single-chamber water-flow pyrheliometer.

With the three pyrheliometers mentioned above, solar irradiance is *calculated* in true heat units from physical parameters of the instrument. For this reason they are called *absolute* pyrheliometers. Unfortunately, these instruments are difficult to operate and are therefore no longer manufactured. They have been described here to facilitate understanding of the pyrheliometric scales in Section 12.4.

B. *Abbot Silver-Disk Pyrheliometer*

Because the water-flow pyrheliometer is not portable, at the turn of this century Abbot [5] constructed the silver-disk pyrheliometer, a sketch of which is given in Fig. 12.3.2. In this instrument the sensor consists of a silver disk a with a radial bore into which is inserted the bulb of a thermometer b. The thermometer is bent at a right angle to make the instrument more compact. The space between the thermometer bulb and the radial bore is filled with mercury for good heat conduction. The silver disk is coated with a highly absorbent black paint. The disk is enclosed in a cylindrical copper box c. To protect the instrument from ambient temperature fluctuations, a wooden box d encloses the copper box.

Solar radiation enters through the collimating tube e fitted with a number of circular diaphragms f_1, f_2, and f_3, which form a small solid angle to view the sun and a small annular area of the sky. A hole through the thermometer support admits a guiding beam of light i to indicate that the pyrheliometer is pointed toward the sun. The rotatable shutters h, made of polished metal plates, provide for cutting off the solar beam in a highly regulated manner.

With the silver-disk pyrheliometer the amount of solar radiation received cannot be computed from first principles. Each instrument is calibrated against an absolute pyrheliometer.

The silver-disk pyrheliometer is mechanically simple and rugged. With reasonable care it continues to give reliable readings of direct solar radiation indefinitely.

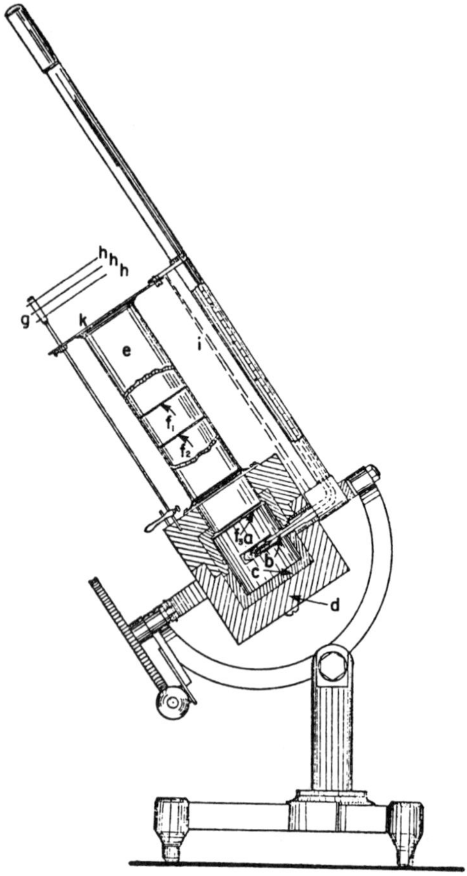

Figure 12.3.2 Abbot silver-disk pyrheliometer. Adapted from Abbot [5].

C. Ångström Electrical Compensation Pyrheliometer

Almost parallel to the development of pyrheliometry at the Smithsonian Institution, Knut Ångström[1] [6] in Sweden developed the first instrument for very precise measurement of direct normal irradiance. This instrument is based on the calorimetric principle. It uses two identical, thin, long Manganin strips approximately 2 mm wide and 20 mm long, coated with a highly absorbent black paint. The strips are placed at the lower end of a collimating tube. A reversible shutter at the mouth of the collimating tube can shade either strip and expose the other to sunlight. The shaded strip is heated by electric current until it attains the temperature of the insolated strip. In this manner the electrical energy supplied to the shaded strip equals the solar energy absorbed by the exposed strip after losses have been taken into account.

The electrical compensation arrangement is shown in Fig. 12.3.3. L and M are the Manganin strips. Thermojunctions T_1 and T_2 attached to the back of the strips are connected to a null galvanometer G whose zero reading indicates equivalence of temperature of the two strips. Rheostat R_1 regulates current to equalize the strip temperatures. C is a commutator switch for the strips. The solar irradiance is proportional to the square of the electric current; in other words,

$$\dot{I}_{\mathrm{n}} \propto i^2, \tag{12.3.1}$$

where $\dot{I}_{\mathrm{n}}$ is the direct normal irradiance and i the electric current. Let K be the constant of proportionality in the preceding equation such that

$$\dot{I}_{\mathrm{n}} = Ki^2. \tag{12.3.2}$$

In practice K is determined from first principles for one instrument by the manufacturer and is obtained by calibration for other instruments. The two original absolute instruments are Nos. 70 and 158 and are still in good condition.

Because the strips are rectangular, the aperture of the instrument is rectangular. The original design had an aperture angle of approximately $24^\circ \times 6^\circ$. This instrument has gone through a number of improvements by the Swedish manufacturer (Swedish Meteorological and Hydrological Institute, SMHI), the Smithsonian Institution, and the Eppley Laboratory (United States). Aperture angles have been reduced and the electrical system has been improved. In the earlier instruments the exposed strip could not be

[1] Knut Ångström (1857–1910) was the son of the famous Swedish astronomer and physicist Anders Jonas Ångström (1814–1874) after whom the Ångström unit of wavelength was named. Anders Ångström (1888–1981), developer of the turbidity formula (Chapter 6) and the sunshine correlation (Chapter 8), was the son of Knut Ångström.

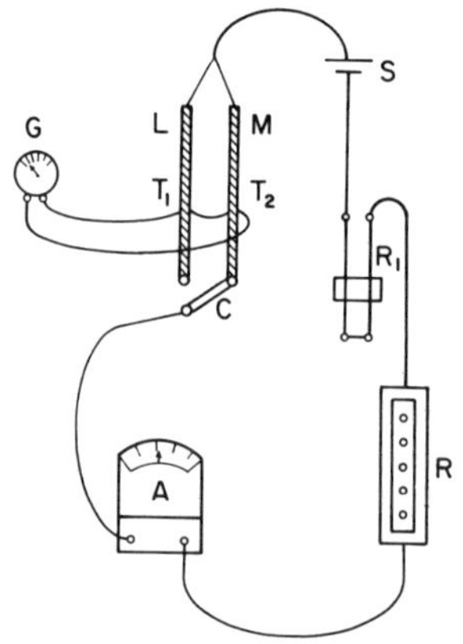

Figure 12.3.3 Electric circuit arrangement of the Ångström electrical compensation pyrheliometer. Adapted from Ångström [6].

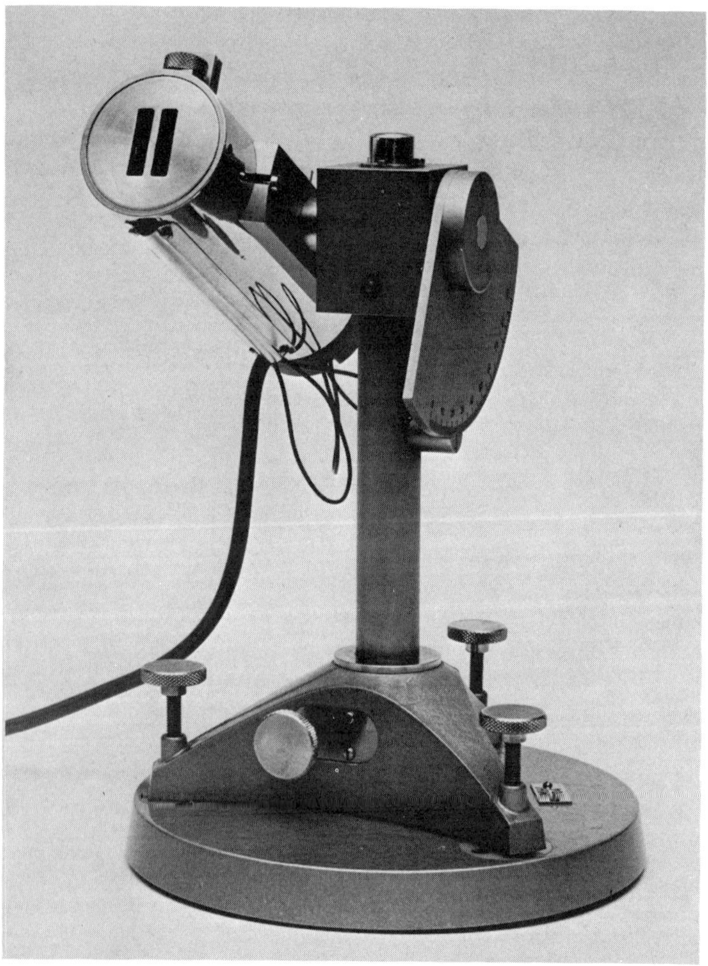

Figure 12.3.4 Ångström electrical compensation pyrheliometer, Stockholm design (photograph by Gunnar Larsson, SMHI).

uniformly irradiated, causing the so-called edge effects. This has been rectified. The Stockholm model of the Ångström electrical compensation pyrheliometer is shown in Fig. 12.3.4.

D. *Cavity-Type Absolute Pyrheliometers*

Since the mid 1960s, a second generation of absolute pyrheliometers has been developed for accurate measurement of solar irradiance. These are electrically self-calibrated, blackened cavity radiometers developed in the United States and in Europe. In the United States, the radiometer was developed by Kendall, Willson, and colleagues [7–12] at the Jet Propulsion Laboratory (JPL), and by Geist [13] at the National Bureau of Standards (NBS). The European designers were Brusa and Fröhlich [14] at the Physico-Meteorological Observatory (PMO) of the World Radiation Centre, Switzerland, and Crommelynck [15] at the Royal Meteorological Institute, Belgium. The present-day value of the solar constant, 1 367 W m^{-2}, is based on measurements by this new generation of cavity-type radiometers. These radiometers absorb radiation on a conical receiver and determine solar radiant heat flux by electrical substitution. The advantage of a conical absorber is described below.

Let us consider a plane surface having an absorptance α_s (Fig. 12.3.5a). The energy reflected from the flat surface is lost forever. We now construct a hollow cone with this surface (Fig. 12.3.5b) and irradiate it with the same quality of radiation as before. Depending on the opening angle of the cone, radiation reflected after the first impingement will strike the walls before escaping. At each strike a fraction α_s will be absorbed. It is obvious, and can be proven, that the mouth of the cavity will have an effective absorptance α_c greater than α_s. The value of α_c increases as the cone angle becomes smaller. This is a well-known laboratory technique of making almost perfectly black surfaces. Further, in a conical cavity a specular black paint will result in a superior absorptance compared to a matte black paint.

At the JPL, one series of instruments is designated Standard Absolute Cavity Radiometers (SACRAD). Modification of these instruments with a

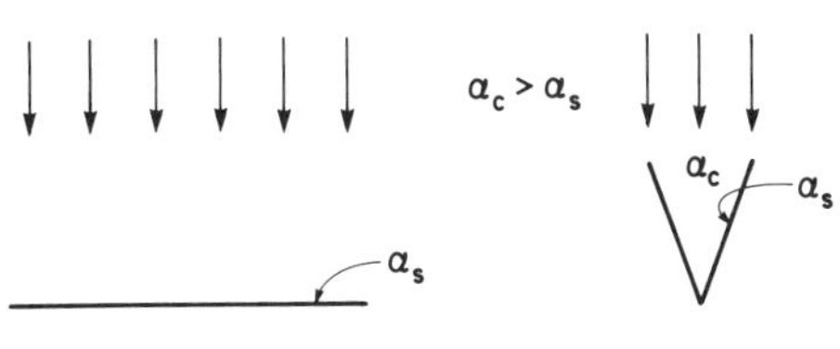

Figure 12.3.5 Effective absorptance of (a) flat surface and (b) conical surface.

view-limiting aperture provided a radiometer for measuring normal solar irradiance [7, 8]. This series is called Primary Absolute Cavity Radiometers (PACRAD).[2] From these radiometers, an instrument called the Active Cavity Radiometer (ACR) has evolved [9]. From these instruments, irradiance can be determined in absolute heat units. Their calibration depends only on dimensions, arrangement of components, and electrical measurements. The accuracy of all these instruments is $\pm 0.3\%$ or better.

The PACRAD instruments were developed principally for defining the absolute radiation scale in laboratory and ground-based solar irradiance experiments. The primary purpose of the ACRs has been the measurement of solar constant [10–12]. The ACRs have gone through a series of developments. The latest version, ACR IV, is capable of measuring solar constant with an uncertainty of $\pm 0.1\%$ in SI units and is described below.

E. *Active Cavity Radiometer ACR IV*

A cross-sectional sketch of the ACR IV [11] is shown in Fig. 12.3.6. The design is based on two circular, conical cavity detectors thermally connected to a heat sink through their respective thermal impedances. The cavities have a 30° opening angle. The interior of the cavities is coated with a specular black paint yielding an effective cavity absorptance of 0.999 999 with an uncertainty of $\pm 0.000\ 3\%$. In the region corresponding to solar irradiance of the cavity interior, a low-temperature-coefficient heater winding is bonded to the back of each cavity. Thin-film temperature sensors are bonded to each thermal impedance near the cavity apertures. The cavities are enclosed in a heat sink, which is insulated from the outside. The instrument has a 5° view-limiting aperture. Through the electric heaters, the primary cavity is accurately maintained at all times at a slightly higher (1°C) temperature than the heat sink. A shutter alternately shades and exposes the primary cavity to solar radiation while the electronic servo-system maintains the required temperature difference. When the sensor is exposed to solar radiation, the cavity heating power is automatically decreased by the electronics in an amount proportional to the absorption of solar irradiance by the cavity. The measured direct normal irradiance $\dot{I}_n$ is given by the following working

[2] In Chapter 2, we have given a value of the Stefan–Boltzmann constant (5.6866×10^{-8} W m^{-2} K^{-4}). It was measured by one of the absolute radiometers at the Jet Propulsion Laboratory. This is believed to be the most accurate measurement thus far obtained and differs from the theoretical value by only 0.3%—an indication of the accuracy obtainable with these instruments.

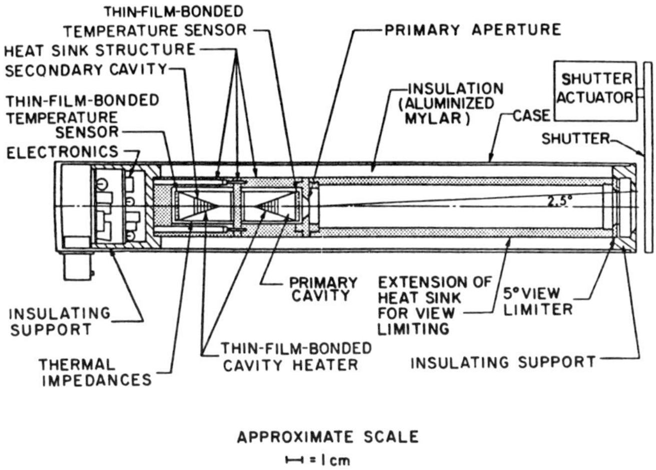

Figure 12.3.6 Active cavity radiometer, model ACR IV. Adapted from Willson [10], copyrighted by the American Geophysical Union.

equation:

$$\dot{I}_n = K[P_c - P_o], \tag{12.3.3}$$

where K is the constant of proportionality, P_c the cavity heater power without an external radiation field (shutter closed), and P_o the cavity heater power under solar irradiance. The constant K contains instrumental parameters such as the detector area, cavity absorptance, and heater resistance. It is determined through fundamental analysis of heat exchange in the radiometer. However, because heat exchange during the solar heating period is not identical to that during the electric heating period, uncertainties remain regarding the accuracy of the constant K.

F. *Field Instruments*

The pyrheliometers described thus far are either absolute instruments (Abbot water-flow and water-stir, Ångström electric compensation, ACR) or reference standard (Abbot silver-disk). Although the Ångström electric compensation is an absolute instrument, it is usually employed as a reference standard. The term *reference standard* is used for radiometers whose instrument constant has been obtained through calibration against an absolute instrument.

Absolute or reference standard instruments are not used for routine measurement of direct normal irradiance. Rather, *field instruments* are employed for this purpose. Among the field instruments, pyrheliometers manufactured by the Eppley Laboratory (United States) and Kipp & Zonen (The Netherlands) are very popular, the former in North America and the latter in Europe, Africa, and Asia.

The Eppley instrument, popularly called the normal-incidence pyrheliometer (NIP), is shown mounted on a solar tracker in Fig. 12.3.7. The design is based on a multijunction thermopile as its sensing element. The thermopile is coated with 3M Velvet Black paint. The collimating tube has a 5°43′30″ field of view and is blackened from the inside. The tube is filled with dry air at atmospheric pressure. The viewing end of the tube is sealed by a removable insert carrying a crystal quartz window 1 mm thick. The sensor response is stabilized through a temperature-compensating circuit. The instrument is fitted with a manually rotatable filter holder having four openings, three for

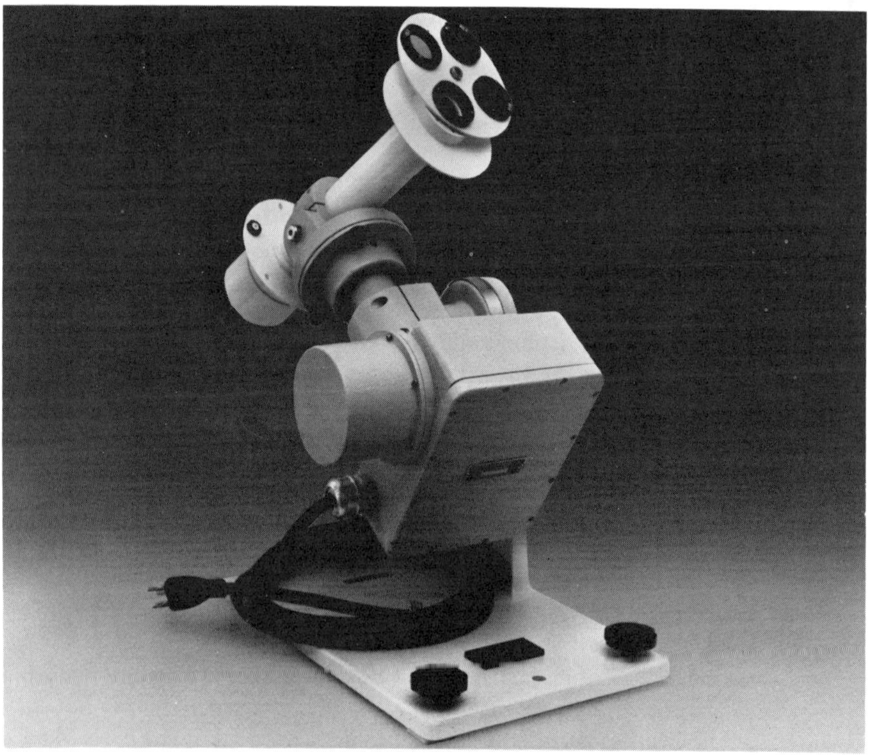

Figure 12.3.7 Eppley normal-incidence pyrheliometer (NIP).

attachment of cutoff filters and the fourth for measurement of complete spectrum.

The pyrheliometer manufactured by Kipp & Zonen, popularly called the *actinometer*, is based on the Moll thermopile as sensing element. A photograph of the instrument is shown in Fig. 12.3.8. The present model, based on the Linke–Feussner design, consists of a thermopile of 46 Constantan–Manganin strips arranged in two arch-shaped opposite groups of 23. This instrument can be used for short-wave and long-wave radiation. A quartz window is used to cut off long-wave radiation. The collimating tube, with $10°12'$ aperture angle, is made of a massive copper block with internal diaphragms. The massiveness of the tube body minimizes the effect of ambient temperature fluctuations. At the front end of the instrument a filter holder is

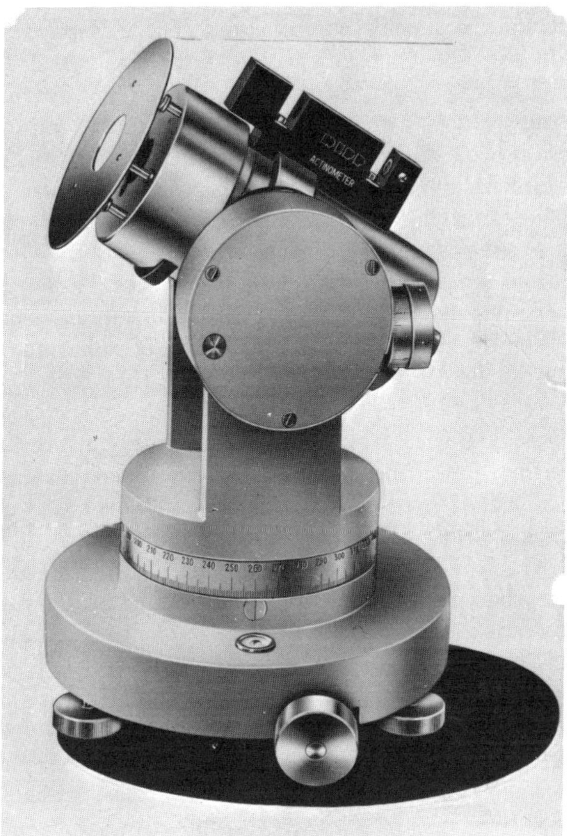

Figure 12.3.8 Kipp & Zonen actinometer.

provided for inserting different cutoff filters to measure solar irradiance within certain bandwidths.

Pyrheliometers based on the thermomechanical (bimetallic) and photovoltaic principles have also been built, but have not received wide acceptance. However, the latter are now gaining some popularity for use as bright sunshine recorders.

12.4 Pyrheliometric Scales

In simple words, the radiation scale of a pyrheliometer is its calculated performance under direct solar irradiance, taking into account all thermal losses. Two radiometers of different design may perform differently. Even two radiometers of the same design but of different dimensions may perform differently. An instrument whose calculated performance has been determined with a high degree of precision is called an *absolute pyrheliometer.* Generally, a group of specimen instruments is calibrated against the absolute instrument and is used as a standard to keep a check on the stability of the absolute instrument. Such instruments are called *standard* or *reference standard* pyrheliometers, and they may or may not be of the same type as the absolute instrument. For example, originally the silver-disk pyrheliometers were standardized against the water-flow instruments. For meteorological data it is necessary to standardize solar radiation measurements. For this purpose it is essential to adopt a standard scale of radiation. Before 1956, the two standard scales of radiation were the *Ångström scale* and the *Smithsonian scale.*

At the International Meteorological Conference at Innsbruck in 1905, the electrical compensation pyrheliometer of K. Ångström was adopted as the standard instrument. The Ångström scale (ÅS 1905) is based on the calculated performance of instrument No. Å 70 kept at the Institute of Physics, Uppsala, Sweden [16]. The surface area of the thin, long Manganin strip in the Ångström pyrheliometer is not uniformly illuminated. On the other hand the companion strip carrying electric charge dissipates heat uniformly over its length. This gives rise to the so-called *edge effects* [17]. According to Lindholm [18], the magnitude of the required correction is slightly above 2%, measurements of the uncorrected scale being too low. Prior to 1956, these corrections were never applied.

The Smithsonian scale, called the Smithsonian revised scale of pyrheliometry of 1913 (SS 1913), is based on several comparisons of a group of instruments, including two absolute instruments: single-chamber water-flow pyrheliometer No. 3 (Fig. 12.3.1) and water-stir pyrheliometer No. 4. These comparisons were spread over a number of years [2, 3]. About two decades after the declaration of SS 1913 when double-chamber water-flow absolute pyrheliometer No. 5 was built and tested, it was noted that the scale of 1913 had an error, measurements in this scale being high by 2.5% [4]. Although this error has been confirmed by several studies since [19], this correction was not applied in meteorological practice prior to 1956.

For almost half a century, the two scales were applied independently. However, during this time a number of individual attempts were made to compare the two scales. Some of the difficulties lie in the unequal and dissimilar aperture angles of the instruments used in the two scales. Drummond [20], through compilation of several comparisons, had shown an inequality of the two scales. Finally, in 1956 at Davos, the International Radiation Commision recommended the adoption of a new scale, known as the International Pyrheliometric Scale of 1956 (IPS 1956), which is based on the following changes in the original scales.

(a) Measurements made according to the original uncorrected Ångström scale are to be increased by 1.5%. In other words,

$$\text{IPS 1956} = 1.015(\text{ÅS 1905}). \tag{12.4.1}$$

(b) Measurements made according to the Smithsonian Scale 1913 are to be decreased by 2.0%, or

$$\text{IPS 1956} = 0.98(\text{SS 1913}). \tag{12.4.2}$$

The new scale was based on the assumption that the total difference between the two scales was 3.5%. Consequently, as a compromise, +1.5% represented the edge-effect correction for the Ångström scale and −2.0% the correction found by measurements with the double-chamber water-flow instrument for the Smithsonian scale.

Following the first comparisons in 1959 (IPC I), International Pyrheliometric Comparisons have been held every five years, all at Davos, in 1964 (IPC II), in 1970 (IPC III), in 1975 (IPC IV), and in 1980 (IPC V). Among these comparisons, the IPC III [21] is the most important, because it was there that the new generation of electrical cavity-type absolute instruments were employed for the first time. It was noted that IPS 56 is approximately 2% too low. Subsequently, from the 1975 comparisons a new scale [22], known as the *World Radiometric Reference* (WRR), has been established. The new scale is 2.2% above the IPS 56. In other words,

$$\text{WRR} = 1.022(\text{IPS 1956}). \tag{12.4.3}$$

This scale is based on five modern pyrheliometers, PMO-2, CROM-2, PACRAD III, ACR 310, and ACR 311. These instruments, called the *World*

Standard Group, have an accuracy better than 0.3%. It may be recalled from Chapter 3 that the current accepted value of the solar constant 1 367 W m^{-2}, is based on WRR.

The development of a third generation of absolute pyrheliometers has now started with the objective of attaining accuracies of 0.02–0.05%. These developments might form a base to establish a new scale valid for terrestrial and extraterrestrial measurements of direct solar irradiance.

12.5 Pyrheliometer Aperture Angle

The various pyrheliometers have varying degrees of field of view. Accordingly they receive varying degrees of the circumsolar radiation, and this has been one of the difficulties in carrying out comparisons of these instruments. The amount of circumsolar radiation varies with air mass and atmospheric turbidity and can lead to errors of 1–3% in measurement of direct normal irradiance—depending on the pyrheliometer used.

Because of practical difficulties of orientation, it is impossible to design a pyrheliometer that will measure radiation from the sun's disk alone. However, it is possible to standardize the instruments by regulating the geometry of their collimating tubes. For this purpose, the World Meteorological Organization [23] defines a set of normalized parameters. Let us consider a circular collimating tube (Fig. 12.5.1). Let

r be the radius of receiver (also called radius of aperture),
R be the radius of limiting aperture (also called radius of diaphragm or radius of field stop), and
L be the distance between the above two circles.

From the above we define the following normalized parameters and angles:

$$a = R/r, \qquad b = L/r,$$

$$z_0 = \text{opening angle} = \tan^{-1}(a/b),$$

$$z_p = \text{slope angle} = \tan^{-1}[(a-1)/b],$$

$$z_1 = \text{limit angle} = \tan^{-1}[(a+1)/b],$$

z_0 is called the aperture angle. The WMO recommends the following limitations for the pyrheliometers:

$$1^\circ \le z_p \le 2^\circ \qquad \text{and} \qquad b \ge 15.$$

From these two conditions, we have

$$z_0 \le 4^\circ.$$

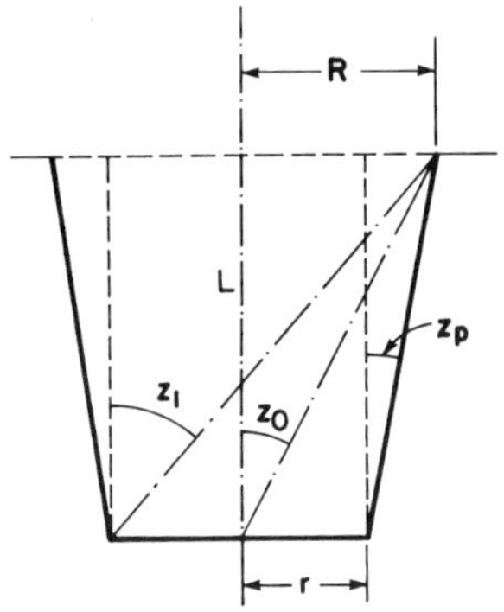

Figure 12.5.1 Opening angles of pyrheliometers.

Table 12.5.1

Opening Angles of Some Well-Known Pyrheliometers

Pyrheliometer	PACRAD PMO ACR	Abbot silver-disk	Ångström electrical compensation	Linke-Fuessner Kipp & Zonen	Eppley NIP
Opening angle z_0	Circular 5°	Circular 6°	Rectangular 5° × 10°	Circular 10°12′	Circular 5°43′30″

The above limits apply to pyrheliometers with circular diaphragms. The Ångström instrument has a rectangular receiver with a rectangular diaphragm. The WMO does not specify any limits for rectangular apertures.

Opening angles of some well-known pyrheliometers are listed in Table 12.5.1.

12.6 Pyrheliometer Classification

The WMO [23] classifies pyrheliometers according to their accuracy and the accuracy of their auxiliary equipment. The various criteria for accuracy are sensitivity, stability of the calibration factor, maximum error due to variations in ambient temperature, errors due to spectral response of the receiver, nonlinearity of response, opening angle, time constant of the system, and effect of auxiliary equipment. In terms of the hierarchy of the instruments, they carry the following designations:

Absolute Pyrheliometer
Reference Standard Pyrheliometer

Secondary Instruments
- (a) First-class pyrheliometer
- (b) Second-class pyrheliometer

The World Radiation Center at Davos maintains a group of absolute pyrheliometers. Against this group, the reference standard pyrheliometers are calibrated and maintained at different designated regional centers around the world and also in some national centers. The secondary instruments are in turn calibrated against the reference pyrheliometers. The calibration constant of a reference standard instrument is the ratio of radiative flux measured by the absolute instrument to the signal response of the reference standard instrument. Such a constant is determined from a number of average readings of each instrument and can vary with the level of radiant flux.

12.7 Measurement of Global Irradiance: Pyranometers

Global solar irradiance is measured by radiometers with hemispherical fields of view, called pyranometers. Sensing elements of most common pyranometers are based on thermoelectric, thermomechanical, or photovoltaic principles. Unlike the conical absorbers of some of the pyrheliometers, the sensing elements of the pyranometers are flat surfaces. In routine meteorological measurements, pyranometers are always placed in a horizontal position.

There are a number of manufacturers of pyranometers. The Eppley Laboratory manufactures two types of instruments, the *black and white* and the *spectral precision* pyranometer, both based on the thermoelectric principle. The *black and white* pyranometer is a development of their well-known "light bulb"-type instrument, which is no longer in production. The detector of this radiometer is a differential thermopile of radial wirewound-plated construction, with three black segments coated with 3M black, and three white segments coated with barium sulphate—hence the name "black and white" pyranometer. The instrument is fitted with a built-in temperature compensation circuit to minimize the effects of ambient temperature fluctuations. This pyranometer, shown in Fig. 12.7.1, has a hemispherical cover of precision-ground optical Schott glass WG295,[3] which transmits uniformly from 0.285 to 2.8 μm. An opening is provided on the side of the instrument to place a dessicator to absorb humidity inside the glass dome. The main body of the instrument carries adjustable leveling screws and a circular

[3] See Section 12.12.

spirit level. A calibration certificate traceable to the World Radiometric Reference is provided with the instrument.

The *Eppley precision spectral pyranometer* comprises a circular multijunction thermopile. The sensor is coated with Parson's black lacquer, which has high absorptance at all wavelengths. The instrument is fitted with two hemispherical covers of WG295 Schott glass. The outer cover can be replaced by Schott glass hemispherical filters, which transmit radiation within specified bandwidths—hence the name *spectral pyranometer*. This instrument is more accurate than the "black and white" pyranometer.

The pyranometers manufactured by Kipp & Zonen, popularly called solarimeters, have Moll thermopile detectors. Two models are manufactured with slightly different accuracies.

A number of models of pyranometers built on the thermomechanical principle are commercially available. This bimetallic pyranometer is generally called the Robitzsch actionograph or pyranograph. In this instrument,

Figure 12.7.1 Eppley black and white pyranometer.

the movement of the free end of the bimetallic strip is magnified by mechanical linkages and recorded by means of a moving arm and pen on a drum. The drum is spring powered and thus the instrument is completely independent of any external power source and is suitable for installation in remote areas. The different models come in various rotational speeds, ranging from one rotation in 24 h to one rotation in several days. A photograph of a typical bimetallic actionograph is shown in Fig. 12.7.2. All models are similar in appearance. A hemispheric glass dome covers the bimetallic strip. The mechanical linkage and the drum are encased in a weather-resistant metal box. A window on the side of the box is provided for viewing the drum and the pen. The instrument should be placed on a horizontal

Figure 12.7.2 Robitzsch (bimetallic) pyranograph.

surface, the strip lying in an east–west direction and the window facing north in the northern hemisphere.

Bimetallic actinographs are popular because of their simplicity, portability, and sturdiness. However, their accuracy ranges from 5 to 10%. They are sensitive to the level of solar flux as well as ambient temperature. Most models have auxiliary shielded bimetallic plates to compensate the effect of environmental temperature variations. Because of the thermal inertia of the bimetallic strip, the instrument response to sudden changes in radiative flux is slow, and consequently it is not suitable for measurements involving periods shorter than a day.

There are a number of commercial suppliers of pyranometers built on silicon photovoltaic cells. Although the accuracy attainable with these radiometers is not high, their response is virtually instantaneous, they can provide an integrated output signal without external power source, and they cost only a fraction of the price of a thermopile instrument. Inaccuracies in the measurements results from the spectral-selective characteristics of the solar cells. For example, the spectral response of a typical photovoltaic sensor is shown in Fig. 12.7.3. It exhibits a relatively sharp peak at about

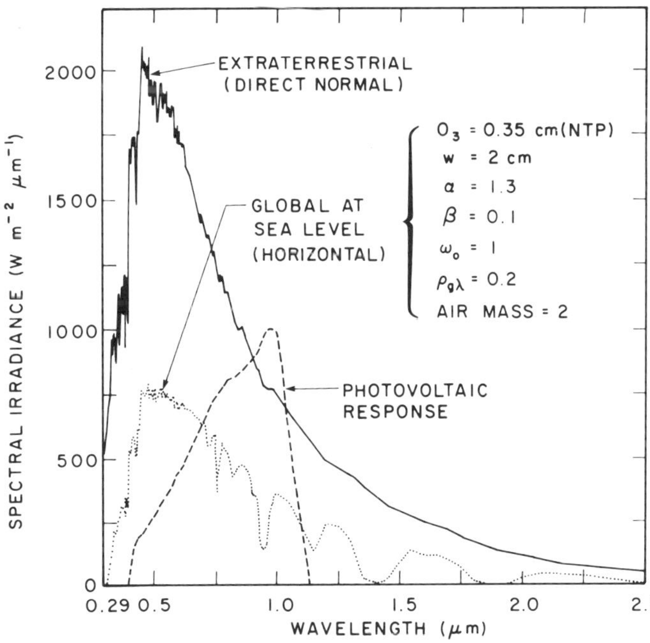

Figure 12.7.3 Spectral response of silicon cell against spectral irradiance.

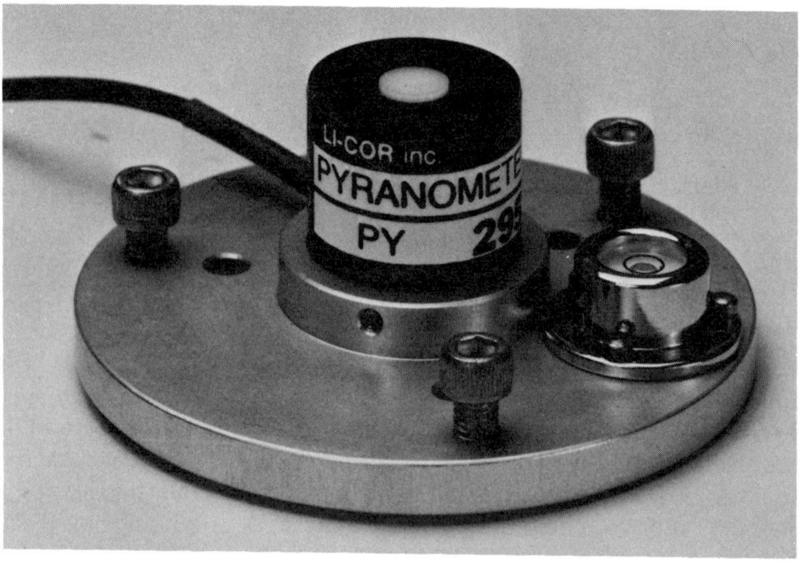

Figure 12.7.4 Li-Cor photovoltaic pyranometer.

1.0 μm, and the sensitivity range is only between 0.4 and 1.1 μm. On the same diagram, global spectral irradiance under a set of typical atmospheric parameters is also drawn. It shows that full solar cell response occurs only in a narrow spectral band. For this reason, calibration of photovoltaic pyranometers poses a problem. Furthermore, cell response deviates from the ideal cosine law.[4] This is partly corrected (for a price) by the use of a plastic diffuser, antireflection coating, or by texturing the surface. A photograph of a Li-Cor silicon cell pyranometer mounted on a base with a spirit level is shown in Fig. 12.7.4.

12.8 Pyranometer Characteristics

In designing pyrheliometers and pyranometers, there are a number of common considerations in attaining high accuracy in measurement. As mentioned earlier, these considerations are, for example, sensitivity, stability of calibration factor, linearity, change in response due to variation in ambient temperature, spectral response of sensor, response time, and effect of auxiliary equipment. However, the effect of circumsolar radiation, which depends on the aperture angle, is a unique consideration in the case of pyrheliometers.

[4] See Section 12.8 regarding the cosine effect.

On the other hand, the directional response of the sensor on the elevation and azimuth of the sun and the effect of inclination of a pyranometer are problems specific to this instrument.

A. *Cosine Effect*

The sensing element of pyranometers is invariably coated with some form of highly absorbent black paint. Although it is easier to achieve high absorptance in the spectral sense, it is difficult to achieve the same in the directional sense. That is, absorptance varies with the angle of incidence. Most surfaces exhibit an angular dependence of absorptance. This variation is not identical for electric nonconductors and electric conductors. For most electric conductors, absorptance remains almost constant until the incident angle exceeds about 70°. Beyond this point absorptance increases considerably and then drops rapidly when the angle of incidence approaches 90°. Figure 12.8.1 shows this characteristic qualitatively. This angular dependence of absorptance is called the *cosine effect*. Fortunately, at low solar elevations, the energy contained in the solar beam is very small, and a small departure from the ideal condition is not critical. The cosine effect varies with the ratio of direct radiation to diffuse radiation. The cosine effect is more pronounced when this ratio is high.

B. *Azimuth Effect*

Under ideal conditions the response of a receiver is proportional to the cosine of the angle of incidence and is constant at all azimuth angles. However, real surfaces contain some imperfections arising from the manufacturing processes and consequently small variations in azimuthal absorptance do occur.

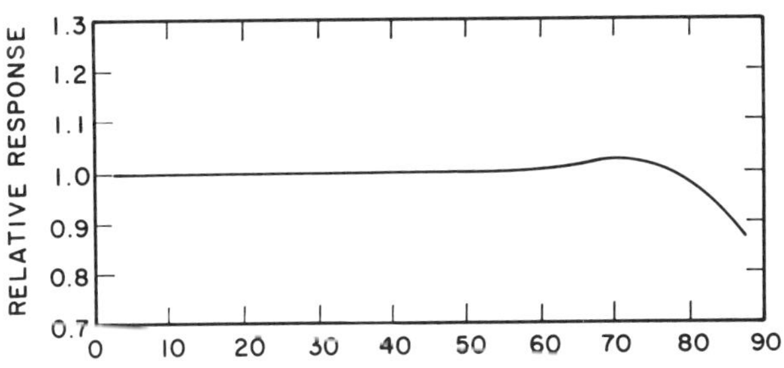

Figure 12.8.1 Pyranometer response as a function of the angle of incidence—cosine effect.

C. *Effect of Inclination*

In many solar energy applications it is necessary to measure insolation on inclined planes. A pyranometer mounted on an inclined plane receives from a solid angle of 2π direct, sky diffuse, and ground-reflected radiation. Since pyranometers are calibrated in a horizontal upward direction, they may show a change in response when mounted in an inclined or downward direction. This change occurs mainly because of a change in the heat transfer pattern inside and outside the hemispherical glass dome. This tilt effect varies with the angle of inclination as well as with the level of irradiance. For some models, such as Kipp & Zoren pyranometers, it also varies with the orientation of the thermopile vis à vis tilt angle. Deviation from the horizontal position increases with the tilt angle and with the level of radiant flux. In a well-designed pyranometer this variation should not exceed 2%.

Many authors [24–27] have investigated tilt effects on pyranometers. However, because of continuous developments and modifications in pyranometry, manufacturers' catalogs should be consulted in order to carry out any corrections necessary. For example, the Eppley Laboratory recommends no tilt correction for its "black and white" and also for precision spectral pyranometers. Kipp & Zonen provide a correction plot for one of their pyranometers.

12.9 Pyranometer Calibration

Unlike pyrheliometers, there are no absolute pyranometers. All pyranometers are calibrated instruments. There are a number of methods of calibrating pyranometers using the sun or laboratory sources. Two methods commonly used are calibrating directly against a reference standard pyrheliometer using the sun as source and calibrating against a reference pyranometer that has already been calibrated as above.

A. *Calibration against a Reference Standard Pyrheliometer*

A perfectly clear day should be chosen to conduct this calibration. The test pyranometer should be fitted with an occulting disk to temporarily eliminate the direct component. Figure 12.9.1 shows such a device. The disk should be large enough to shade the glass dome at all zenith angles. A disk 10 cm in diameter and held 1 m from the pyranometer will approximate a 6° aperture angle of the reference standard pyrheliometer. This arrangement will cut off direct plus circumsolar radiation.

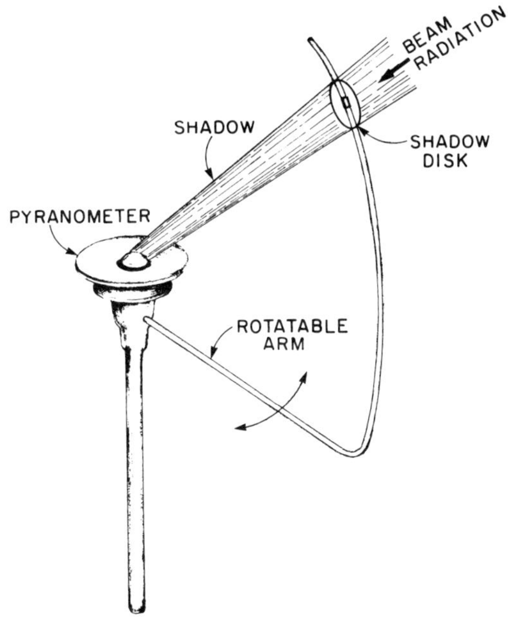

Figure 12.9.1 Pyranometer with a shading disk. Adapted from Latimer [41].

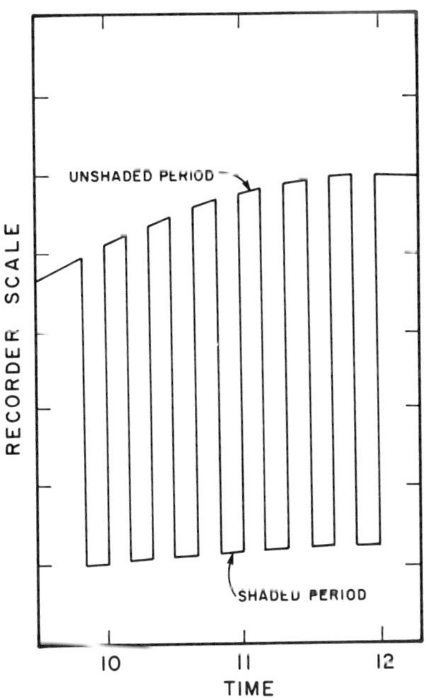

Figure 12.9.2 Pyranometer response with and without a shading disk. Adapted from Latimer [41].

The test should begin by keeping the pyrheliometer pointed directly at the sun. At the same time the occulting disk should alternately shade and expose the pyranometer for durations of about 10 min. Figure 12.9.2 shows qualitatively the pyranometer recorder plot obtained from this procedure. The pyranometer calibration factor K is obtained from the following:

$$K = \frac{\text{Pyranometer reading averaged over the exposed period minus that averaged over the shaded period}}{\dot{I}_{\rm n} \cos \theta_{\rm z}}, \quad (12.9.1)$$

where $\dot{I}_{\rm n}$ is the direct normal irradiance calculated from the pyrheliometer reading.

B. *Calibration against a Reference Pyranometer*

The test pyranometer and the reference pyranometer are both placed in a horizontal position and operated simultaneously. Again, a clear day should be chosen for this test. The calibration factor is a ratio of the measurement from the test pyranometer to the measurement from the reference pyranometer.

12.10 Pyranometer Classification

The WMO [23] classifies pyranometers as *first class*, *second class*, and *third class*. This classification is based on the accuracy of these instruments in terms of the various criteria we have already mentioned in Section 12.8.

12.11 Measurement of Diffuse Irradiance: Pyranometer with Shadow Band

When global radiation is measured by a pyranometer and direct radiation is measured separately by a pyrheliometer, diffuse irradiance can be found by calculation. Thus

$$\dot{I}_{\rm d} = \dot{I} - \dot{I}_{\rm n} \cos \theta_{\rm z}. \quad (12.11.1)$$

Because pyrheliometry is expensive, diffuse radiation is measured by a pyranometer equipped with a special shading device to exclude direct radiation from the sun. Such a pyranometer is sometimes called a diffusograph.

In Fig. 12.9.1 we have shown a diffusograph with a shading disk. However, to operate this device a tracking mechanism is required to follow the sun continuously. Again, this equipment is expensive. It is more customary to install a shadow band. The band is installed parallel to the equatorial plane and is inclined from the vertical at an angle equal to the latitude. It is necessary to adjust the center of the band periodically as the solar declination changes. An adjustment should be made daily or every other day when the sun is not obscured by clouds. The inner surface of the band should be painted black to minimize the effect of multiple reflections. A photograph of a pyranometer installed with an Eppley shadow band is shown in Fig. 12.11.1. The stand carries engraved markings for latitudes and declination. The arms can slide to adjust for solar declination. In this design it is not necessary to wait for clear-sky conditions to adjust the band position.

Because a shadow band screens the sensor from a portion of the diffuse radiation coming in from the sky, a correction must be made to the measurement. However, because of the anisotropy of diffuse radiation, with its maximum being close to the sun, it is difficult to compute an exact theoretical

Figure 12.11.1 Eppley model of a shadow band mounted on a pyranometer.

correction. Therefore, a combined theoretical and experimental approach is followed.

The theoretical component is based on the assumption of isotropic sky conditions. Let i_d be the isotropic diffuse intensity. From Chapter 11, we have [Eq. (11.4.3)]

$$\dot{I}_d = \pi i_d.$$

For the type of shadow band described above, and for isotropic conditions, Drummond [28] computed the fraction of diffuse radiation blocked by the shadow band. This fraction is given by

$$X = \frac{2b}{\pi r}\cos^3\delta\left[\left(\frac{\pi}{180}\psi_s\right)\sin\phi\sin\delta + \cos\phi\cos\delta\sin\psi_s\right] \quad (12.11.2)$$

where

b is the width of the shadow band,
r is the radius of the shadow band,
ψ_s is the azimuth angle of the sun at sunrise (degrees),
ϕ is the latitude (degrees),
δ is the declination of the sun (degrees).

For isotropic conditions, K', the correction to be applied, is

$$K' = 1/(1 - X) \quad (12.11.3)$$

and depends on three constants, b, r, and ϕ, and the two variables δ and Ψ_s.

Robinson and Stoch [29] developed an expression slightly different from (12.11.2) for a nonadjustable shadow band.

The blockage effect under real-sky conditions can be determined by simultaneous measurements from two pyranometers, one fitted with a shadow band and the other with a shadow disk. Any such comparison is limited to a particular site and a particular size and type of shadow band. From a large mass of data, and for a particular site and size of band, Drummond [30] observed that a further correction of +7% is necessary for cloudless skies, +3% for overcast skies, and +4% for average partly cloudy skies. Schüepp [31], in a similar study carried out at a different site and with a different size band, found that an additional correction of 6–9% was required for clear skies and no additional correction for overcast skies. LeBaron *et al.* [32] and Painter [33] studied the effects of additional factors, such as pyranometer type, cosine response, reflectance of inner surface of the shadow band, and ratio of diffuse to global irradiance. It seems very difficult to develop a single unified treatment of this subject.

Because of the complexity of the problem and because Drummond carried out his investigations with an Eppley shadow band, his analysis can be adopted by users of this model. Let K be the total correction factor (by which the diffusograph measurement is to be multiplied); then

$$\begin{aligned} K &= K' + 0.07 \qquad \text{(cloudless skies)},\\ &= K' + 0.03 \qquad \text{(overcast skies)},\\ &= K' + 0.04 \qquad \text{(partly cloudy skies)}. \end{aligned}$$

For the Eppley shadow band ($r = 31.7$ cm, $b = 7.6$ cm), K varies from 1.24 to 1.05 depending on latitude and month. At any given north latitude, the correction is maximum during March and September, and minimum during December.

□ EXAMPLE 12.11.1. The correction factor K by which the diffusograph measurement is to be multiplied for the month of June for the Eppley shadow band (width 7.6 cm, radius 31.7 cm) at 40° N latitude.

Solution. $b = 7.6$ cm, $r = 31.7$ cm, $\phi = 40°$, and $\delta_c = 23.04°$ for June (Table 4.2.1).

From Eq. (1.5.2a), the solar azimuth at sunset is given by

$$\psi_s = \cos^{-1}[-\sin(23.04)/\cos(40)] = 120.7°.$$

From Eq. (12.11.2), we obtain

$$X = 0.135\,1.$$

Therefore

$$K' = 1/(1 - 0.135\,1) = 1.156.$$

We add 4% for anisotropic sky radiation and obtain

$$K = 1.156 + 0.04 = 1.196 \simeq 1.2.$$ □

12.12 Optical Filters

In many areas of engineering and meteorological research, such as turbidity measurements, determination of radiation in well-defined spectral regions of sunlight is required. This can be measured either by expensive spectrophotometers or by aid of rather simple and comparatively inexpensive instruments—pyrheliometers or pyranometers in combination with glass filters.

In meteorological applications, two types of optical filters are used—absorption filters or interference filters. The first is a group of colored glasses manufactured by Schott Glaswerke (Mainz, Germany). Both types are available as plane or hemispherical specimens.

An extensive series of the Schott colored-glass filters is available for various cutoff wavelengths. Manufacturers' catalogs [34] should be consulted for full details of their series and for spectral transmittance of individual filters. The wavelengths are usually expressed in millimicrons (nm or mμ). For meteorological studies, the three most common filters used internationally are Nos. OG530, RG630, and RG695 (old numbers OG1, RG2, and RG8.) Their colors are orange, red, and black, respectively. Ideal characteristics of the three filters are given below:

OG530 opaque up to 530 nm; transparent 530–2 800 nm,
RG630 opaque up to 630 nm; transparent 630–2 800 nm,
RG695 opaque up to 695 nm; transparent 695–2 800 nm.

However, in reality, between the lower cutoff wavelength and the beginning of almost uniform transmittance there is a short transitional range (see Fig. 12.12.1). The position of the center of lower cutoff of Schott filters varies

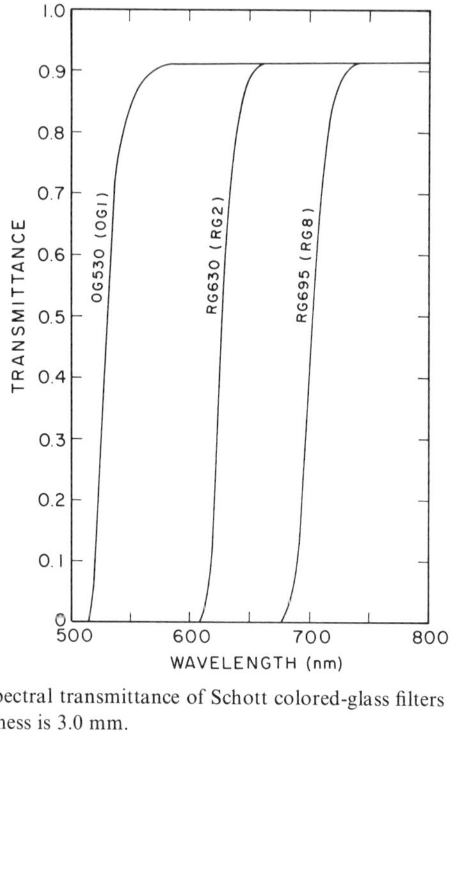

Figure 12.12.1 Spectral transmittance of Schott colored-glass filters OG 530, RG 630, and RG 695. Filter thickness is 3.0 mm.

with temperature and thickness. These glasses also exhibit a variation with thickness of transmittance in the main region of transmission. It may be noted that for colorless glass, the region of high transparency is 350–2800 nm, and 250–4000 nm for quartz.

The direct normal irradiance, say in the range 630–695 nm, can be determined through a series of subtractions of pyrheliometric measurements as follows.

Through the use of RG630 and RG695, the following integrals are evaluated:

$$\int_{630}^{\infty} \dot{I}_{n\lambda}\, d\lambda, \qquad \int_{695}^{\infty} \dot{I}_{n\lambda}\, d\lambda.$$

Combining these measurements with measurements of total solar radiation, we determine the integrals

$$\int_{0}^{630} \dot{I}_{n\lambda}\, d\lambda, \qquad \int_{0}^{695} \dot{I}_{n\lambda}\, d\lambda.$$

The difference between the above two quantities gives us the following:

$$\int_{630}^{695} \dot{I}_{n\lambda}\, d\lambda = \int_{0}^{695} \dot{I}_{n\lambda}\, d\lambda - \int_{0}^{630} \dot{I}_{n\lambda}\, d\lambda. \tag{12.12.1}$$

The radiation defined by the band 630–695 nm may be considered monochromatic and consequently turbidity β can be evaluated by assigning a certain value to the wavelength exponent α (see Chapter 6 for examples).

The Schott glass filters cut off energy in the unwanted wavelength bands, mainly through absorption. The absorbed energy elevates their temperature. In pyrheliometric applications, where a filter is freely exposed in front of the aperture, the filter temperature has a minimal effect on the radiative exchange with the sensor. However, in pyranometric applications, the infrared emission effects will influence the detector through radiative exchange with the inner glass dome. For the Eppley precision spectral pyranometer, Drummond and Roche [35] recommend the use of the following multiplicative correction factors to reduce values in measured energy units to true energy units:

OG530, 0.94; RG630, 0.925; RG695, 0.91.

The heating effect is only one of the problems introduced by placement of Schott colored glasses. Gulbrandsen [36] studied the effect of radiative cooling and zero-point deviation of pyranometers. However, adjustments to account for these factors are difficult to incorporate in routine measurements.

12.13 Measurement of Albedo

Ground albedo can be determined in a very simple manner as a ratio of measurement of two horizontally placed identical pyranometers. One pyranometer faces the sky and the other the ground. It is common to attach the two pyranometers back to back on a horizontal rod which is attached to a mast or a stand. The pyranometers are placed several meters above ground.

12.14 Measurement of Bright Sunshine Duration

The number of hours of sunshine is a very useful indicator of the amount of solar radiation arriving on the earth. Data on hours of sunshine are needed by meteorological services as well as by the tourism industry and local chambers of commerce. In addition, a time record of the exact time of day at which the sun may have been visible at a particular place is also required by local courts to substantiate or refute the testimony of a witness. In Chapter 8, we have studied the use of hours of bright sunshine in developing methods of predicting global and diffuse radiation arriving on the ground. Because most sunshine measuring instruments are cheaper to buy and operate than pyranometers, they form an essential part of meteorological networks.

Two main types of sunshine duration sensors in use are (i) the focusing type and (ii) the photovoltaic type. These instruments are described here.

The focusing type includes the Campbell–Stokes sunshine recorder shown in Fig. 12.14.1. It consists of a solid polished sphere of precision optical glass, about 10 cm in diameter, supported in a spherical bowl so that the sun's image is brought to focus on a chemically treated thin card held in a groove inside the bowl. The sphere is mounted on an axis parallel to that of the earth. The bowl is part of a spherical shell and surrounds the lower half of the focusing sphere. The bowl is cut in such a manner that at all times of the day the sun's image can fall on some part of the inner surface containing the card. The card lies in an east–west direction and is printed with hour lines so that the duration as well as the exact time of sunshine can be measured.

The altitude of the sun changes with the season, causing the sun's image to travel up and down the bowl from north to south. To accommodate this effect, three sets of the card-retaining grooves are provided in the bowl's surface, and three different sets of cards are used. Long, curved cards are used for summer; short, curved cards for winter; and straight cards for spring and autumn. The grooves are arranged so that sufficient overlap is present to keep the burn away from the edges of the card.

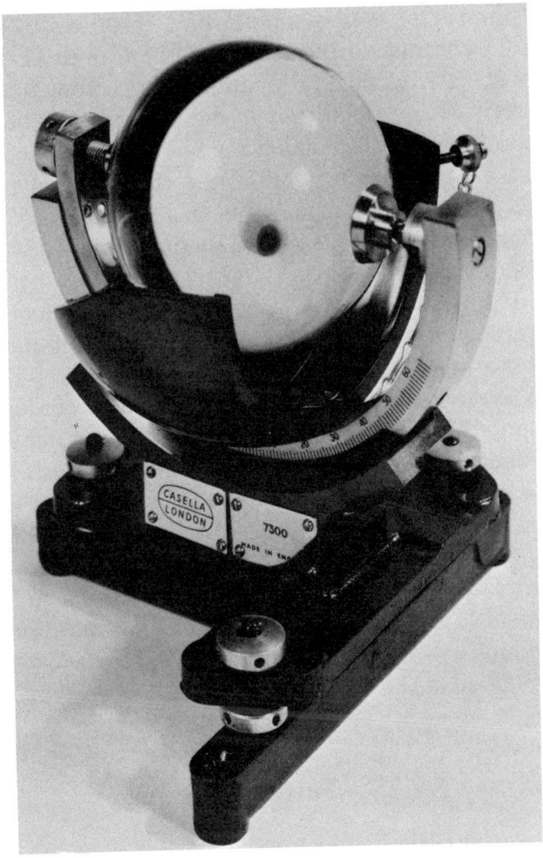

Figure 12.14.1 Campbell–Stokes sunshine recorder.

Solar rays passing through the glass sphere are focused on the card. As the sun moves through the sky a brown line is burned into the paper whenever the radiation is above a certain threshold level. When the solar radiation is below threshold level, the burned line is interrupted by an unburned length. The number of hours of bright sunshine during a day is determined by measuring the total length of the burned segments.

A major problem with this instrument is the effect of humidity on the card and its ability to establish a fixed threshold level of radiation. Under very humid conditions, burn may not begin until the threshold level is 280 W m^{-2}, whereas in a very dry climate, it may begin at 70 W m^{-2}. Moreover, assessment of total daily sunshine relies on personnel training, and measured sunshine totals are in fact estimates which can vary from observer to observer.

There are a number of designs of sunshine recorders based on photovoltaic sensors. In contrast to the record of the Campbell–Stokes instrument, which must be read manually, observations from the photovoltaic devices can be recorded automatically in digital form. Further, they do not require the inconvenience of changing the cards daily. The U.S. National Weather Service uses the Foster sunshine recorder in its meteorological network. This recorder was designed by Foster and Foskett [37] of the NWS. However, it is not commercially available. It consists of two selenium photovoltaic cells, one exposed to direct sunshine and the other shaded by a shadow band. The two cells are connected in a differential form so that when the skies are overcast there is no signal output, and under direct sunlight a signal is recorded.

Examination of records from the various sunshine recorders shows large differences of up to 20% in monthly totals. Apart from human factors, these differences are due to the different threshold values of sunshine. To provide a uniform standard, the WMO has adopted for many years the Campbell–Stokes recorder as a standard with 210 W m^{-2} as an average lower limit for threshold of radiant flux. The Commission for Instruments and Methods of Observation of the WMO during its Eighth Session in October 1981 revised this standard. The new standard for sunshine duration is the period of time during which a threshold of 120 W m^{-2} is exceeded, using a standard pyrheliometer pointed at the solar disk.

12.15 Further Reading

The literature on solar radiation measuring instruments has been well documented in a number of specialized publications. One delightfully written document is by CSAGI [38] (Comité Spéciale de l'Année Géophysique Internationale). The texts by Coulson [39] and Robinson [40] also treat this subject in depth. The WMO publication [23] on measurement of radiation and sunshine provides us with the internationally accepted guidelines regarding instrument and measurement standards (a revised edition of this document is currently under press). Latimer [41] has provided details regarding instrument calibration, installation, and measurement techniques. Brief descriptions of a series of new sunshine recorders are given by Riches [42].

In this chapter, we have been concerned with individual instruments. However, to assess the total availability of solar radiation over a geographic region, the instruments form part of an integrated network. Design of such networks has been reported by Carter *et al.* [43].

Nomenclature

a, b Constants (dimensionless)
$\dot{I}$ Global irradiance on a horizontal surface (W m^{-2})
$\dot{I}_d$ Diffuse irradiance on a horizontal surface (W m^{-2})
$\dot{I}_n$ Direct normal irradiance (W m^{-2})
i Electric current (A)
i_d Intensity of diffuse irradiance (W m^{-2} sr^{-1})
K Instrument constant (dimensionless)
L Distance between receiver and diaphragm of a pyrheliometer (cm)
P_c Cavity heater power with shutter closed (W)
P_o Cavity heater power with shutter open (W)
R Radius of limiting aperture of a pyrheliometer (cm)
z_1 Limiting angle (degrees)
z_o Opening angle or aperture angle (degrees)
z_p Slope angle (degrees)
α_c Cavity absorptance (dimensionless)
α_s Cavity surface absorptance (dimensionless)
δ Solar declination (degrees)
θ_z Zenith angle (degrees)
λ As a subscript, monochromatic value
τ Cutoff filter transmittance (dimensionless)
ϕ Latitude (degrees)
ψ_s Solar azimuth at sunrise (degrees)

References

1. W. J. H. Moll, A thermopile for measuring radiation. *Proc. Phys. Soc. London, Ser. B.* **35,** 257–260 (1923).
2. C. G. Abbot and L. B. Aldrich, Pyrheliometric apparatus. *Ann. Astrophys. Obs., Smith. Inst.* **2,** 34–49 (1908).
3. C. G. Abbot, Standard pyrheliometers. *Ann. Astrophys. Obs. Smith. Inst.* **3,** 52–72 (1913).
4. C. G. Abbot and L. B. Aldrich, An improved water-flow pyrheliometer and the standard scale of solar radiation. *Smith Misc. Collect.* **87** (15), 1–8 (1932).
5. C. G. Abbot, The silver disk pyrheliometer, *Smith Misc. Collect.* **56** (19), 1–11 (1911).
6. K. Ångström, The absolute determination of the radiation of heat with the electric compensation pyrheliometer and examples of the application of this instrument. *Astrophys. J.* **9,** 332–346 (1899).
7. J. M. Kendall and C. M. Berdahl, Two blackbody radiometers of high accuracy. *Appl. Opt.* **9** (5), 1082–1091 (1970).
8. M. S. Reid, C. M. Berdahl, and J. M. Kendahl, Calibration standards and field instruments for precision measurement of insolation. *Sol. Energy* **20** (4), 357–358 (1978).
9. R. C. Willson, Active cavity radiometer. *Appl. Opt.* **12** (4), 810–817 (1973).
10. R. C. Willson, Accurate solar "constant" determinations by cavity pyrheliometers. *J. Geophys. Res.* **83** (C8), 4003–4007 (1978).
11. R. C. Willson, Active cavity radiometer type IV. *Appl. Opt.* **18** (2), 179–188 (1979).
12. R. C. Willson, Solar total irradiance observations by active cavity radiometer. *Sol. Phys.* **74,** 217–229 (1981).

13. J. Geist, Optical radiation measurements. National Bureau of Standards, T.N., 594-1 (1972).
14. R. W. Brusa and C. Fröhlich, Realization of the absolute scale of total irradiance. Scientific Discussions International Pyrheliometer Comparisons IV. W.R.C., Davos (1975).
15. D. Crommelynck, Théorie instrumentale en radiométrie absolue. Pub. Ser. A, No. 81. Institut Royal Météorologique de Belgique, Brussels (1973).
16. C. Fröhlich, The relation between the IPS now in use and Smithsonian scale 1913, Ångström Scale and Absolute Scale. *Proc. Symp. Solar Rad. Measurements Instrum.* pp. 61–77 (1973).
17. A. Ångström, On pyrheliometric measurements. *Tellus* **10,** 342–354 (1958).
18. F. Lindholm, on the Ångström absolute pyrheliometric scale *Tellus* **10,** 249–255 (1958).
19. L. B. Aldrich and C. G. Abbot, Smithsonian Pyrheliometry and the Standard Scale of Solar Radiation, *Smith Misc. Collect.* **110** (5), 1–4 (1948).
20. A. J. Drummond, A contribution to absolute pyrheliometry. *Q. J. R. Meteorol. Soc.* **82,** 481–493 (1956).
21. C. Fröhlich, J. Geist, J. Kendall, and R. M. Marchgraber, The Third International Comparisons of Pyrheliometers and a comparison of radiometric scales. *Sol. Energy* **14** (2), 157–166 (1973).
22. World Meteorological Organization, Commission for Instruments and Methods of Observation, Abridged Final Report of the Seventh Session, WMO No. 490 (1978).
23. World Meteorological Organization, Measurement of Radiation and Sunshine. Guide to meteorological instruments and observing practices. 4th ed., W.M.O., No. 8, T.P.3, Geneva, Switzerland (1971). A revised edition is in press.
24. D. J. Norris, Calibration of pyranometers. *Sol. Energy* **14** (2), 99–108 (1973).
25. D. J. Norris, Calibration of pyranometers in inclined and inverted positions. *Sol. Energy* **16** (1), 53–55 (1974).
26. E. Flowers, Test and evaluation of the performance of solar radiation sensors at inclinations from the horizontal under laboratory and field conditions. ERDA-NOAA, IAA, E(49-26)-1041, T-003 (1977).
27. B. Goldberg, Errors associated with precision pyranometers on a tilted surface. *Sol. Energy* **25** (3), 285–286 (1980).
28. A. J. Drummond, On the measurement of sky radiation. *Arch. Meteorol. Geophys. Bioklimatol. Ser. B* **7** (3/4) (1956).
29. N. Robinson and L. Stoch, Sky radiation measurement and correction. *J. Appl. Meteorol.* **3,** 179–181 (1964).
30. A. J. Drummond, Comments on sky radiation measurements and corrections. *J. Appl. Meteorol.* **13,** 810–811 (1964).
31. W. Schüepp, Enregistrement séparé des composantes du rayonnement solaire. *Bull. Serv. Metéorol. Congo Belge* **8,** 11–20 (1952) (as cited by Reference 33).
32. B. A. LeBaron, W. A. Peterson, and I. Dirmhirn, Corrections for diffuse irradiance measured with shadowbands. *Sol. Energy* **25** (1), 1–13 (1980).
33. H. E. Painter, The shade ring correction for diffuse irradiance measurements. *Sol. Energy* **26** (4), 361–363 (1981).
34. Schott, Colour Filter Glass, Optics Division, Schott Glaswerke, D-6500, Mainz, Federal Republic of Germany (1982).
35. A. J. Drummond and J. J. Roche, Corrections to be applied to measurements made with Eppley (and other) spectral radiometers when used with Schott colored glasses. *J. Appl. Meteorol.* **4,** 741–744 (1965).
36. A. Gulbrandsen, On the use of pyranometers in the study of spectral solar radiation and atmospheric aerosols. *J. Appl. Meteorol.* **17,** 899–904 (1978).

37. N. B. Foster and L. W. Foskett, A photoelectric sunshine recorder, *Bull. Am. Meteorol. Soc.* **34,** 212–215 (1953).
38. C.S.A.G.I., Radiation Instruments and Measurements. "Manual of the International Geophysical Year," Vol. 5, Part VI, pp. 367–467. Program Press, New York, (1958).
39. K. L. Coulson, "Solar and Terrestrial Radiation." Academic Press, New York, 1975.
40. N. Robinson (ed.), "Solar Radiation." Elsevier, Amsterdam, 1966.
41. J. R. Latimer, Radiation Measurement, International Field Year for the Great Lakes, Technical Manual Series, No. 2 (1978).
42. M. R. Riches, Duration of Sunshine. *in* An Introduction to Meteorological Measurements and Data Handling for Solar Energy Applications, Chapter 9 DOE/ER-0084 (1980).
43. E. A. Carter, D. L. Christensen and B. B. Williams, Solar radiation data sources, applications and network design. The Univ. of Alabama in Huntsville, Huntsville, Alabama (1978).

Appendix A

LIST OF BASIC CONSTANTS AND DIMENSIONS

Table A.1

Constant	SI units	Nonmetric units
AU, astronomical unit	1.496×10^{11} m	9.3×10^{7} mile
c_0, speed of light in vacuum	2.998×10^{8} m s^{-1}	9.836×10^{8} ft s^{-1}
C_1, first radiation constant	3.7427×10^{8} W μm^{4} m^{-2}	1.19×10^{8} Btu μm^{4} ft^{-2} h^{-1}
C_2, second radiation constant	1.4388×10^{4} μm K	2.59×10^{4} μm °R
I_{sc}, solar constant	1367 W m^{-2}	433.3 Btu ft^{-2} h^{-1}
	4921 kJ m^{-2} h^{-1}	1.96 cal cm^{-2} min^{-1}
r_e, radius of earth	6.37×10^{6} m	3.96×10^{3} mile
r_s, radius of sun	6.96×10^{8} m	4.3×10^{5} mile
σ, Stefan–Boltzmann constant		
theoretical value	5.6697×10^{-8} W m^{-2} K^{-4}	0.1714×10^{-8} Btu ft^{-2} h^{-1} °R^{-4}
experimental value	5.6866×10^{-8} W m^{-2} K^{-4}	0.1719×10^{-8} Btu ft^{-2} h^{-1} °R^{-4}

Appendix B

CONVERSION FACTORS

Table B.1

Conversion Factors for Quantity of Radiation per Unit Area

Units	$J\ m^{-2}$	$W\ h\ m^{-2}$	$cal\ cm^{-2}$ [a]	$Btu\ ft^{-2}$
$1\ J\ m^{-2}$	1	2.778×10^{-4}	2.39×10^{-5}	8.81×10^{-5}
$1\ W\ h\ m^{-2}$	3.60×10^{3}	1	0.0860	0.317
$1\ cal\ cm^{-2}$	4.187×10^{4}	11.63	1	3.69
$1\ Btu\ ft^{-2}$	1.136×10^{4}	3.155	0.271	1

[a] It is useful to note that in the United States, cal cm^{-2} is also called *Langley*, named after Samuel Pierpont Langley (1834–1906) of the Smithsonian Institution, a pioneer of solar radiation studies, astronomer, physicist, inventor and engineer. He built the first steam-driven pilotless airplane.

Table B.2

Conversion Factors for Radiant Flux per Unit Area

Units	$W\ m^{-2}$	$cal\ cm^{-2}\ min^{-1}$	$Btu\ ft^{-2}\ h^{-1}$
$1\ W\ m^{-2}$	1	1.433×10^{-3}	0.317
$1\ cal\ cm^{-2}\ min^{-1}$	698	1	221.2
$1\ Btu\ ft^{-2}\ h^{-1}$	3.155	4.521×10^{-3}	1

Appendix C

SPECTRAL DISTRIBUTION OF THE EXTRATERRESTRIAL SOLAR IRRADIANCE

Table C.1

Spectral Distribution of the Extraterrestrial Solar Irradiance from 0.25 to 25 μm at Mean Sun–Earth Distance[a]

λ	Value	λ	Value	λ	Value	λ	Value	λ	Value	λ	Value
0.0	0.0	0.368	2.46	0.488	3.54	0.608	3.52	0.728	2.73	0.920	4.28
0.250	2.51	0.370	2.35	0.490	3.88	0.610	3.39	0.730	2.70	0.925	4.13
0.252	0.13	0.372	2.31	0.492	3.80	0.612	3.47	0.732	2.72	0.930	4.16
0.254	0.12	0.374	1.89	0.494	3.84	0.614	3.42	0.734	2.71	0.935	4.14
0.256	0.17	0.376	2.07	0.496	3.95	0.616	3.42	0.736	2.71	0.940	4.03
0.258	0.27	0.378	2.62	0.498	3.98	0.618	3.36	0.738	2.69	0.945	3.98
0.260	0.27	0.380	2.51	0.500	3.82	0.620	3.49	0.740	2.61	0.950	3.93
0.262	0.22	0.382	2.33	0.502	3.63	0.622	3.37	0.742	2.59	0.955	3.85
0.264	0.32	0.384	1.44	0.504	3.84	0.624	3.38	0.744	2.57	0.960	3.87
0.266	0.57	0.386	2.04	0.506	3.93	0.626	3.24	0.746	2.58	0.965	3.84
0.268	0.57	0.388	1.93	0.508	3.85	0.628	3.40	0.748	2.58	0.970	3.84
0.270	0.55	0.390	2.24	0.510	3.88	0.630	3.32	0.750	2.57	0.975	3.80
0.272	0.55	0.392	2.56	0.512	3.93	0.632	3.23	0.752	2.52	0.980	3.84
0.274	0.44	0.394	1.49	0.514	3.74	0.634	3.33	0.754	2.50	0.985	3.85
0.276	0.37	0.396	2.53	0.516	3.77	0.636	3.32	0.756	2.50	0.990	3.82
0.278	0.52	0.398	1.72	0.518	3.40	0.638	3.32	0.758	2.48	0.995	3.80
0.280	0.28	0.400	3.21	0.520	3.45	0.640	3.31	0.760	2.45	1.000	3.73
0.282	0.37	0.402	3.60	0.522	3.76	0.642	3.18	0.762	2.44	1.005	3.71
0.284	0.67	0.404	3.36	0.524	3.84	0.644	3.28	0.764	2.44	1.010	3.69
0.286	0.45	0.406	3.27	0.526	3.88	0.646	3.26	0.766	2.41	1.015	3.65
0.288	0.72	0.408	3.25	0.528	3.56	0.648	3.20	0.768	2.41	1.020	3.57
0.290	0.90	0.410	3.39	0.530	3.84	0.650	3.14	0.770	2.38	1.025	3.54
0.292	1.24	0.412	3.45	0.532	3.95	0.652	3.25	0.772	2.37	1.030	3.50
0.294	1.13	0.414	3.52	0.534	3.76	0.654	3.21	0.774	2.36	1.035	3.46
0.296	1.14	0.416	3.45	0.536	3.92	0.656	2.97	0.776	2.35	1.040	3.44
0.298	1.07	0.418	3.56	0.538	3.81	0.658	2.81	0.778	2.34	1.045	3.42
0.300	0.99	0.420	3.32	0.540	3.76	0.660	3.07	0.780	2.40	1.050	3.37
0.302	1.12	0.422	3.67	0.542	3.67	0.662	3.15	0.782	2.38	1.055	3.29
0.304	1.06	0.424	3.29	0.544	3.78	0.664	3.16	0.784	2.35	1.060	3.24
0.306	1.09	0.426	3.50	0.546	3.81	0.666	3.13	0.786	2.35	1.065	3.22
0.308	1.22	0.428	3.25	0.548	3.76	0.668	3.07	0.788	2.33	1.070	3.21
0.310	1.10	0.430	3.05	0.550	3.84	0.670	3.04	0.790	2.30	1.075	3.18
0.312	1.31	0.432	2.92	0.552	3.77	0.672	2.98	0.792	2.27	1.080	3.14
0.314	1.36	0.434	3.31	0.554	3.77	0.674	2.93	0.794	2.26	1.085	3.09
0.316	1.42	0.436	3.54	0.556	3.82	0.676	3.03	0.796	2.33	1.090	3.07
0.318	1.44	0.438	3.70	0.558	3.67	0.678	2.97	0.798	2.31	1.095	3.04
0.320	1.43	0.440	3.38	0.560	3.67	0.680	2.94	0.800	2.29	1.100	3.03
0.322	1.56	0.442	3.64	0.562	3.69	0.682	2.95	0.805	5.72	1.105	3.03
0.324	1.36	0.444	3.90	0.564	3.76	0.684	2.93	0.810	5.61	1.110	3.02
0.326	1.39	0.446	3.78	0.566	3.67	0.686	2.89	0.815	5.52	1.115	2.99
0.328	2.12	0.448	3.92	0.568	3.70	0.688	2.79	0.820	5.39	1.120	2.93
0.330	1.99	0.450	4.03	0.570	3.70	0.690	2.80	0.825	5.31	1.125	2.88
0.332	2.00	0.452	4.37	0.572	3.57	0.692	2.86	0.830	5.26	1.130	2.85
0.334	1.90	0.454	3.89	0.574	3.74	0.694	2.85	0.835	5.15	1.135	2.83
0.336	1.87	0.456	4.06	0.576	3.72	0.696	2.92	0.840	5.12	1.140	2.81
0.338	1.61	0.458	4.13	0.578	3.61	0.698	2.92	0.845	5.08	1.145	2.79
0.340	1.90	0.460	3.98	0.580	3.64	0.700	2.87	0.850	5.11	1.150	2.76
0.342	1.94	0.462	4.15	0.582	3.72	0.702	2.84	0.855	4.83	1.155	2.75
0.344	1.89	0.464	4.17	0.584	3.69	0.704	2.81	0.860	5.06	1.160	2.74
0.346	1.66	0.466	3.96	0.586	3.62	0.706	2.83	0.865	4.96	1.165	2.68
0.348	1.81	0.468	3.91	0.588	3.61	0.708	2.76	0.870	4.80	1.170	2.67
0.350	1.80	0.470	4.02	0.590	3.35	0.710	2.77	0.875	4.92	1.175	2.65
0.352	2.02	0.472	3.90	0.592	3.62	0.712	2.84	0.880	4.86	1.180	2.61
0.354	2.04	0.474	4.04	0.594	3.54	0.714	2.78	0.885	4.80	1.185	2.57
0.356	2.27	0.476	4.03	0.596	3.57	0.716	2.75	0.890	4.75	1.190	2.56
0.358	1.78	0.478	4.03	0.598	3.60	0.718	2.74	0.895	4.70	1.195	2.54
0.360	1.72	0.480	4.12	0.600	3.44	0.720	2.71	0.900	4.62	1.200	2.50
0.362	2.04	0.482	4.10	0.602	3.44	0.722	2.71	0.905	4.51	1.205	2.48
0.364	2.16	0.484	4.01	0.604	3.55	0.724	2.81	0.910	4.39	1.210	2.47
0.366	2.19	0.486	3.83	0.606	3.47	0.726	2.79	0.915	4.37	1.215	2.48

***Table C.1** (Continued)*

λ	Value	λ	Value	λ	Value	λ	Value	λ	Value	λ	Value
1.220	2.46	1.520	1.44	1.820	0.80	2.480	1.02	3.680	0.24	4.880	0.08
1.225	2.41	1.525	1.44	1.825	0.79	2.500	0.98	3.700	0.23	4.900	0.08
1.230	2.42	1.530	1.43	1.830	0.78	2.520	0.95	3.720	0.23	4.920	0.07
1.235	2.41	1.535	1.39	1.835	0.77	2.540	0.93	3.740	0.22	4.940	0.07
1.240	2.39	1.540	1.38	1.840	0.76	2.560	0.90	3.760	0.22	4.960	0.07
1.245	2.36	1.545	1.38	1.845	0.76	2.580	0.87	3.780	0.21	4.980	0.07
1.250	2.35	1.550	1.37	1.850	0.75	2.600	0.85	3.800	0.21	5.000	0.07
1.255	2.35	1.555	1.37	1.855	0.74	2.620	0.83	3.820	0.21	5.100	0.33
1.260	2.27	1.560	1.36	1.860	0.72	2.640	0.80	3.840	0.20	5.200	0.31
1.265	2.21	1.565	1.34	1.865	0.72	2.660	0.78	3.860	0.20	5.300	0.29
1.270	2.20	1.570	1.31	1.870	0.70	2.680	0.76	3.880	0.19	5.400	0.27
1.275	2.23	1.575	1.30	1.875	0.68	2.700	0.74	3.900	0.19	5.500	0.25
1.280	2.23	1.580	1.29	1.880	0.69	2.720	0.72	3.920	0.19	5.600	0.23
1.285	2.19	1.585	1.27	1.885	0.70	2.740	0.70	3.940	0.18	5.700	0.22
1.290	2.21	1.590	1.25	1.890	0.69	2.760	0.68	3.960	0.18	5.800	0.20
1.295	2.21	1.595	1.23	1.895	0.69	2.780	0.66	3.980	0.18	5.900	0.19
1.300	2.20	1.600	1.24	1.900	0.68	2.800	0.65	4.000	0.17	6.000	0.18
1.305	2.17	1.605	1.23	1.905	0.68	2.820	0.63	4.020	0.17	6.100	0.17
1.310	2.12	1.610	1.22	1.910	0.69	2.840	0.61	4.040	0.17	6.200	0.16
1.315	2.09	1.615	1.22	1.915	0.69	2.860	0.60	4.060	0.16	6.300	0.15
1.320	2.09	1.620	1.21	1.920	0.68	2.880	0.58	4.080	0.16	6.400	0.14
1.325	2.07	1.625	1.22	1.925	0.67	2.900	0.57	4.100	0.16	6.500	0.13
1.330	2.04	1.630	1.22	1.930	0.66	2.920	0.55	4.120	0.15	6.600	0.12
1.335	2.02	1.635	1.20	1.935	0.66	2.940	0.54	4.140	0.15	6.700	0.11
1.340	2.00	1.640	1.18	1.940	0.65	2.960	0.53	4.160	0.15	6.800	0.11
1.345	1.99	1.645	1.17	1.945	0.64	2.980	0.51	4.180	0.15	6.900	0.10
1.350	1.96	1.650	1.17	1.950	0.64	3.000	0.50	4.200	0.14	7.000	0.10
1.355	1.93	1.655	1.17	1.955	0.62	3.020	0.49	4.220	0.14	7.100	0.09
1.360	1.91	1.660	1.17	1.960	0.62	3.040	0.48	4.240	0.14	7.200	0.09
1.365	1.88	1.665	1.16	1.965	0.63	3.060	0.47	4.260	0.13	7.300	0.08
1.370	1.85	1.670	1.14	1.970	0.63	3.080	0.45	4.280	0.13	7.400	0.08
1.375	1.85	1.675	1.12	1.975	0.64	3.100	0.44	4.300	0.13	7.500	0.07
1.380	1.84	1.680	1.10	1.980	0.64	3.120	0.43	4.320	0.13	7.600	0.07
1.385	1.82	1.685	1.10	1.985	0.62	3.140	0.42	4.340	0.12	7.700	0.07
1.390	1.81	1.690	1.10	1.990	0.61	3.160	0.41	4.360	0.12	7.800	0.06
1.395	1.79	1.695	1.08	1.995	0.61	3.180	0.40	4.380	0.12	7.900	0.06
1.400	1.78	1.700	1.08	2.000	0.60	3.200	0.40	4.400	0.12	8.000	0.06
1.405	1.76	1.705	1.07	2.020	2.25	3.220	0.39	4.420	0.12	8.100	0.05
1.410	1.74	1.710	1.04	2.040	2.20	3.240	0.38	4.440	0.11	8.200	0.05
1.415	1.73	1.715	1.04	2.060	2.09	3.260	0.37	4.460	0.11	8.300	0.05
1.420	1.72	1.720	1.05	2.080	1.97	3.280	0.36	4.480	0.11	8.400	0.05
1.425	1.73	1.725	1.01	2.100	1.91	3.300	0.35	4.500	0.11	8.500	0.04
1.430	1.72	1.730	0.97	2.120	1.81	3.320	0.34	4.520	0.11	8.600	0.04
1.435	1.67	1.735	0.95	2.140	1.75	3.340	0.34	4.540	0.10	8.700	0.04
1.440	1.65	1.740	0.95	2.160	1.65	3.360	0.33	4.560	0.10	8.800	0.04
1.445	1.62	1.745	0.94	2.180	1.60	3.380	0.32	4.580	0.10	8.900	0.04
1.450	1.61	1.750	0.93	2.200	1.47	3.400	0.32	4.600	0.10	9.000	0.04
1.455	1.58	1.755	0.94	2.220	1.52	3.420	0.31	4.620	0.10	9.100	0.03
1.460	1.56	1.760	0.94	2.240	1.42	3.440	0.30	4.640	0.09	9.200	0.03
1.465	1.57	1.765	0.92	2.260	1.44	3.460	0.30	4.660	0.09	9.300	0.03
1.470	1.56	1.770	0.90	2.280	1.35	3.480	0.29	4.680	0.09	9.400	0.03
1.475	1.55	1.775	0.88	2.300	1.32	3.500	0.28	4.700	0.09	9.500	0.03
1.480	1.53	1.780	0.86	2.320	1.21	3.520	0.28	4.720	0.09	9.600	0.03
1.485	1.51	1.785	0.85	2.340	1.12	3.540	0.27	4.740	0.09	9.700	0.03
1.490	1.51	1.790	0.85	2.360	1.23	3.560	0.27	4.760	0.09	9.800	0.03
1.495	1.51	1.795	0.86	2.380	1.19	3.580	0.26	4.780	0.08	9.900	0.02
1.500	1.49	1.800	0.86	2.400	1.16	3.600	0.26	4.800	0.08	10.000	0.02
1.505	1.48	1.805	0.84	2.420	1.10	3.620	0.25	4.820	0.08	25.000	0.73
1.510	1.47	1.810	0.82	2.440	1.10	3.640	0.25	4.840	0.08		
1.515	1.46	1.815	0.80	2.460	1.00	3.660	0.24	4.860	0.08		

[a]By C. Frölich and C. Wehrli. Solar irradiance values in watts per square meter. Wavelength λ in micrometers. Integrated values between the two wavelengths. For example for 0.0 to 0.250 μm the value is 2.51 W m^{-2}. The sum of all the value columns is 1367 W m^{-2}, the solar constant.

Appendix D

METHODS OF STATISTICAL TESTS

In this book we have studied a number of models for predicting radiation arriving on a horizontal or an inclined surface. Comparison of an individual calculated value against a measured value is not a sufficient test of accuracy of a predictive model. Rather, it is necessary to analyze a large body of data. For solar radiation models, a number of useful statistical tests are based on calculation of *mean bias error* (MBE), *root mean square error* (RMSE), *correlation coefficient* (CC), and *time of day bias* (TODB). These statistical tests are defined below.

D.1 Mean Bias Error

The MBE is an indication of the average deviation of the predicted values from the measured values. It is defined by

$$\mathrm{MBE} = \frac{\sum_{i=1}^{N} (y_i - x_i)}{N}, \tag{D.1}$$

where y_i is the ith predicted value, x_i the ith measured value, and N the number of observations. Ideally a zero value of MBE should be obtained.

D.2 Root Mean Square Error

The RMSE is a measure of the variation of predicted values around the measured values. It is defined as follows:

$$\text{RMSE} = \left\{ \frac{\sum_{i=1}^{N} (y_i - x_i)^2}{N} \right\}^{1/2}. \tag{D.2}$$

The RMSE is always positive; however, a zero value is ideal. It may be noted that a few large variations of the calculated amount of radiation from the measured radiation can substantially increase RMSE.

In some statistical tests, the *standard deviation* (SD), which is a measure of the dispersion or spread of the data from the mean value, is also used. However, in analyzing the radiation models, RMSE test is applied.

D.3 Correlation Coefficient

The CC is a test of the linear relationship between the calculated and measured values. It is defined by

$$\text{CC} = \frac{\sum_{i=1}^{N} (y_i - \bar{y})(x_i - \bar{x})}{\{[\sum_{i=1}^{N} (y_i - \bar{y})^2][\sum_{i=1}^{N} (x_i - \bar{x})^2]\}^{1/2}},$$

where $\bar{y}$ is the predicted mean value and $\bar{x}$ is the measured mean value. Ideally CC should be 1.

D.4 Time of Day Bias

In Chapter 8 we have seen that approximately 90% of the day's radiation is received during the middle two thirds of the day. Therefore, the foregoing statistical tests should be carried out over different periods of the day, such as:

(a) peak radiation period, 11–13,
(b) moderate radiation periods 9–11, 13–15, and
(c) low radiation periods, 7–9, 15–17.

D.5 Tests of Models

Let us assume two models, A and B, are to be tested against measured hourly radiation data. A number of possibilities may exist. The ideal will be that the calculated and the measured values are identical. However, let us assume a hypothetical case in which one model uniformly overpredicts and the second uniformly underpredicts, such as shown in Fig. D.1a. The differences between calculated and measured values are plotted in Fig. D.1b and are called plot residuals. Model A has less bias than model B. The bias of A is positive and that of B is negative. The RMSE will be greater for model B than for model A. However, the two models will have identical correlation coefficients, and each will be nearly equal to one. Consequently, in this instance, RMSE will be the best test.

Let us now consider two models with a particular type of difference plotted in Fig. D.2. Model A shows very good correspondence with measured values during two hours around solar noon; however, correspondence is poor during the rest of the day. Model B has good correspondence with measured radiation during early morning and late afternoon hours and overestimates during the middle third of the day. Both models will have

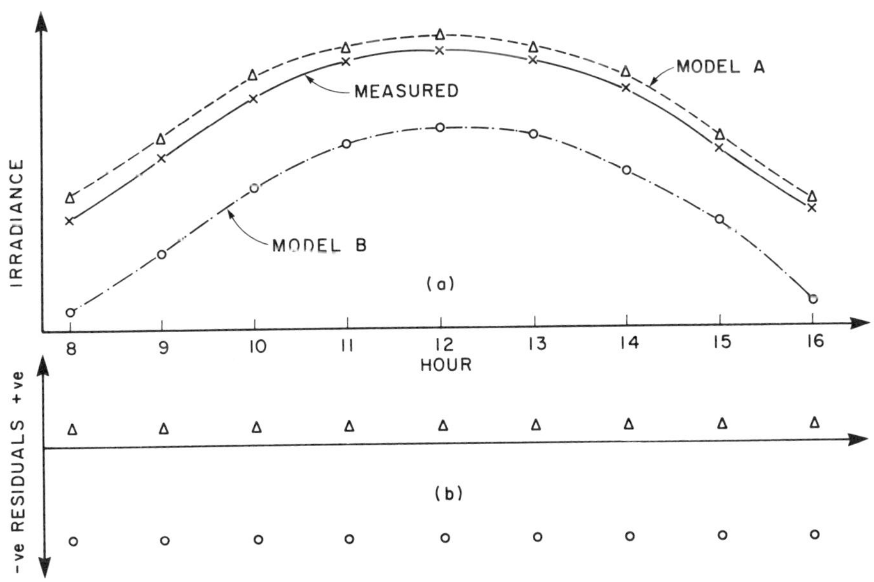

Figure D.1 (a) Diurnal variation of the measured against the predicted insolation. (b) Plot residuals of (a).

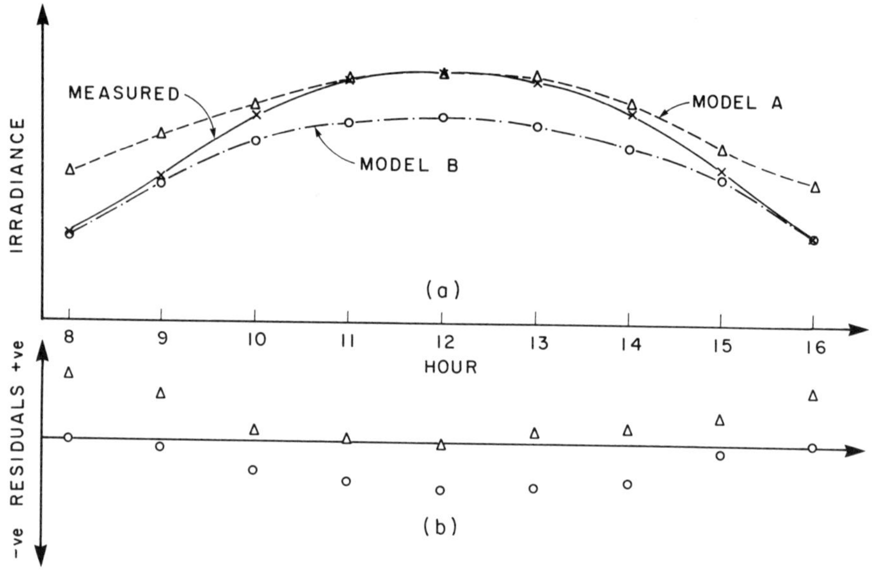

Figure D.2 (a) Diurnal variation of the measured against the predicted insolation. (b) Plot residuals of (a).

approximately the same value of MBE, RMSE, and CC. Consequently, these tests will not distinguish between the two models. However, these tests carried out during, say, the middle third of the day will distinguish between the two models.

What has been said above regarding hourly values applies also to daily or monthly average daily values. The three tests, MBE, RMSE, and CC, should be carried over different seasons. For solar heating applications, tests should cover only winter months. On the other hand, for solar cooling applications, summer months are more important. MacLaren *et* al. [1] have shown applications of these tests and methods of comparing various insolation models.

Reference

1. J. F. MacLaren Ltd., Hooper and Angus Associates Ltd., J. Hay and H. Davies. Define, develop and establish a merged solar and meteorological computer data base. Atmospheric Environment Service, Downsview, Ontario (1980).

INDEX

PRINTED IN THE UNITED STATES OF AMERICA